Minitab Handbook

Third Edition

Barbara F. Ryan
Minitab Inc.
State College, Pennsylvania

Brian L. Joiner
Joiner Associates Incorporated
Madison, Wisconsin

Duxbury Press
An Imprint of Wadsworth Publishing Company
Belmont, California

Computational Statistics Editor: Stan Loll
Editorial Assistant: Claire Masson
Production Editor: Deborah Cogan
Managing Designer: Carolyn Deacy
Print Buyer: Randy Hurst
Permissions Editor: Peggy Meehan
Copy Editor: Judith Abrahms
Compositor: Meredith Flynn
Cover Designer: Craig Hanson
Printer: Courier

MINITAB is a registered trademark of Minitab, Inc.
Macintosh is a registered trademark of Apple Computer, Inc.

International Thomson Publishing
The trademark ITP is used under license

Duxbury Press
An Imprint of Wadsworth Publishing Company
A division of Wadsworth, Inc.

2 3 4 5 6 7 8 9 10—98 97 96 95 94

Printed in the United States of America

Library of Congress Cataloging-in-Publication Data
Ryan, Barbara F.
      Minitab handbook. — 3rd ed. / Barbara F. Ryan, Brian L. Joiner.
          p.   cm.
      Includes bibliographical references and index.
      ISBN 0-534-21240-9
      1. Minitab.   2. Statistics—Data processing.   3. Mathematical statistics—Data
processing.   I. Joiner, Brian L.   II. Title.
QA276.4.R88   1994
519.5′0285′5369—dc20   93-46752

Preface

Minitab is a powerful statistical software program that provides a wide range of basic and advanced data analysis capabilities. Currently in use at over 5,000 sites, including 2,000 universities around the world, more college students are trained on Minitab than on any other statistics package. Additionally, over 250 statistics textbooks and supplements support Minitab-aided instruction. Minitab is also used by a large number of commercial and government analysts; in fact, three-quarters of the top 50 companies in the Fortune 500 depend on Minitab for its unique combination of ease-of-use and powerful data analysis capabilities.

The *Minitab Handbook* is designed to teach students how to do data analysis using Minitab. It is primarily written for use as a supplementary text for first- or second-level courses in statistics (pre- or post-calculus), as well as for researchers new to Minitab. The *Handbook* emphasizes aspects of statistics that are particularly appropriate to computer use, such as the creative use of plots, application of standard methods to real data, in-depth exploration of data, simulation as a learning tool, screening data for errors, manipulating data, transformations, plus multiple regression and analysis of variance.

About the Third Edition

The third edition reflects the changes made in the Minitab software and in statistics teaching that have occurred since the second edition was published. Most of the chapters have been revised and some of the material has been rearranged. Additionally, there are four completely new chapters: Chapter 2, *Versions of Minitab;* Chapter 14, *Control Charts;* Chapter 15, *Additional Topics in Regression;* and Chapter 16, *Additional Topics in Analysis of Variance.*

This edition is designed to be used with Releases 7, 8, and 9, as well as later releases of Minitab. Chapter 2, *Versions of Minitab,* gives a brief overview of the user interface for each currently available version of Minitab.

A new release of Minitab, Release 10 for Windows, will be available shortly after this book is published. We indicate, in footnotes, the main differences between Releases 9 and 10. Most of the *Handbook* emphasizes commands, since they are the same on all versions of Minitab. We do, however, show where to find the dialog box associated with each command and provide a few sections that show how to create high-resolution graphs.

Examples and Exercises

The *Handbook* includes numerous examples and exercises that demonstrate step-by-step how to use the computer to explore and analyze data. Many of these examples and exercises have been obtained directly from the authors, while consulting with researchers from a wide variety of application areas. They are presented in sufficient detail to provide most anyone with an idea of the scope and importance of problems amenable to statistical treatment. Obviously, the computer does not replace hand work, so the *Handbook* emphasizes a mixture of both hand and computer work. The chapters and sections are designed to be as independent of each other as possible. Once the main points of Chapters 1 (*Introduction to Minitab*), 2 (*Versions of Minitab*), 3 (*One Variable: Displays and Summaries*), and 4 (*Plotting Data*) have been covered, the remaining chapters can be covered in almost any order.

Data Sets Used in This *Handbook*

We have incorporated over fifty data sets in this book. Many of them are distributed along with the Minitab software. The complete set is available directly from Duxbury Press.

History of Minitab

Minitab Statistical Software was originally developed in 1972 for students in introductory statistics courses at The Pennsylvania State University. Since Minitab was developed as a portable program, that is, one that could run on many different types of computers, it was subsequently widely distributed to other universities. The first edition of the *Minitab Handbook* was published in 1976 and was the first widely used book of its kind. It helped instructors, especially those new to computers, to integrate the use of computers into teaching statistics. Since then, Minitab's use has expanded beyond introductory-level coursework and its capabilities have been extended beyond just basic statistics.

Originally, Minitab was developed as a command-based system, using simple English-like commands to allow the user to "speak" to the computer.

We find that most of our users still do much of their work using commands. However, with Release 8 we added menus to the DOS and Macintosh versions, and with Release 9 we provided a full Windows interface. Commands, however, remain a common denominator and are the same on all versions of Minitab, with the later versions offering additional commands and options.

Minitab is currently available on a wide variety of computers, including microcomputers, minicomputers, and mainframes. For further information about the software contact:

Minitab, Inc.
3081 Enterprise Drive
State College, PA 16801-3008
phone: (814) 238-3280
fax: (814) 238-4383

Acknowledgments

We are grateful for the many suggestions we have received from users of Minitab and earlier editions of this *Handbook.* Our colleagues and students at The Pennsylvania State University and the University of Wisconsin deserve special mention. In writing and revising the *Handbook,* we are especially grateful to Thomas P. Hettmansperger, who provided extensive suggestions (particularly in the nonparametric statistics chapter); Paul Velleman, who always made insightful and thought-provoking comments; and Lawrence Klimko for his careful and extensive criticisms. Reference is made throughout the book to those who kindly made their data available to us. We are also grateful to The Pennsylvania State University for their partial support while writing the first edition of the *Handbook.* Finally, we want to acknowledge appreciation for the thoughtful advice we received from the many reviewers of the manuscripts for each edition, including Sung K. Ahn, Washington State University; Nicholas R. Farnum, California State University, Fullerton; Michael Martin, Stanford University; David Metcalf, University of Virginia; Richard Moreland, University of Pittsburgh; David Rogosa, Stanford University; David Viglienzoni, Cabrillo College; Vern C. Vincent, University of Texas—Pan American; and W. T. Woodard, University of Hawaii.

Minitab began as a student-oriented adaptation of the National Bureau of Standards' "Omnitab" system, originated by Joseph Hilsenrath and developed further by David Hogben and Sally Peavy. The name "Minitab" was conceived from that origin.

Contents

PART ONE
Overview of Minitab

1

Introduction to Minitab

Minitab is an easy-to-use general-purpose statistical computing system. It is a flexible and powerful tool, designed especially for those who have no previous experience with computers. In this chapter we give a brief introduction to Minitab. In the next chapter, we will discuss the various versions of Minitab in detail. In the remaining chapters, we will show how Minitab can be used to do various statistical analyses.

1.1 Versions of Minitab

Minitab runs on many computers. Several releases are currently in use:

Release 7, available for most computers

Release 8 for DOS computers

Release 8 for Macintosh computers

Release 9 for Windows

Release 9 for VAX/VMS and some UNIX workstations

We will base this book on the features in Release 7 because these are available on all releases currently in use.

Different versions of Minitab have different user interfaces, which determine how Minitab looks on your screen and how you tell Minitab which operations to perform. All versions allow you to type simple commands, and we will use commands in most of our examples. Release 8, both for DOS and for Macintosh, and Release 9 for Windows also have menus. On these versions you can use menus or commands, whichever you wish.

Release 8, for DOS and Macintosh, and Release 9 for Windows have a data editor. This is a simple spreadsheet-like capability that allows you to enter, view, and edit data. Commands can be used to enter, view, and edit data in all versions.

All versions have simple character-based graphs, and we will mostly use these. Releases 7 and 8 also have simple high-resolution graphs, and Release 9 has a sophisticated high-resolution graphics capability.

Finally, versions of Minitab with higher release numbers have more statistical commands. However, these are mostly advanced commands and will not be needed for this book.

Thus your first task is to find out what version of Minitab you will be using. Chapter 2 includes a separate section for each version of Minitab. After going through this chapter, read the material in Chapter 2 that corresponds to the version you will be using.

1.2   A Simple Example

Table 1.1 shows data for 11 women. Each woman took her resting pulse rate, then ran in place for one minute, then took her pulse again. In addition, whether or not the woman smokes regularly was recorded, using 1 = yes and 2 = no.

We want to calculate how much the pulse rate of each woman changed after exercise, and to find the average pulse rates for the 11 women before

Table 1.1   Pulse Rate Data

Pulse Rate Before Exercise	Pulse Rate After Exercise	Whether Woman Smokes
96	140	2
62	100	2
78	104	1
82	100	2
100	115	1
68	112	2
96	116	2
78	118	2
88	110	1
62	98	1
80	128	2

and after exercising and the average difference between them. Minitab commands that do this are shown in Exhibit 1.1.

Minitab stores data in a worksheet of rows and columns similar to a spreadsheet. The columns of the worksheet are designated as C1, C2, C3, C4, ... The first command,

```
READ   C1 C2 C3
```

tells Minitab to put the lines of data that follow into columns C1, C2, and C3. The command

```
END
```

signals the end of the data. The next command is

```
NAME C1 = 'Before' C2 = 'After' C3 = 'Smokes' C4 = 'Change'
```

The NAME command assigns names to the three columns already filled and to a fourth column we plan to create. We can use the command NAME any time we want to assign or change the name of a column.

The command

```
LET C4 = C2 - C1
```

Exhibit 1.1   Minitab Commands for the Pulse Rate Data

```
READ   C1 C2 C3
    96    140    2
    62    100    2
    78    104    1
    82    100    2
   100    115    1
    68    112    2
    96    116    2
    78    118    2
    88    110    1
    62     98    1
    80    128    2
END
NAME C1 = 'Before'    C2 = 'After'   C3 = 'Smokes' C4 = 'Change'
LET C4 = C2 - C1
PRINT C1 C2 C4 C3
DESCRIBE C1 C2 C4
```

Exhibit 1.2   Minitab's Worksheet After Using READ and LET

Before	*After*	*Smokes*	*Change*		
C1	*C2*	*C3*	*C4*	*C5*	*C6*
96	140	2	44		
62	100	2	38		
78	104	1	26		
82	100	2	18		
100	115	1	15		
68	112	2	44		
96	116	2	20		
78	118	2	40		
88	110	1	22		
62	98	1	36		
80	128	2	48		

calculates the changes. LET can do simple calculations like this and very complex calculations we will learn about later. Exhibit 1.2 shows Minitab's worksheet at this point.

Next we display all four columns, putting the column for smoking last, by using

```
PRINT C1 C2 C4 C3
```

The last command is

```
DESCRIBE C1 C2 C4
```

It calculates many descriptive statistics, including the averages for C1, C2, and C4.

Exhibit 1.3 shows a Minitab session using these commands. This is how your screen would look if you were to type the commands on your computer. Minitab uses two prompts. The first, MTB >, says Minitab is ready for a new command. If we type the READ command, Minitab asks for data with the DATA> prompt, and continues asking for data until we type END, at which point it tells us it has read 11 rows of data. Next, Minitab asks for another command. We type NAME and then LET. These display no output on the screen. When we use PRINT, Minitab prints the columns we requested, in the order we requested, using their names. Finally we use the command DESCRIBE to get many statistics.

Exhibit 1.3   Input and Output for the Commands in Exhibit 1.1

```
MTB > READ   C1 C2 C3
DATA>    96    140    2
DATA>    62    100    2
DATA>    78    104    1
DATA>    82    100    2
DATA>   100    115    1
DATA>    68    112    2
DATA>    96    116    2
DATA>    78    118    2
DATA>    88    110    1
DATA>    62     98    1
DATA>    80    128    2
DATA> END
      11 ROWS READ

MTB > NAME C1 = 'Before'   C2 = 'After'   C3 = 'Smokes'   C4 = 'Change'
MTB > LET C4 = C2-C1
MTB > PRINT C1 C2 C4 C3

 ROW  Before  After  Change  Smokes

   1      96    140      44       2
   2      62    100      38       2
   3      78    104      26       1
   4      82    100      18       2
   5     100    115      15       1
   6      68    112      44       2
   7      96    116      20       2
   8      78    118      40       2
   9      88    110      22       1
  10      62     98      36       1
  11      80    128      48       2

MTB > DESCRIBE C1 C2 C4

                N      MEAN    MEDIAN    TRMEAN     STDEV    SEMEAN
Before         11     80.91     80.00     80.89     13.31      4.01
After          11    112.82    112.00    111.44     12.83      3.87
Change         11     31.91     36.00     32.00     11.94      3.60

               MIN       MAX        Q1        Q3
Before       62.00    100.00     68.00     96.00
After        98.00    140.00    100.00    118.00
Change       15.00     48.00     20.00     44.00
```

Exercises

1-1   The DESCRIBE command in Exhibit 1.3 printed 10 values for each of C1, C2, and C4. These will all be discussed in Chapter 3. Now, however, see whether you can guess what any of them are (that is, try to define MEAN, MIN, and so on).

1-2   Interpret the output in Exhibit 1.3. On the average, did the pulse rates increase or decrease after exercise? Who had the greatest change? The smallest?

1-3   The following Minitab commands do some simple calculations.
```
READ C1 C2 C3
  15 28 30
  14 30 31
  16 30 34
  13 27 31
END
PRINT C1 C2 C3
LET C4=C1+C2+C3
READ C10 C11
  32 26
  34 27
  32 25
END
LET C12=C10-C11
PRINT C4 C10 C12
```

Pretend you are Minitab and carry out the commands. Keep a worksheet as you go along. What does the worksheet contain after each command? What is printed out?

1.3   A Brief Overview of Minitab

Minitab consists of a worksheet where data are stored and commands that operate on the data. In the worksheet, you can store columns of data and single-number constants. The columns are denoted by C1, C2, C3, C4, . . . and may have names. The stored constants are denoted by K1, K2, K3. . . Constants can have names in Release 9, but not in earlier releases. The total worksheet size and the number of columns and stored constants available to you depend on your particular computer. On some computers, the worksheet is enormous; on others, it is quite restricted. The total worksheet size available to you is displayed whenever you start Minitab.

When you want Minitab to analyze data, you type the appropriate commands. There are commands to read, edit, and print data, to do plots and histograms, to do arithmetic and transformations, and to do various statistical analyses such as tests, regression, and analysis of variance. A brief summary of all Minitab commands is given in Appendix B.

Some Basic Rules

1. Every command starts with a command name, such as READ or DESCRIBE. In most commands, this name is followed by arguments. An argument is either a column number (such as C10), a column name (such as Change), a constant (such as 75.34), a stored constant (such as K15), or the name of a computer file.

2. Only the first four letters of the command name and the arguments, which must be in the proper order, are used by Minitab. Other text may be added for annotation if you wish. However, we recommend that you use only letters and commas for this extra text. Never use numbers or symbols (such as ; : − * & or +), because these are used in special ways by Minitab.

   Using these rules, the command

   ```
   READ C1 C2 C3
   ```

   could be written as

   ```
   READ the following data into C1 C2 and C3
   ```

   We could not have written

   ```
   READ 1st pulse into C1, 2nd into C2 and smokes into C3
   ```

   because Minitab would try to use the 1 in 1st and the 2 in 2nd as arguments.

3. You may add any annotation you wish if you first type the symbol #. Minitab ignores everything on the line after it sees the #. A # can even be the first character on a line. For example,

   ```
   # Now I will enter my data
   READ  C1 C2 C3  # 1st pulse is in C1 and 2nd is in C2
   ```

4. A list of consecutive columns or stored constants may be abbreviated using a dash. For example, you could use

Minitab Command Box Format

Minitab commands are described throughout this book in boxes like this one. In these descriptions, the symbol C can be replaced by any column name or number, K by any constant, and E by either a column or constant. Arguments in square brackets are optional. Any extra text used to help explain the commands is written in lowercase. Required text is in boldface. Here are some examples.

PRINT the data in C

says you may print the data in a column. To print the data in column C18, you could replace C by C18, omit the extra text, and type

```
PRINT C18
```

PRINT E...E

says you may display any list of columns and constants. For example,

```
PRINT C1 C4 C2 K1-K4
```

SUM C [store in K]

calculates the sum of the numbers in the column. Storage in K is optional. If you type

```
SUM C10
```

Minitab calculates and prints the sum of all the numbers in C10. If you type

```
SUM C10 K1
```

Minitab calculates, prints, and stores the sum in K1.

In menu versions of Minitab, you can use the menus to execute a command instead of typing it. At the bottom of this box, we will indicate how to find a command in the menus. Most commands are found in the same location in all menu versions of Minitab. In such cases, you will see just one line at the bottom of a command box. In a few cases, there are differences among the versions, and you will see separate references for the different versions. Chapter 2 will explain this more fully.

For example, to find the DESCRIBE command, you look under the Stat menu for the submenu Basic Statistics, then under this submenu for the entry Descriptive Statistics. Thus at the bottom of the box for DESCRIBE, you will see

Stat > Basic Statistics > Descriptive Statistics

```
READ C2-C4
```

instead of

```
READ C2 C3 C4
```

5. If you type a number, do not use commas within the number. Thus, type 1041 instead of 1,041.

6. Each command must start on a new line. You need not start in the first space, however. If the entire command will not fit on one line, end the first line with the continuation symbol, &, and continue the command on the next line. For example,

```
PRINT C2, C4-C20, C25, C26, C30, C33 &
C35-C40, C42, C50
```

7. You may use uppercase or lowercase letters for command names and arguments. Thus the following are equivalent:

```
READ C1 C2 C3
read c1 c2 c3
ReAd C1 c2 C3
```

How to NAME Columns

Any column may be given a name. The name serves two purposes: First, the column may be referenced by its name. It's often easier to remember the name of a variable than the number of a column. Second, all output is labeled with the name. Many users find that naming columns takes a little extra time, but pays off in the long run by making output easier to read.

Stored Constants

Any calculation that results in a single number can store that number in a stored constant. The stored constant may then be used in place of the number in any command. SUM is a command that results in a one-number answer. If C1 contains the numbers 5, 6, and 2, then SUM C1 calculates $5 + 6 + 2 = 13$. Since the answer is one number, we can store it in a constant. The LET commands can do arithmetic using both columns and constants. Here is an example:

NAME   C = 'name'   ...   C = 'name'

A name may be from one to eight characters long. Any characters may be used, with two exceptions: A name may not begin or end with a blank, and a name may not contain the symbol ' (called an apostrophe or single quote).

Either the column number or its name may be used in any succeeding command. When a column name is used, it must be enclosed in single quotes (apostrophes). When you type a name in a command, Minitab does not distinguish between uppercase and lowercase. Thus PRINT 'Height' and PRINT 'height' are equivalent. On printed output, Minitab uses the uppercase and lowercase letters you used when you named the column. The name of a column can be changed by using another NAME command. Here are some examples,

```
NAME C1 = 'Height' C2 = 'L + W'
PRINT 'Height' C2
NAME C1 = 'Ht'
```

In menu versions, use the data editor to name columns.

```
READ C1
   5
   6
   2
END
SUM C1  K1
LET K2  = 4
LET K3  = K1+K2
LET C2  = K1+C1
```

Subcommands

Some Minitab commands have subcommands. These allow more control over the way the command works. The DESCRIBE command we used in Exhibit 1.2 has a subcommand, BY. We could use

```
DESCRIBE   'Change';
  BY 'Smokes'.
```

This calculates descriptive statistics for Change, separately for those who smoke and those who do not. Notice the punctuation: The main command ends in a semicolon. This tells Minitab there will be a subcommand on the next line. The subcommand ends with a period. The period tells Minitab you are finished.

In general, if you use subcommands, end the main command line with a semicolon. End each subcommand line with a semicolon until you are finished typing subcommands. End the last subcommand line with a period. Minitab waits until it sees the period to start doing the calculations. If you accidentally forget to end the last subcommand with a period, you can put the period on the next line, all by itself.

Always begin each subcommand on a new line. You can type the subcommand anywhere on the line. We often indent subcommands for clarity.

Occasionally you will discover an error after entering a subcommand. In this case, type ABORT as the next subcommand. This will cancel the whole command. You can then start over again with the main command.

Missing-Data Code

Many data sets are missing one or more observations. When you enter the data, type an asterisk (*) in place of the missing value. For example,

```
READ   C1   C2
  28     5.6
  24     5.2
  25     *
  24     51
END
```

If you ask Minitab to do a calculation that is impossible, such as taking the square root of a negative number, Mintab sets the answer to *. All Minitab commands automatically take asterisks into account when they do an analysis.

If after you have entered data, you discover a value that is clearly wrong, you can change that value to *. In Chapter 2 we will show you how.

Numeric and Alpha Data

Worksheet columns can store two types of data: numbers (or numeric data) and words (or alpha data). Most of your work will be done with numeric data. Most Minitab commands work only with numeric data, so you may not use alpha data at all.

You can input alpha data with the READ command (and with some other commands), but you must use the special subcommand FORMAT (see Section 17.5 if you ever want to do this). If a Minitab saved worksheet already has alpha data in it, you can easily bring it into Minitab (saved worksheets are described in Chapter 2). In Releases 8 and 9 Windows, you can input alpha data very easily using the data editor, described in Chapter 2.

Note: If you try to store numeric data in a column that has alpha data in it, you will probably get an error message. In that case, ERASE the column containing alpha data first, then store your numbers (the ERASE command is described in Section 1.4).

1.4   Some Basic Minitab Capabilities

This section introduces the basic capabilities of Minitab and shows you how to write simple Minitab programs of your own. More information on basic capabilities is given in Chapter 2.

Inputting Data

The commands READ and SET let you type data into Minitab's worksheet. In most examples in this book, we will use READ and SET to enter data. If you are using Release 8 or Release 9 Windows, you can also enter data using Minitab's data editor (see Chapter 2). In most cases, especially if you are not a very good typist, you will probably find it easier to enter data by using the data editor than by using READ and SET.

Both READ and SET allow you to enter data. The difference between them is the way you type the data: READ expects data one row at a time; SET expects it one column at a time. When you enter data, use whichever command you prefer. Sometimes SET will be easier; sometimes READ will be.

SET has special features that allow you to easily enter data that follow a pattern. You can use a colon to abbreviate a list of numbers. For example, to enter the integers from 5 to 12 into C2, type

```
SET   C2
   5:12
END
```

A slash is used to indicate an increment other than 1. For example,

```
SET   C5
   1:3/.2
END
```

puts the numbers 1, 1.2, 1.4, 1.6, 1.8, 2.0, 2.2, 2.4, 2.6, 2.8, 3 into C5.

Parentheses can be used to repeat a list of numbers. For example,

READ the following data into **C...C**

This command is followed by lines of data. Each line is put into one row of the worksheet.

You may select any columns for storage. The following example puts data into C2 and C5:

```
READ  C2   C5
   3   71
   5   78
   4   81
END
```

Numbers can be put anywhere on a line as long as they are in the correct order.

File > Import ASCII Data

END of data

Type this command after typing the data for READ or SET.
END tells Minitab you are finished typing data.

SET the following data into **C**

This command is followed by lines of data. You can put as many numbers as you want on one line, and you can use as many lines as you want. All the numbers will go into the same column. The following example puts 11 numbers into C2:

```
SET C2
   140   100   104   100   115   112
   116   118   110    98   128
END
```

Numbers can be put anywhere on a line as long as they are in the correct order. Numbers must be separated by blanks or commas.

Edit > Set Patterned Data

```
SET C2
  3(1, 2)   2(9)
END
```

repeats the list 1, 2 three times, then adds two 9s. Thus, C2 contains 1, 2, 1, 2, 1, 2, 9, 9. The number before the parentheses is called a repeat factor. Do not put any space between it and the open parenthesis. If you use a repeat factor after the close parenthesis, each number is repeated. For example,

```
SET C2
  (1, 2)4
END
```

puts 1, 1, 1, 1, 2, 2, 2, 2 into C2.

Printing Data on Your Screen

If you want to see what is in your worksheet, or if you want to include a listing of your data with your output, you can use the PRINT command.

PRINT the data in E...E

This command may be used to display columns or constants that have been stored in the worksheet. You may display any columns or constants you want in any order you want. For example,

```
PRINT C1-C4 C10
PRINT K2 K4 K6
PRINT C20
PRINT 'Height'
```

Release 8: **Edit > Display Data**
Windows: **File > Display Data**

Doing Arithmetic with LET

LET can do basic arithmetic as well as calculate very complicated expressions. We will give a brief introduction to its capabilities here.

LET uses the following symbols for arithmetic:

+   add
−   subtract
*   multiply
/   divide
**   raise to a power (exponentiation)

Here is an example with several LET commands:

```
LET K1 = 3
LET K2 = 5*13
LET K3 = K1+K2+4
SET C1
 4 6 5 2
END
LET C2 = 2*C1
LET C3 = K1*C1
LET C4 = C2+1
LET C5 = C3+C4
LET C6 = C1**2
```

After these commands, K1 = 3, K2 = 65, and K3 = 72. The following table shows what C1–C6 contain.

C1	C2	C3	C4	C5	C6
4	8	12	9	21	16
6	12	18	13	31	36
5	10	15	11	26	25
2	4	6	5	11	4

Parentheses may be used for grouping. For example:

```
READ C1 C2
 8 1
 6 3
 4 4
END
LET C3 = 10*(C1+C2)
LET C4 = 10*C1+C2
LET C5 = C1/C2+1
LET C6 = C1/(C2+1)
```

At this point the worksheet contains

C1	C2	C3	C4	C5	C6
8	1	90	81	9	4.0
6	3	90	63	3	1.5
4	4	80	44	2	0.8

Note that the expressions for C3 and C4 look similar, as do those for C5 and C6, but the results are not the same. LET follows the usual precedence rules of arithmetic; that is, operations within parentheses are always performed first, then **, then * and /, and finally + and −. If you are not sure of the sequence of a calculation, you can always use parentheses to make sure it is done right.

LET can also be used to calculate many functions. For example, LET C2 = SQRT(C1) will calculate the square root of every number in C1 and store the answers in C2.

LET can be used to calculate single-number answers. For example, LET K1 = SUM(C1) will calculate the sum of all the numbers in C1 and store the answer in K1. As another example, LET C3 = C1 − MEAN(C1) will first calculate the mean of C1, then subtract that mean from each number in C1.

LET can access one number in a column. For example, C5(4) = 72.1 puts 72.1 into the fourth row of C5, and LET K1 = C2(5) puts the number in row 5 of C2 into K1.

LET can also use comparison operators and Boolean operators. If you need to use any of these features, you can use Minitab's on-line help facility (see Chapter 2 to find out about Help) or read the Minitab manual.

LET E = arithmetic expression

An arithmetic expression can be made up of columns and constants, arithmetic symbols (+ − * / **), and parentheses.

LET can calculate the functions ABSOLUTE, SQRT, LOGTEN, LOGE, EXPONENTIATE, ANTILOG, ROUND, SIN, COS, TAN, ASIN, ACOS, ATAN, SIGNS, NSCORES, PARSUMS, and PARPRODUCTS.

LET can calculate the one-number answers COUNT, N, NMISS, SUM, MEAN, STDEV, MEDIAN, MINIMUM, MAXIMUM, SSQ, SORT, RANK, and LAG.

LET can calculate the comparisons <, >, <=, >=, =, and ~= and the Booleans AND, OR, and NOT. Answers that are true are set equal to 1; those that are false are set equal to 0.

Unlike other Minitab commands, no extra text may be used on a LET line (except after a #, as discussed in Section 1.3).

Release 8: **Calc > Functions and Statistics > General Expressions**
Windows: **Calc > Mathematical Expressions**

Subsetting Data with COPY

The COPY command, along with its subcommand, USE, allows you to select a subset of data. Here is a simple example. Suppose we've entered sex (coded 1 = male, 2 = female), height, and weight for seven people into C1–C3, which we have named HT, WT, and SEX. Then the following commands put the data for women into C12–C13:

```
COPY C2 C3  C12 C13;
  USE 'SEX' = 2.
NAME C12 = 'HT.F' C13 = 'WT.F'
```

All rows of C2 and C3 where SEX = 2 are copied into C12 and C13. Now the worksheet contains:

C1	C2	C3		C12	C13
SEX	HT	WT		HT.F	WT.F
2	66	130		66	130
1	70	155		64	125
2	64	125		65	115
2	65	115		63	108
2	63	108			
1	66	145			
1	69	160			

In general, you may specify any values for the variable on USE. For example, to copy all rows of C1–C3 that have any of the values 64, 65, or 69 for HT, use

```
COPY C1-C3  C21-C23;
  USE 'HT' = 64 65 69.
```

You can use a colon to indicate a range of values. For example, to copy the data for people who are 64 inches tall or shorter, use

```
COPY C1-C3  C31-C33;
  USE 'HT' = 0:64.
```

You can subset in place. If you plan to use the full data set again, first save it in a file. (Chapter 2 shows how to save data.) Then create the subset and store it in the original columns. For example:

```
COPY C1-C3  C1-C3;
  USE 'SEX' = 2.
```

COPY C...C into C...C

USE rows where C = K...K

If you do not use the subcommand, COPY makes a copy of all the data in the first list of columns.

USE allows you to copy selected rows, based on the values in a column. An interval can be abbreviated with a colon. Thus, USE C1 = 5:8, 20 says to copy all rows in which C1 is either in the range of 5 through 8 or equal to 20.

COPY has several other features. These are explained in Section 17.1.

Manip > Copy Columns, use buttons for USE

Managing the Worksheet

Occasionally you may forget what you have in the worksheet. The INFO command can help. Users of Release 8 for the Macintosh and Release 9 for Windows can also use the INFO Window.

The ERASE command allows you to erase columns and constants you no longer need. Sometimes you may need to do this because you run out of space. On other occasions, you will want to do it to reduce unnecessary clutter.

INFORMATION on status of the worksheet

Prints a list of all columns used, the number of values in each, the name of each (if they have been named), and a list of all stored constants used.

Release 8: **Edit > Get Worksheet Info**
Release 9 Windows: Use the info window or type this command

ERASE E...E

You may erase any combination of columns and stored constants. For example,

```
ERASE C2 C5-C9 K1-K7 C20
```

Release 8: **Calc > Erase Variables**
Windows: **Manip > Erase Variables**

Ending a Session

STOP

Type STOP when you are finished using Minitab. After typing STOP, you
will be in your computer's operating system.

Release 8 Macintosh: **File > Quit**
Release 8 DOS and Release 9 Windows: **File > Exit**

Exercises

1-4   Find the errors in each of the following (there may be more than one error
in each part).

(a) ```
     READ C1-C3
        5   2
        6   14
        2   18
     END
     ```

(b)  ```
     READ C2 C3
         983 2
       1,102 5
         992 7
     END
     ```

(c) ```
     READ C2 C5
        2   16
        4   12
        5   11
        6   12
     END
     PRNT C2
     DESCRIBE THE 4 NUMBERS IN C2
     LET C2 = C2 + C5
     LET C10 = C5 + C6
     LET C11 = C1+C2+C5
     ```

(d)  ```
     READ INTO C1 AND C2
        5.6    23.0
        5.5    23.1
        5.4    23.3
     END
     DESCRIBE DATA IN COLUMN 1
     LET C5=C1+C2 LET C6=C1-C2
     ```

1-5 (a) In the following commands, SET was used to enter data. Show how
          to use READ to do the same job.
```
SET C1
   2   3   5   7   11   15
END
SET C2
   11   18   9   16   12   10
END
SET C3
   1:6
END
SET C4
   (1 2)3
END
```
     (b) In the following commands, READ was used to enter data. Show how
          to use SET to do the same job.
```
READ C1-C4
   5   6   3   1
   1   2   8   2
   5   1   1   3
   6   2   3   4
END
```

1-6 What values does C1 have after the following SET?
```
SET C1
   980   992   1,140   801
   963   1,002
END
```

1-7 What's wrong with each of the following sets of commands?
     (a)
```
SET C1-C2
   5    2
   13   6
END
```
     (b) `LET C3=C1+C2 THIS IS TOTAL SCORE`
     (c) `NAME C1=SCORE1 C2=SCORE2`
     (d)
```
READ C1 C2
   5   3
   6   1
END
LET K1=C1+C2
```
     (e) `NAME C4='EDUCATION'   C2='1ST JOB'`
     (f) `PRINT LENGTH`
     (g) `NAME C1 = 'length' C2 = 'Width   C3 = 'Temperature'`

1-8   Write a Minitab program that produces a temperature table in degrees Fahr-
       enheit (from $20°$ to $80°$ in steps of $5°$) and the equivalent temperatures in
       degrees Celsius. To convert from Fahrenheit to Celsius, use the formula
       Celsius = (Fahrenheit − 32) * (5/9).

1-9   What will the worksheet contain at the end of this program?
```
READ C1 C2 C3
     1    24    132
     1    22    141
     3    29    153
     4    22    114
     3    35    155
     4    33    130
END
COPY C1 C11;
   USE 3 5.
COPY C1-C3   C21-C23;
   USE C1 25:40.
```

1-10   Suppose C1 is named x and contains the eleven integers $0, 1, 2, \ldots, 10$.
       Show how to use LET to calculate each of the following formulas. Will any
       of the values in y be set to the missing value code?
       (a) $y = 2x + 1/3$
       (b) $y = x^2 / 5$
       (c) $y = (x + 5)^2 / (2x − 1)$
       (d) $y = \sqrt{x^3 − 1}$
       (e) $y = \dfrac{x^2 + 5}{x − 3}$

2

Versions of Minitab

This chapter gives a brief introduction to five versions of Minitab. Read the section that applies to the version you will be using. To find out more about a version, read the manual for that version and look at the appropriate on-line help information.

Section 2.1 describes commands that are available in all versions of Minitab. Read this section if you are using Release 7 on any computer, or Release 9 on a VAX or UNIX workstation. Section 2.5 provides additional information for Release 9 VAX and UNIX users, so read this section if you are using one of these computers. If you are using Release 8 or Release 9 Windows, you need read only the section that applies to your computer. However, if you want to learn more about commands, you might read Section 2.1 as well.

Versions of Minitab that have menus and a data editor often have easier ways to do the operations described in Section 2.1. However, if you are using a menu version of Minitab and want to learn about these commands, read Section 2.1. Otherwise, just read the section that applies to your version.

This chapter describes the versions of Minitab that are available as this book goes to press. Minitab, Inc. periodically releases new versions of its software, so you may be using a more recent version. Changes usually consist of added features and perhaps some changes to the menu interfaces. See the on-line help and the manuals that came with your software for the most up-to-date information.

We assume you know the basics of the computer you are using. In particular, you should know how to start an application program such as Minitab, and how to use files and directories. If you have difficulties with these, see the user documentation for your computer or ask your instructor or system administrator for help.

2.1   Command Versions

Starting Minitab

Start your computer or terminal if necessary. Then type **minitab** and press Enter (or Return on some keyboards). The opening screen displays some information about Minitab, followed by the prompt MTB >. You can now type Minitab commands, as we did in Chapter 1.

Help on Minitab

Information about Minitab is stored in the computer. If you forget how to use a command or subcommand, or need general information, you can ask Minitab for help. For example, to find out about the command DESCRIBE, type

```
HELP DESCRIBE
```

In general, to get help on a command, type HELP followed by the command name. To get help on a subcommand, type HELP followed by the command and then the subcommand. For example, to get help on the BY subcommand of DESCRIBE, type

```
HELP DESCRIBE BY
```

If you type HELP OVERVIEW, Minitab gives you a list of topics that describe various aspects of Minitab. If you type HELP COMMANDS, Minitab gives you a summary of Minitab's commands. If you type HELP HELP, Minitab gives you general information on how to use the Help system.

Correcting Data in the Worksheet

Minitab has three commands that are useful for correcting numbers you have already entered in the worksheet: LET, DELETE, and INSERT.

Exhibit 2.1 shows a worksheet as it should be and as it was actually typed with two errors: one number is wrong and one line is omitted. The number in row 3 of C1 should be 1.3, not 3.1. To correct it, we use a special feature of LET.

```
LET C1(3) = 1.3
```

Exhibit 2.1   Correcting Data in the Worksheet

Worksheet Should Be		But It Was Typed As		After LET		After INSERT	
C1	C2	C1	C2	C1	C2	C1	C2
1.5	102	1.5	102	1.5	102	1.5	102
1.7	106	1.7	106	1.7	106	1.7	106
1.3	120	3.1	120	1.3	120	1.3	120
1.4	118	1.4	118	1.4	118	1.4	118
1.5	101	1.5	101	1.5	101	1.5	101
2.1	130	1.1	124	1.1	124	2.1	130
1.1	124					1.1	124

says to change the number in row 3 of C1 to 1.3. Exhibit 2.1 shows the worksheet after LET was used.

To correct the omission of the line, we need to insert a row containing the numbers 2.1 and 130 between rows 5 and 6. To insert this line, we use

```
INSERT 5 6   C1 C2
   2.1 130
END
```

Sometimes you enter data into the worksheet, then discover that a value is wrong, and have no way to find out the correct value. You can change this value to *, using LET. For example, if the wrong value is the fifth number in C18, use

```
LET C18(5) = '*'
```

You must enclose the asterisk in single quotes (apostrophes) when using it with LET.

DELETE rows K...K of C...C

Deletes the indicated rows of data and closes up the worksheet. You may abbreviate a list of consecutive rows by using a colon. For example,

```
DELETE   1:10 25:30 C1
```

deletes rows 1 through 10 and rows 25 through 30 from C1.

Calc > Delete Rows, or use the data editor to delete rows

INSERT between rows **K** and **K** of **C...C**

This command inserts rows of data into the worksheet. The data are typed following the INSERT line. The row numbers K and K must be consecutive integers. (The second K is redundant, but helps avoid errors.)

To add rows of data at the tops of columns, use

```
INSERT 0 1  C...C
```

To add rows of data to the ends of columns, omit the row numbers and use

```
INSERT C...C
```

If you insert data into just one column, then you can string the data across the data lines, as you can with the SET command, and you can use the patterned-data feature of SET.

In menu versions, you can use the data editor to insert data.

Saved Worksheets

Once you have entered data into Minitab, you usually want to store it in a computer file so you will have it available for use in another session on another day. The most convenient type of file for use with Minitab is a Minitab saved worksheet, created by the SAVE command. The file will contain all columns, stored constants, and column names you have. This file can very easily be input by Minitab's RETRIEVE command.

Many of the data sets used in this book are on your computer as Minitab saved worksheets. The columns have already been given names. You can access these data sets by using Minitab's RETRIEVE command.

Using files does require some knowledge of your computer's operating system. If you have any difficulty using SAVE and RETRIEVE, check with someone who knows your local computer system.

Only Minitab can use Minitab saved worksheets. You cannot use your computer's editor to print or modify them and you cannot use them with other programs. If you want to transfer data in a Minitab worksheet to another program, use Minitab's WRITE command. If you want to transfer a Minitab saved worksheet from one type of computer to another, you can use the PORTABLE subcommands SAVE and RETRIEVE.

SAVE the worksheet in 'filename'

PORTABLE

Puts the entire Minitab worksheet into a computer file. The file will contain all columns, stored constants, and column names from the worksheet. This file can be input by Minitab's RETRIEVE command. The file name must be enclosed in single quotes (apostrophes). Minitab automatically adds the file extension .MTW, unless you supply an extension.

If you want to save your worksheet on a floppy disk or in a directory other than your current directory, include the full path name within the single quotes.

The subcommand PORTABLE says to use a format that will allow you to move the saved worksheet to Minitab on a different type of machine, for example from a VAX computer to a Macintosh.

File > Save Worksheet As

RETRIEVE the worksheet stored in 'filename'

PORTABLE

Inputs data from a Minitab saved worksheet. After using RETRIEVE, the worksheet will contain the same columns, stored constants, and column names it had when SAVE was used to create the file. Anything you have in the worksheet when you type RETRIEVE is erased before the saved worksheet is brought in.

If you want to retrieve a worksheet that is not in your current directory, include the full path name within the single quotes.

The subcommand PORTABLE says the worksheet was saved in a portable format.

File > Open Worksheet

ASCII Data Files

You can transfer data between Minitab and other applications by using a data, or ASCII, file. Almost every application—every editor, spreadsheet, and word processor—can input and output data in ASCII format.

To import an ASCII data file into Minitab, use either READ, SET, or INSERT. For example, suppose the file myfile.DAT contains 10 lines of data, each with 4 numbers. To import the file into C11–C14, use

```
READ 'myfile' C11-C14
```

Each line of data will be put into one row of the worksheet, into the columns listed on READ. When the file extension is .DAT, you can leave it off and type just the filename, as we have done here.

You can save the columns of your current worksheet in an ASCII file using the WRITE command. This file will contain only columns of data. Stored constants and column names will be lost. Suppose your worksheet contains 20 rows of data in each of C1–C10 and you want to output C1–C5 and C9 to a new file called myfile2. You can use

```
WRITE 'myfile2' C1-C5 C9
```

READ   '**filename**' into **C...C**
SET   '**filename**' into **C**
INSERT '**filename**' between rows **K** and **K** of **C...C**

    FORMAT (**Fortran format statement**)

These three commands input an ASCII data file. They are the same commands described in Chapter 1, except that now the data set comes from the specified file rather than being typed after the command. The name of the file must be enclosed in single quotes.

If the ASCII file contains only numbers, with at least one space or comma between numbers, then you will not need the FORMAT subcommand. This is usually the case. However, if the file contains alpha data, if there are no spaces or commas between numbers, if blanks are used for missing values, if two or more lines of data are to go into one row of Minitab's worksheet, or if the file has any other unusual properties, you will need to use the FORMAT subcommand to describe the data format. Formats are described in Section 17.5.

File > Import ASCII Data

WRITE 'filename' C...C

FORMAT (Fortran format statement)

WRITE stores the specified columns in the specified file. Minitab adjusts the format to make the data as compact as possible. If the specified columns do not all fit on one line, Minitab puts the continuation symbol, &, at the end of the line and continues the data on the next line. If the columns are of different lengths, Minitab makes them all the same length by adding missing-value codes (*) to the shorter columns.

Minitab automatically adds the file extension .DAT if you do not specify an extension.

File > Export ASCII Data

Using files does require some knowledge of your computer's operating system. If you want to WRITE a file, you must know what a legal file name can be on your computer. You also need to know where to store the file. If you want to READ a file, you must know where the file is located on your computer. All this depends on what type of computer you are using. If you have any difficulty using READ and WRITE, check with someone who knows your local computer system.

Saving and Printing the Minitab Session

Minitab does not automatically save your commands and output; once you have seen them on the screen, they are gone. It is therefore a good practice to begin each session by starting a Minitab outfile with the OUTFILE command. All subsequent commands you type, as well as output Minitab prints, are automatically put into this file. If you want a copy of your data set on paper, use Minitab's PRINT command to display the data on the screen and thereby include it in your outfile.

After you leave Minitab, you can use operating system commands to view or print the file, or you can import the file into an editor or word processor to create a report. Minitab saves everything—input commands, error messages, output you really don't want anymore—in an outfile, so you might want to edit the outfile before you print it out.

Suppose you want to save your output in a file called myoutput. Type

```
OUTFILE 'myoutput'
```

OUTFILE 'filename'
NOOUTFILE

When you type OUTFILE, all output on your screen is saved in the file you specify. This continues until you type NOOUTFILE, or until you exit Minitab. If you specify a file name on OUTFILE that has already been used for output, on most computers the new output is appended. On a VAX/VMS computer, a new version of the file is created.

File > Other Files > Start Recording Session, for OUTFILE
File > Other Files > Stop Recording Session, for NOOUTFILE

From this point on, all commands and output that appear on the screen are stored in the file myoutput.LIS. Minitab adds the extension .LIS for these files. When you want to stop recording your session, type NOOUTFILE. If at a later time you want to record more of your session, you can append it to this file by typing OUTFILE with the same file name. If you want to put your output into a new file, then use a new file name.

Graphs

There are two types of graphs. Character graphs appear along with printed output. High-resolution graphs appear on a separate screen. For most analyses, character graphs are good enough, although they do not look as pretty as the high-resolution graphs and lack a few of their options.

In order to use high-resolution graphs, you must have a graphics terminal that Minitab supports, and you must tell Minitab what type of terminal you have with the command GOPTIONS. For example,

```
GOPTIONS;
  DEVICE = 'TEK4010'.
```

tells Minitab you have a Tektronics 4010 terminal, or a terminal that emulates a Tektronics 4010. From this point on, Minitab formats graphs for this terminal. Some of the more common devices Minitab supports are:

Screens: TEK4010, TEK4014, TEK4015, TEK4105, TEK4112, VT125, VT240
Printers: DECLN03, HPLASERP for the HP Laser Jet II
Plotters: HP7470A, HP7475A, HP7550A

Let's look at Minitab's histogram. Retrieve the TREES data set. To get a character histogram of C1, type HISTOGRAM C1. Output is shown in Exhibit 2.2. If you want a high-resolution histogram, type GHISTOGRAM C1. (Session commands for high-resolution graphs all start with G.) Output is in Exhibit 2.3. Press [Enter] to return to typing commands.

If you want to print a graph, you must follow three steps: First, use the

Exhibit 2.2 Character Histogram of C1, Using Trees Data

```
MTB > HISTOGRAM C1

Histogram of DIAMETER    N = 31

Midpoint    Count
       8       1   *
       9       2   **
      10       0
      11      10   **********
      12       2   **
      13       3   ***
      14       4   ****
      15       1   *
      16       2   **
      17       1   *
      18       4   ****
      19       0
      20       0
      21       1   *
```

Exhibit 2.3 High-Resolution Histogram of C1, Using Trees Data

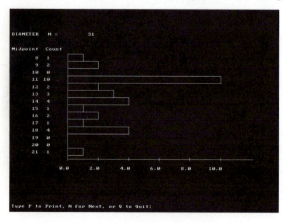

GOPTIONS command to specify the device on which you will print the graph. For example,

```
GOPTIONS;
  DEVICE = 'HPLASERP'.
```

Then type the graph command with the subcommand FILE to save the graph in a file. For example,

```
GHISTOGRAM C1;
  FILE 'myfile'.
```

Notice that the histogram does not appear on your screen. Instead, it is saved in a file named myfile.PLT. Last, exit Minitab and use your operating system's print command to print the file on the printer.

To send a graph to a plotter, you must have a plotter that is cabled directly to your terminal. See the MINITAB reference manual for more information.

If you want to see high-resolution graphs on your screen again, you must respecify your screen device with GOPTIONS.

2.2   Release 8 DOS

Starting Minitab

Start your computer. To start Minitab, type **minitab**. If Minitab is not in your path, type **cd\minitab** and press Enter (or go to the appropriate subdirectory). Then type **minitab**.

You can use Release 8 DOS as if it were a command-only version of Minitab. Start Minitab and you will see what is called the Session window, shown in Exhibit 2.4. You can type all of Minitab's commands here.

In addition to the Session window, the interface for Release 8 DOS includes a Data window (Minitab calls this the Data screen) and various menus and dialog boxes. Release 8 uses many of the same conventions that are found in Microsoft's Windows interface, so if you later use Minitab or any other program on Windows, you will already be familiar with the Windows conventions.

Session Window

The Session window is where you can type Minitab commands and where Minitab displays output. It has a menu bar on top and the usual Minitab prompt, MTB >. If you want to type a command, type it after the MTB > prompt and press Enter, and the command will be executed.

Exhibit 2.4   Opening Screen for Minitab

Exhibit 2.5   Session Window with READ, NAME, and DESCRIBE Commands

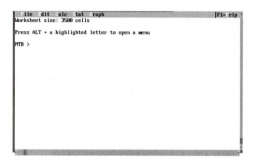

In Exhibit 2.5, we use the READ command to enter seven rows of data into C1–C3, name the columns, and DESCRIBE C1 C2. The Session window scrolls as more output goes into it. You can scroll up and down in the window by using the [PgUp] and [PgDn] keys and the up-arrow and down-arrow keys.

You can also use a mouse. If your mouse has more than one button, use only the left button in Minitab. The thin strip on the right side of the window is called a scroll bar. Position the mouse on the up arrow located at the top of this strip. Hold down the left mouse key and the window will scroll up. Do the same on the down arrow located at the bottom of this strip and the window will scroll down. Move the small rectangle located inside the strip and the window will scroll to the spot you choose. Click in the strip and the window will move one page up or down, depending on whether you click above or below the small rectangle.

The Session window is quite limited in size. The number of screens you can scroll back through depends on how much is on each line. Usually it is between 5 and 15 screens. If you want to save more of your output, you can create an outfile. (See "Saving and Printing the Session Window," later in this section.)

Editing the Session Window. You can use the Edit menu to copy, edit, and paste commands in the Session window. First scroll so that the complete block is visible. Then highlight the block. There are two ways to do this:

Keyboard. Move the cursor to the beginning of the block. While holding down the [Shift] key, move to the end of the block.

Exhibit 2.6   File Menu with Submenu

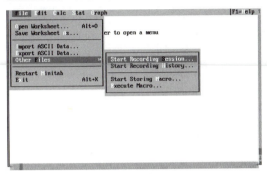

Mouse. Point to the beginning of the block, press the mouse button and hold it down, drag to the end of the block, and release the mouse button.

Choose Edit, then Copy Commands Only, to copy the block. Choose Edit, then Paste, to paste it at the Minitab prompt. Minitab does not read the pasted text until you press ⏎Enter, so you have a chance to edit it first. You cannot edit any of Minitab's output.

Menus

Most of Minitab's commands can be executed through menus and dialog boxes as well as by typing them in the Session window. The menu bar at the top shows the menus: File, Edit, Calc, Stat, Graph, and Help.

To open a menu, press ⎇Alt and the highlighted letter in the menu name. For example, to open the File menu, press ⎇Alt in combination with the key for the letter F. If you have a mouse, you can also click on the word File. In either case, a list drops down. To access Help, you can also press ⎘F1. Exhibit 2.6 shows the File menu and a submenu.

Look at the items in the File menu. Most items end with an ellipsis (...). If you choose one of these, a dialog box will appear. One item, Other Files, ends in >>. If you choose this, a second list, called a submenu, will appear. This submenu is shown in Exhibit 2.6. Two items in the File menu do not end in an ellipsis or in >>. If you choose one of these, Minitab will execute the corresponding Session command immediately. For example, if you choose Exit, Minitab will exit (the corresponding Session command is STOP). A few menu entries have keyboard shortcuts, shown to the right of the entry. For example, Open Worksheet is followed by Alt+O. Any time you type ⎇Alt in combination with the letter O, you will get the Open Worksheet dialog box directly.

Exhibit 2.7   Dialog Box for DESCRIBE

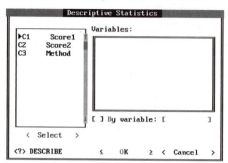

There are three ways you can choose an item in a menu:

You can use the arrow key to move to it, then press Enter.

You can press the highlighted letter. For example, to choose Save Work-sheet As, press the A key (no need to use Alt here).

You can click on the entry with a mouse.

Pressing Esc will always take you back: from a dialog box to the entry that opened the dialog, from a submenu to the main menu it came from, and from a main menu to no opened menus.

Dialog Boxes

Suppose you want to use the menus to execute the Minitab command DE-SCRIBE. First open the Stat menu (press Alt +S or use your mouse). Then open the Basic Statistics submenu (press the down arrow once, then press Enter, or press B, or use your mouse). Finally, open the Descriptive Statistics entry (press Enter, or press D, or use your mouse). This gives the dialog box shown in Exhibit 2.7.

Note: In most of this book we will abbreviate a menu/submenu selection in the following style: **Stat > Basic Statistics > Descriptive Statistics**.

Let's look at the parts of this dialog box. The box on the left lists all the worksheet columns you currently have. It is called a list box. The button under this box, Select, is used to select variables. The box on the right is a text box, where you enter the variables (columns) you want to use in DESCRIBE. The entry "By variable" is an option. It corresponds to the BY subcommand of DESCRIBE. The last line has three buttons. The first is a

?. Choose this, and you will get help on this dialog box. The word next to the ? is the name of the corresponding Minitab Session command. This is just for your information; it does not do anything. Next is the OK button. Select this and the command is executed. Last is the Cancel button. Select this and the command is canceled.

Notice that the Select button is "grayed out." That is because you cannot use it at this point. An item can be "grayed out" for any number of reasons. If it is, you cannot use it until you make some change in the dialog box. In this case, you need to highlight some variables.

Moving Around. You can move from one part of the dialog box to another with the [Tab] key. Pressing [Tab] alone moves you forward through the items; pressing [Shift]+[Tab] moves you backward. To move among items within a group—for example, the different column statistics shown in Exhibit 2.8—use the arrow keys.

If an item has a highlighted letter, you can go directly there by using [Alt] plus the letter. When the current (or active) selection is a list box or a Variables box, the box has a double-line border. In all other cases, the current selection is highlighted.

If you have a mouse, you can go directly to any item you want by just clicking it.

Selecting Variables. Suppose you want to enter Score1 and Score2 into the Variables box. You have three choices:

Typing. First use [Tab] to move to the Variables box if you are not already there. Type the variables you want, using either column numbers, such as C1 C2, or names, such as Score1 Score2. If you type a name, you do not have to enclose it in quotes (unless the name contains blanks or special characters, or you are using the dialog box for the LET command). You can use a dash to abbreviate a list of consecutive columns. If you make a typing mistake, edit your text in the usual way, with arrow keys, the delete key, and retyping.

Selecting from the list box. First use [Tab] to move to the Variables box if you are not already there. Then press [F2]. This moves you over to the list box. Use the arrow keys to move up and down the list. The current location is indicated by a small triangle. Press the spacebar to highlight it. If you change your mind, press the spacebar again to unhighlight the variable. Continue highlighting variables until you have what you want, then press [F2] again. The highlighted variables appear in the Variables box.

Using a mouse. First click in the Variables box to select it. Then click the variables in the list box to highlight them. Clicking a second time will high-light a variable. Finally, click the Select button underneath the list box to transfer all the highlighted variables into the Variables box. You can also select variables one at a time. If you double-click (that is, click the mouse button twice) on a variable, it will be written in the Variables box. Continue double-clicking until you have all the variables you want.

When you are finished, select the OK button. Use [Tab] to get to it, then press [Enter], or press [Alt]+O to go directly to it and press [Enter], or click OK with the mouse. Minitab writes the corresponding session command into the Session window, just as if you had typed it there yourself. It then does the calculations and displays output.

Check Boxes. These are small boxes used for checking certain options. For example, the Descriptive Statistics dialog box contains a check box, "By variable", corresponding to the BY subcommand. Some check boxes, such as this one, have a text box that goes with them. There are two ways to use this option:

Keyboard. Use [Tab] to move to the check box. Press the spacebar to put an x in the box. If you change your mind and want to uncheck the box, press the spacebar again. Now move to the box to the right. This is another text box. You can type the column number or name of the variable you want. You can also select it from the list box. Press [F2], move to the variable you want, then press [F2] again. Since you can select only one entry for this box, you need not use the spacebar to highlight the entry before pressing [F2].
Mouse. Click the check box to put an x in it. If you change your mind and want to uncheck the box, click it again. Now click in the box to the right. This is another text box. Next, click the variable you want, then click the Select button (or just double-click the variable you want).

Radio Buttons. These are another element of a dialog box. Exhibit 2.8 shows a dialog box that calculates one statistic (just one number) for one column. We got to this dialog box by starting with the Calc menu, then selecting the Functions and Statistics submenu, then selecting the entry for Column Statistics. In abbreviated form, this is written **Calc > Functions and Statistics > Column Statistics**.

Look at the items listed under Statistic. Each is preceded by a pair of parentheses. These are called radio buttons. You select one and only one of

Exhibit 2.8   Example with Radio Buttons

Exhibit 2.9   Data Screen

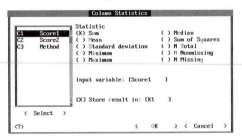

these. The currently selected button is Sum. Suppose you want the minimum value. You have two choices:

Keyboard. First use ⌨Tab⌨ to move to the Statistic group. Then use the arrow keys to move around until you reach the desired item, Minimum. You could also press ⌨Alt⌨ plus the highlighted letter to get there—in this case, ⌨Alt⌨+I. Now use ⌨Tab⌨ to move out of the Statistic group and finish the dialog box.
Mouse. Just click the item you want.

Data Window (Data Screen)

The Data window, or Data screen, shows the columns in your worksheet. Here you can enter, edit, and view your data. To get to the Data screen, open the Edit menu and select Data Screen. There is also a keyboard shortcut, ⌨Alt⌨+D.

Exhibit 2.9 shows the Data screen, containing the numbers that are currently in the worksheet. Recall that these were entered using the READ command in Exhibit 2.5. You could also have entered them directly into the Data screen.

The current cell is highlighted. You can move around the screen using the arrow keys and the ⌨PgUp⌨ and ⌨PgDn⌨ keys. In addition, ⌨Ctrl⌨+⌨Home⌨ takes you to the first row and first column of your data set, and ⌨Ctrl⌨+⌨End⌨ takes you to the last row and last column. There are no scroll bars in this window, so you cannot use a mouse to scroll; you must use the keyboard. You can, however, use a mouse to select a cell by clicking in it.

Entering and Editing Data. Move to the cell in row 8 of C1, the beginning of the first empty row. Make sure this cell is highlighted. Type the number 45. The number appears in the cell. If you press ⌨Enter⌨, the number will be put into that cell permanently and the cursor will move to the right. If you press any arrow key, the number will also be put into the cell permanently and the cursor will move in the direction of the arrow you pressed. If you press (Esc), the cell will be returned to its original value. Press (Enter).

Now you are in C2. Type 49 and press (Enter). Finally, enter 1 in C3. You have now reached the end of row 8. Press [F6] to go to the beginning of the next row. Now enter the numbers 27, 42, and 1.

If at any time you realize the number in a cell is incorrect, move to that cell and type the correct number.

Column Names. You can enter or change the name of a column. Move into the area above a column number. Either press the up arrow until you get there, use the menu described below, or click with the mouse. Then type the name for the column.

Data Screen Menu. Now look at the entry on the bottom right of the screen. It says F10=Menu. Press (F10) and a menu pops up, as shown in Exhibit 2.10. Press (Esc) and the menu disappears. All the menu items can be selected in the usual ways: press arrow keys then (Enter), press the high-lighted letter, press the keyboard shortcut, or click with the mouse. Remember, if you use a keyboard shortcut, you do not have to first open this menu with (F10). We will mention some of the items in this menu. You can learn about the rest through Help.

Suppose you want to delete the next-to-last row of data (row 8). High-light any cell in that row and choose Delete Row (select this from the menu or press (Shift)+(F8)). The row is deleted.

Exhibit 2.10 Data Screen, New Data, and Menu

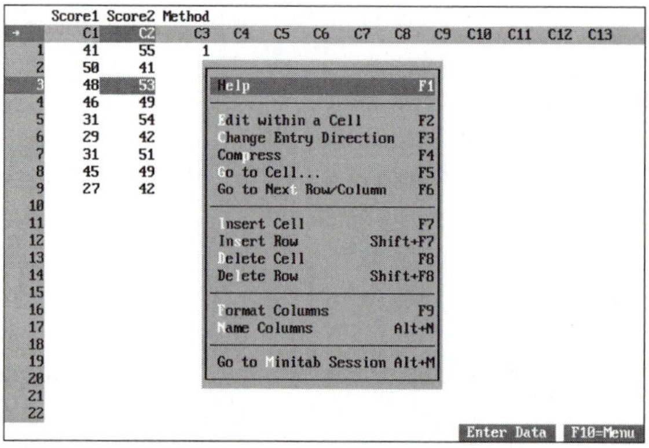

Suppose you want to insert a row between rows 7 and 8. Highlight any cell in row 8 and choose Insert Row. The data set is opened up and a row of stars is filled in for row 8. You can now type the data you want in that row.

You can also delete a cell from just one column. For example, highlight the cell in row 8 of C2 and choose Delete Cell. The cell is deleted and now C2 is shorter than the other columns. Similarly, you can insert a cell in just one column. Highlight the cell you want to insert and choose Insert Cell; the column opens up and the cell contains a star. Now type your data.

Notice the little arrow in the upper-left part of the Data screen. Selecting Change Entry Direction from the menu changes the way this arrow points (you can also click the arrow with a mouse). The arrow's direction controls the following keys:

Key	Action When Arrow Points Right	Action When Arrow Points Down
Home	Moves to beginning of current row.	Moves to top of current column.
End	Moves to end of current row.	Moves to bottom of current column.
Enter	Enters number and moves right.	Enters number and moves down.
F6	Enters number, moves to beginning of next row.	Enters number, moves to beginning of next column.

Help on Minitab

Help is available in the Session window, in the Data screen, and in dialog boxes. In all cases, pressing F1 gives you Help. In the Session window, you can also select Help from the menu bar. In the Data screen, you can also select Help from the menu. In dialog boxes, you can also select Help by choosing the ? button.

Selecting Help from the Session window menu gives a long list of categories. Select one and you will get the relevant information. Selecting Help from the Data screen gives a short list of categories. Again, select one and you will get the relevant help. In a dialog box, selecting ? gives you information on that dialog box.

Help information is displayed in a window of its own. If the help text is long, you can scroll the window just as you scroll the Session window. When you are finished, press Esc to leave Help. You can also press Alt+D or click Done with the mouse.

In this chapter, we give an overview of special features of Release 8. You can learn other things by reading through the Help information included with the software.

Exhibit 2.11 First Dialog Box for
Save Worksheet As

Exhibit 2.12 Second Dialog Box for
Save Worksheet As

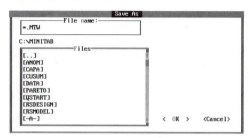

Saved Worksheets

Once you have entered data into Minitab, you usually want to store it in a
computer file so you will have it available for use in another session on
another day. The most convenient type of file for use with Minitab is a
Minitab saved worksheet. This file contains all columns, stored constants,
and column names you have.

The session commands used with saved worksheets, SAVE and RE-
TRIEVE, are described in Section 2.1. Here we will show you how to use
the dialog boxes for SAVE and RETRIEVE.

Saving Worksheets. Suppose you want to save the data currently in the
worksheet on a floppy disk in drive B. Select **File > Save Worksheet As**.

The dialog box, shown in Exhibit 2.11, gives you a choice of three
formats. The default is Minitab worksheets saved in DOS format. These
have the extension .MTW. The second is portable format, used when you
want to move a Minitab worksheet from one type of computer to another,
for example from a DOS machine to a Macintosh. These have the extension
.MTP. The third is Lotus format, used to move data between Minitab and
Lotus or Symphony. These have the extensions .WK1, .WKS, .WR1, and
WRK.

To select the default file format, a Minitab worksheet, just choose Select File. This gives the dialog box shown in Exhibit 2.12.

The current directory is C:\MINITAB. The Files box displays a list of
the files in this directory and also lists your drives. Subdirectories and
drives are shown in square brackets. Press [Tab] to move to the Files box,
then scroll down to highlight [-B-] and select OK. The Files box now displays a list of the files on the floppy disk in drive B. This floppy disk happens to contain two files and no subdirectories. Move to the File name box

Exhibit 2.13 Dialog Box to Save
Worksheet as EX1.MTW

Exhibit 2.14 Dialog Box for Open
Worksheet

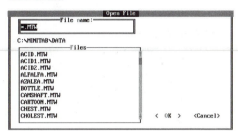

and type EX1 as the name for your file. Exhibit 2.13 shows the dialog box
after the file name has been typed. Select OK and your file is saved as
EX1.MTW on the floppy disk in drive B. Minitab automatically adds the
extension .MTW.

Retrieving Worksheets. Minitab comes with many data sets stored as
saved worksheets. Let's look at the one called Trees. Select **File > Open
Worksheet**. The dialog box that appears gives you a choice of three formats. What you want is the default, a Minitab worksheet. Choose the Select
File button and you will get a dialog box, similar to the one shown in Exhibit 2.13. In general, dialog boxes for files show the directory that you last
used. Since you just stored a file on a floppy disk that is still in drive B, the
contents of this disk are shown.

The file we want is in C:\MINITAB\DATA. Use ⌈Tab⌋ to move to the
Files box, highlight [-C-], and select OK to get back to the Minitab directory on the C drive. Highlight [DATA] and select OK to get to the DATA
subdirectory. This gives the dialog box in Exhibit 2.14. It contains a list
of files. Scroll down until you highlight TREES.MTW and then select
OK. The data set is now in Minitab. You can view it in the Data screen if
you like.

Suppose that when you tried to select [DATA], you accidentally selected [ANOM] instead. Now how do you get to [DATA]? Scroll down and
you will see the entry [..]. This represents the parent directory of the current
directory. Highlight [..] and select OK. This moves you up one level. Now
you can start over.

Note: Anything you have in the worksheet when you retrieve a data set
is erased before the saved worksheet is brought in.

Exhibit 2.15   First Dialog Box for Exporting
ASCII Data

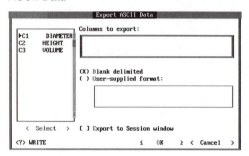

Exhibit 2.16   First Dialog Box for Importing
ASCII Data

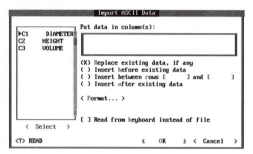

ASCII Data Files

You can transfer data between Minitab and other applications by using a data, or ASCII, file. Almost every application—every editor, spreadsheet, and word processor—can input and output data in ASCII format.

The Session commands used with ASCII files—READ, SET, INSERT, and WRITE—are described in Section 2.1. Here we will show you how to use the dialog boxes corresponding to these commands.

Exporting ASCII Files. You can save the columns of your current worksheet in an ASCII file. Select **File > Export Data** to get the dialog box in Exhibit 2.15. Enter the columns to be exported. Only columns of data can be stored in an ASCII file. Stored constants and column names will be lost. Select OK and you will get a file dialog box, similar to the one shown in Exhibit 2.14. Select the directory for your file, then type a name in the File name box and select OK. Minitab automatically adds the file extension .DAT, if you do not specify an extension.

When Minitab stores the specified columns, it adjusts the format to make the data as compact as possible. If all columns do not fit on one line, Minitab puts the continuation symbol, &, at the end of the line and continues the data on the next line. If the columns are of different lengths, Minitab makes them all the same length by adding missing-value codes (*) to the shorter columns.

Importing ASCII Files. To import an ASCII data file into Minitab, first select **File > Import ASCII**. This gives the dialog box shown in Exhibit 2.16. This box corresponds to two Minitab Session commands, READ and INSERT. The first radio button is for READ. READ replaces any existing data. The other three buttons are for the different cases of INSERT. The first puts the new data at the tops of the existing columns. The second choice requires two consecutive row numbers; the new data are put between these rows. The last choice puts the new data at the bottom of the worksheet.

Next, you select columns where you will store the data. For example, suppose you have a file called mydata.DAT that contains 10 lines of data, each with 4 numbers. Suppose you want to put the data into C1–C4, replacing anything that is in these columns. Select the first radio button, enter C1–C4 in the box titled "Put data in column(s)," and select OK. This gives the usual box for selecting files. Data files have the extension .DAT, so only these are displayed. Now go to the subdirectory that contains the file mydata.DAT, highlight the file name, and select OK.

The dialog box in Exhibit 2.16 contains a Format button. Most ASCII files you use will contain rows and columns of numbers, with at least one space or comma between numbers. In such cases, you do not need this option. However, if the file contains alpha data, if there are no spaces or commas between numbers, if blanks are used for missing values, if two or more lines of data are to go into one row of Minitab's worksheet, or if the file has any other unusual properties, then you will need to use the Format option to describe the data format. Formats are discussed in Section 17.5.

Saving and Printing the Session Window

You can save everything that appears in the Session window in a file, called an outfile. Then, after you have left Minitab, you can print the file or import it into a word processor to create a report.

Minitab's outfile option is basically a switch. When you turn it on, everything in the Session window is put into the outfile. When you turn it off, Minitab stops storing the Session window. When you turn it on again, Minitab stores the contents of the Session window again.

It is a good practice to begin every Minitab session by starting an outfile, so that you will have a permanent record of your session.

The Session commands used to start and stop an outfile, OUTFILE and NOOUTFILE, are described in Section 2.1. Here we will show you how to use the dialog boxes corresponding to these commands.

To start an outfile, choose **File > Other Files > Start Recording Session**. This gives the dialog box shown in Exhibit 2.17. Since the default version of this dialog box is what we want, choose Select File. This gives the dialog box shown in Exhibit 2.18. As usual, the directory that is shown is the last directory you used. If this is not what you want, move to the appropriate drive and directory. Files that contain Minitab output have the

Exhibit 2.17   First Dialog Box
for Recording a Session

Exhibit 2.18   Second Dialog Box for Recording
a Session

extension .LIS, so only these files are displayed. Type a name in the File name box and select OK.

From this point on, all commands and output that appear in the Session window will also be put into this file. (High-resolution graphs, however, will not. You must store these separately.) When you want to stop recording your session, select **Files > Other Files**. You will now see the entry **Stop Recording Session**. Choose this entry.

If, at a later time, you want to record more of your session, select **Files > Other Files > Start Recording Session**. If you choose the same file name as before, your new output will be appended to the old. If you type a new file name, the new output will go into the new file.

If you want a copy of your data on paper, start an outfile if you don't already have one, then use Minitab's PRINT command to display the data on the screen (you can use the corresponding dialog box by selecting **Edit > Display Data**). The data will be put into the outfile. When you leave Minitab, you can print this outfile.

Graphs

There are two types of graphs. Character graphs appear in the Session window. High-resolution graphs appear in a separate window. For most analyses character graphs are good enough, although they do not look as pretty as the high-resolution graphs and lack a few of their options.

Let's look at Minitab's histogram. Retrieve the TREES data set. To get a character histogram of C1, type HISTOGRAM C1 in the Session window. Output is shown in Exhibit 2.19. If you want a high-resolution histogram, type GHISTOGRAM C1. (Session commands for high-resolution graphs all start with G.) Output is shown in Exhibit 2.20. You now have three

Exhibit 2.19 Session Window with Character
Histogram

Exhibit 2.20 High-Resolution
Histogram

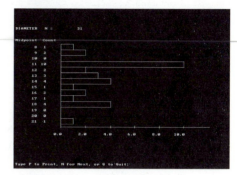

Exhibit 2.21 Dialog Box for a Histogram

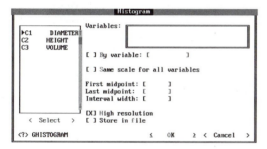

choices: Press P to print the graph on your printer, N to see the next graph
(you would need this if you had asked for GHISTOGRAM C1–C3), or Q
to quit and go back to the Session window.

Exhibit 2.21 shows the dialog box from **Graph > Histogram**. The
"High resolution" box is checked. By default, you will get a high-resolution
graph. If you want a character graph, uncheck this box. Notice the item
"Store in file". You can store a high-resolution graph in a file, which, by
default, will be given the file extension .PRI. Then later, from DOS, print
it out by typing

```
COPY filename.ext PRN:
```

2.3   Release 8 Macintosh

Starting Minitab

Start your computer. Open the Minitab folder and double-click Minitab's icon. You are now in Minitab.

Using Minitab's Windows

Minitab has six types of windows, and they can all be open at the same time (see Exhibit 2.22). The Data window displays your worksheet. You can enter and edit data here. The Session window displays output. You can also type commands here. The History window contains a record of previously executed commands. The Info window summarizes the data in the current worksheet. A graph window is created each time you create a graph. All of these windows are listed in the Windows pull-down menu, and can be made active from there. The Help window gives you Help information on Minitab, and is listed under the Apple menu.

You can use Release 8 as if it were a command-only version of Minitab. Make the Session Window active (see next section) and type Minitab's commands.

Session Window

The Session window is where you can type Minitab commands and where Minitab displays output. To make this window active, select it from the Window menu, or press ⌘+M, or click it if it's visible on your screen. If

Exhibit 2.22   Overview of Minitab's Windows

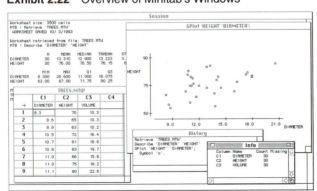

you want to type a command, type it after the MTB > prompt and press return; the command will be executed.

In Exhibit 2.23, we use the READ command to enter seven rows of data into C1–C3, name the columns, and DESCRIBE C1 and C2. The Session window scrolls as more output goes into it. You can scroll up and down to see various parts of your output.

The Session window is quite limited in size. The number of screens you can scroll back through depends on how much is on each line. Usually it's between 5 and 15 screens. If you want to save more of your output, you can create an outfile. (See "Saving and Printing the Session Window," later in this section.)

You can use some editing features in the Session window. You can copy anything from the Session window into another application. You can copy Minitab commands, subcommands, and data from another part of the Session window, or from an editor, and paste them at the current Minitab prompt. When you press return, the pasted material will be executed. You can edit the material before you press return. You cannot edit any of Minitab's output in the Session window.

Menus

The menu bar across the top of Exhibit 2.24 lists Minitab's seven main menus (File, Edit, Calc, Stat, Window, Graph, and Editor) and the Apple menu, which contains special Macintosh functions and Minitab's Help. You can execute nearly all of Minitab's commands through these menus.

Minitab's menus work the same as menus in any other Macintosh application. To open a menu, click its name, hold the mouse button down and drag to the item you want, then release the button. Exhibit 2.24 shows the File menu and a submenu.

Exhibit 2.23 Session Window with READ, NAME, and DESCRIBE

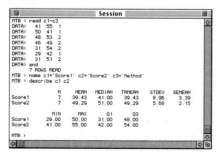

Exhibit 2.24 File Menu with Submenu

Let's look at the items in the File menu. Most end with an ellipsis (...). If you choose one of these, a dialog box will appear. One item ends in ▶. If you drag down to this, a second list, called a submenu, will appear. This submenu is shown in Exhibit 2.24. Two items do not end in an ellipsis or in ▶. If you choose one of these, Minitab will execute the corresponding command immediately. For example, if you choose Quit, then Minitab will quit (the corresponding Session command is STOP). A few menu entries have keyboard shortcuts, shown to the right of the entry. For example, Open Worksheet is followed by ⌘+O. Any time you type ⌘ in combination with the letter O, you will get the Open Worksheet dialog box directly.

Dialog Boxes

Suppose you want to use the menus to execute the Minitab command DE-SCRIBE. First open the Stat menu, then open the Basic Statistics submenu, then select the Descriptive Statistics entry. This gives the dialog box shown in Exhibit 2.25.

Note: In most of this book, we will abbreviate a menu/submenu selection in the following style: **Stat > Basic Statistics > Descriptive Statistics**.

Let's look at the parts of this dialog box. The box on the left lists all the worksheet columns you currently have. It is called a list box. The button under this box, Select, is used to select variables. The box on the right is a text box, where you enter the variables (columns) you want to use in DE-SCRIBE. The entry "By variable" is an option. It corresponds to the BY subcommand of DESCRIBE. The last line has three buttons. The first is ?. Click this, and you will get help on this dialog box. The word next to ? is the name of the corresponding Minitab Session command. This is provided

Exhibit 2.25   Dialog Box for DESCRIBE

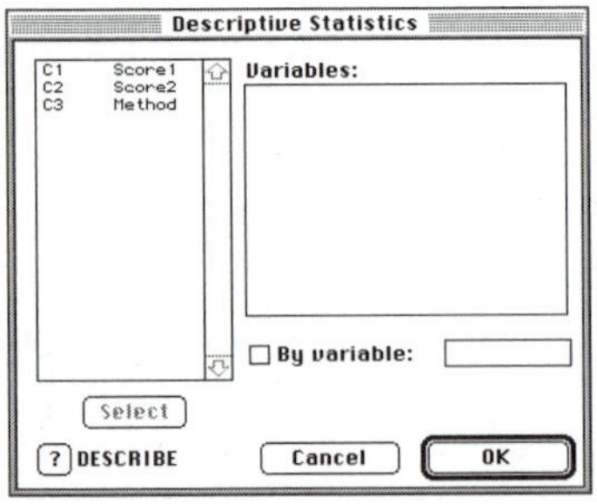

just for your information; it does not do anything. Next is the Cancel button. Click this and the command is canceled. Last is the OK button. Click this and the command is executed.

Notice that the Select button is "grayed out." That is because you cannot use it at this point. An item can be "grayed out" for any number of reasons. If it is, you cannot use it until you make some change in the dialog box. In this case, we need to highlight some variables.

Selecting Variables. Suppose you want to enter the two scores into the Variables box. You have two choices:

Typing. First select the Variables box by clicking in it. Type the variables you want, using either column numbers, such as C1 C2, or names, such as Score1 Score2. If you type a name, you do not have to enclose it in quotes (unless the name contains blanks or special characters, or you are using the dialog box for the LET command). You can use a dash to indicate a list of consecutive columns. If you make a typing mistake, edit your text in the usual way.

Selecting from the list box. First, click the Variables box to select it. Then click the variables in the list box to highlight them. If you change your mind, click a variable again to unhighlight it. Finally, click Select to transfer all the highlighted variables into the Variables box. You can also select variables one at a time. If you double-click a variable, it will be written in the Variables box. Continue double-clicking until you have all the variables you want.

When you are done, click OK. Minitab writes the corresponding session command into the Session window, just as if you had typed it there yourself. It then does the calculations and displays output.

Check Boxes. These are small boxes used for checking certain options. For example, the Descriptive Statistics dialog box contains a check box, "By variable," corresponding to the BY subcommand. Some check boxes, such as this one, have a text box that goes with them.

Let's use the DESCRIBE command with a BY variable. Open the Descriptive Statistics dialog box and put Score1 and Score2 into the Variables box. Next, click the check box before "By variable" to select the entry. To uncheck a box, click it again. Now click in the box to the right. This is another text box. You can type the column number or name of the variable you want. You can also select it from the list box. Select C3. Finally, click the OK button to have Minitab do the calculations.

Radio Buttons. Exhibit 2.26 shows a dialog box that calculates one statistic (just one number) for one column. We got to this dialog box by starting with the Calc menu, then selecting the Functions and Statistics submenu, then selecting the entry for Column Statistics. In abbreviated form, this is **Calc > Functions and Statistics > Column Statistics**.

Look at the items listed under Statistic. Each is preceded by a circle. These are radio buttons. You select one and only one of them. The currently selected button, Sum, has a dot inside it. Suppose you want the minimum value. Click on its radio button.

Data Window

The Data window shows the columns in your worksheet. Here you can enter, edit, and view your data. To make this window active, select it from the Window menu, or press ⌘+D, or click it if it's visible.

Exhibit 2.27 shows the Data window, containing the numbers that are currently in the worksheet. Recall that these were entered using the READ command in Exhibit 2.23. You could also have entered them directly into the Data screen.

The current cell is highlighted. You can move about in the window as you do in other Macintosh applications, using the scroll bars, the arrow keys, and the [page up] and [page down] keys. In addition, [⌘]+[home] takes you to the first row and first column of your data set, and [⌘]+[end] takes you to the last row and last column. Click in a cell to make it the current cell.

Entering and Editing Data. Click in the cell in row 8 of C1, the beginning of the first empty row. Type the number 45. The number appears in the cell. If you press [tab], the number will be put into that cell and the cursor will

Exhibit 2.26   Example with Radio Buttons

```
                     Column Statistics
 ┌──────────────┬──────────────────────────────────────┐
 │ C1  Score1 ⬏ │ Statistic                            │
 │ C2  Score2   │  ◉ Sum              ○ Median          │
 │ C3  Method   │  ○ Mean             ○ Sum of Squares  │
 │              │  ○ Standard deviation ○ N Total       │
 │              │  ○ Minimum          ○ N Nonmissing    │
 │              │  ○ Maximum          ○ N Missing       │
 │              │                                      │
 │              │  Input variable: [           ]       │
 │              │                                      │
 │            ⬎ │  ☐ Store result in: [        ]        │
 ├──────────────┘                                      │
 │  [ Select ]                                          │
 │  [?] SUM                      [ Cancel ]  [  OK  ]   │
 └─────────────────────────────────────────────────────┘
```

Exhibit 2.27   Data Window

	C1	C2	C3	C4	C5	C6
→	Score1	Score2	Method			
1	41	55	1			
2	50	41	1			
3	48	53	2			
4	46	49	2			
5	31	54	2			
6	29	42	1			
7	31	51	2			
8						
9						

move to the right. If you press [return], the number will be put into the cell and the cursor will move down. If you press any arrow key, the number will also be put into the cell and the cursor will move in the direction of the arrow you pressed. If you press [esc], the cell will be returned to its original value. Press [tab].

Now you are in C2. Type 49 and press [tab] Finally, enter 1 in C3. You have now reached the end of row 8. Press [⌘]+[tab] to go to the beginning of the next row. Now enter the numbers 27, 42, and 1. Note that pressing [⌘]+[return] puts the number in the current cell and moves to the top of the next column.

If at any time you realize the number in a cell is incorrect, move to that cell and type the correct number.

Column Names. You can enter or change the name of a column. Click in the cell just below the column number and type the name for the column.

Editor Menu. The Editor menu controls the Data window. It is shown in Exhibit 2.28, along with the extra data we entered.

Suppose you want to delete the next-to-last row of data (row 8). Highlight any cell in that row and choose **Editor > Delete Row**. The row is deleted.

Suppose you want to insert a row between rows 7 and 8. Highlight any cell in row 8 and choose **Editor > Insert Row**. The data set is opened up and a row of stars is filled in for row 8. You can now type the data you want in that row.

You can also delete a cell from just one column. For example, highlight the cell in row 8 of C2 and choose Delete Cell. The cell is deleted and now

Exhibit 2.28 Data Window, New Data, and Menu

	C1	C2	C3	C4	Editor menu
	Score1	Score2	Method		Next Column
					Next Row
1	41	55	1		Go To... ⌘G
2	50	41	1		Go To Active Cell
3	48	53	2		
4	46	49	2		Format Column... ⌘Y
5	31	54	2		Set Column Widths...
6	29	42	1		
7	31	51	2		Compress Display ⌘K
8	45	49	1		Change Entry Direction
9	27	42	2		
10					Insert Cell
11					Insert Row
12					Delete Cell
13					Delete Row
					Repeat: Insert Cell ⌘R

File Edit Calc Stat Window Graph Editor

Untitled Worksheet

C2 is shorter than the other columns. Similarly, you can insert a cell in just one column. Highlight the cell you want to insert and choose Insert Cell, and the column opens up and the cell contains a star. Now type your data.

You can use Cut, Copy, and Paste from the Edit menu as you can in other Macintosh programs, but in Minitab's Data window, you can use them for only one cell at a time.

Notice the little arrow in the upper-left part of the Data window. Selecting **Editor > Change Entry Direction** changes the way this arrow points (you can also click the arrow with a mouse). The arrow's direction controls the following keys:

Key	Action When Arrow Points Right	Action When Arrow Points Down
home	Moves to beginning of current row.	Moves to top of current column.
end	Moves to end of current row.	Moves to bottom of current column.
return	Enters number, moves right.	Enters number, moves down.
⌘+return	Enters number, moves to beginning of next row.	Enters number, moves to beginning of next column.

Help on Minitab

Help is available from the Apple menu. Exhibit 2.29 shows the Help window when you first open it. The box on the left is an "index" of topics. Now it gives you four choices: Overview gives an overview of Minitab; Menus gives information on commands listed in menus; the last two, COMMANDS by Topic and COMMANDS by Name, give information on Session commands.

Exhibit 2.29   Minitab's Help Window When First Opened

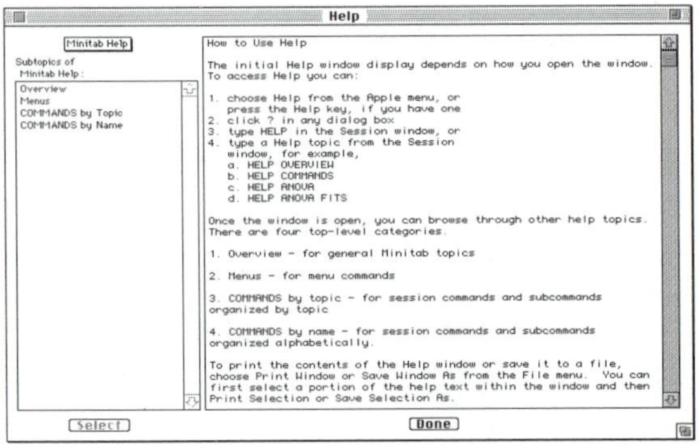

Let's get information on doing descriptive statistics. Select Menu (just) double-click Menu) and you will get a list of Minitab's seven menus. Select Stat and you will get a list of the items in the Stat menu. Select Descriptive Statistics and you will get information on this command. This is shown in the box on the right.

Now suppose you want to go elsewhere in Help. Notice the small box on the left that gives your current location, Descriptive Statistics. Click this and you will see five entries. Your current location, Descriptive Statistics, is on top, followed by your previous location, Basic Statistics, followed by your location previous to that, Stat, and so on. Drag down to the item you want and you will be positioned there.

Help is also available from dialog boxes. Just click the ? button in the lower-left area of the dialog box.

This chapter provides an overview of the user interface and some features of Release 8 of Minitab for Macintosh. You can learn other things by reading through the Help information included with the software.

Saved Worksheets

Once you have entered data into Minitab, you usually want to store it in a computer file so that you have it available for use in another session on another day. The most convenient type of file for use with Minitab is a Minitab saved worksheet. This file contains all columns, stored constants, and column names you have used.

The session commands used with saved worksheets, SAVE and RETRIEVE, are described in Section 2.1. Here we will show you how to use the dialog boxes for SAVE and RETRIEVE.

Saving Worksheets. Suppose you want to save the data currently in the worksheet on a floppy disk. Select **File > Save Worksheet As** to get the dialog box shown in Exhibit 2.30. This is a standard Macintosh dialog box for opening and saving files. You can move among folders and drives in the usual way.

Minitab can save two types of files. The default is Minitab worksheets saved in Macintosh format. These have the extension .MTW. The other is a portable format, used when you want to move a Minitab worksheet from one type of computer to another, for example from a Macintosh to a DOS machine. These have the extension .MTP and are selected with the check box for Portable format.

Now click desktop with System 7 (or Drive with System 6) to go to the floppy drive. This floppy disk happens to contain two Minitab worksheets.

Exhibit 2.30 Dialog Box
for Save Worksheet As

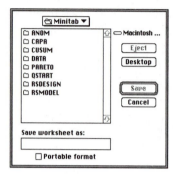

Exhibit 2.31 Dialog Box to
Save Worksheet as ex1.MTW

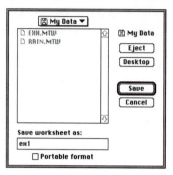

Type ex1 as the name for your file. Exhibit 2.31 shows the dialog box after the file name has been typed. Click Save and your file is saved as ex1.MTW on the floppy disk. Minitab automatically adds the extension .MTW.

You can also save a file by choosing **File > Save Worksheet**. In that case, you will not be given a dialog box. Your worksheet will be saved in the same file from which you opened it.

Retrieving Worksheets. Minitab comes with many data sets stored as saved worksheets. Let's look at the one called Trees. Select **File > Open Worksheet** and you will get a dialog box that looks similar to the one in Exhibit 2.31. In general, dialog boxes for files show the folder you last used. Since you just stored a file on a floppy disk that is still in the floppy drive, the contents of the floppy disk are shown.

Go to the Minitab folder on the hard disk and then to the Data folder. This gives the dialog box in Exhibit 2.32, containing a list of files. Scroll down until you see trees.MTW. Click it, then click Open. The data set is now in Minitab. You can view it in the Data window if you like.

Note: Anything you have in the worksheet when you retrieve a data set is erased before the saved worksheet is brought in.

ASCII Data Files

You can transfer data between Minitab and other applications by using a data, or ASCII, file. Almost every application—every editor, spreadsheet, and word processor—can input and output data in ASCII format.

The session commands used with ASCII files—READ, SET, INSERT, and WRITE—are described in Section 2.1. Here we will show you how to use the dialog boxes corresponding to these commands.

Exhibit 2.32 Dialog Box for Open
Worksheet

Exhibit 2.33 First Dialog Box for
Exporting Data to an ASCII File

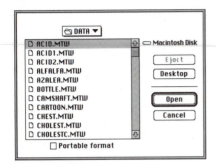

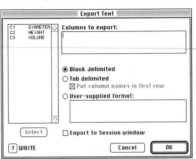

Exporting ASCII Files. You can save the columns of your current work-sheet in an ASCII file. Select **File > Export Data** to get the dialog box shown in Exhibit 2.33. Enter the columns to be exported. Only columns of data can be stored in an ASCII file. Stored constants and column names will be lost. Click OK and you will get a file dialog box, similar to the one shown in Exhibit 2.31. Select a folder for your file, then type a name and click OK. Minitab automatically adds the file extension .DAT, if you do not specify an extension.

When Minitab stores the specified columns, it adjusts the format to make the data as compact as possible. If all columns do not fit on one line, Minitab puts the continuation symbol, &, at the end of the line and continues the data on the next line. If the columns are of different lengths, Minitab makes them all the same length by adding the missing-values code (*) to the shorter columns.

Importing ASCII Files. To import an ASCII data file into Minitab, first select **File > Import ASCII**. This gives the dialog box in Exhibit 2.34. This box corresponds to two Minitab commands, READ and INSERT. The first radio button is for READ. READ replaces any existing data. The other three buttons are for the different cases of INSERT. The first puts the new data at the tops of the existing columns. The second choice requires two consecutive row numbers; the new data are put between these rows. The last choice puts the new data at the bottom of the worksheet.

Next, you select columns where you will store the data. For example, suppose you have a file called mydata.DAT that contains 10 lines of data, each with 4 numbers. Suppose you want to put the data into C1–C4, replacing anything that is in these columns. Select the first radio button, enter C1–C4 in the box titled "Put data in column(s)," and click OK. This gives the usual box for selecting files. Data files have the extension .DAT, so only

Exhibit 2.34   First Dialog Box for Importing
ASCII Data

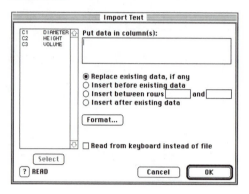

these are displayed. Now go to the subdirectory that contains the file
mydata.DAT, select it, and click OK.

The dialog box in Exhibit 2.34 contains a Format button. Most ASCII
files you will use will contain rows and columns of numbers, with at least
one space or comma between numbers. In such cases, you do not need this
option. However, if a file contains alpha data, if there are no spaces or
commas between numbers, if blanks are used for missing values, if two or
more lines of data are to go into one row of Minitab's worksheet, or if the
file has any other unusual properties, then you will need to use the Format
option to describe the data format. Formats are discussed in Section 17.5.

Saving and Printing the Session Window

You can save everything that appears in the Session window in a file, called
an outfile. Then, after you have left Minitab, you can print the file or import
it into a word processor to create a report.

Minitab's outfile option is basically a switch. When you turn it on,
everything in the Session window is put into the outfile. When you turn it
off, Minitab stops storing the Session window. When you turn it on again,
Minitab stores what appears in the Session window again.

It is a good practice to begin every Minitab session by starting an outfile, so that you will have a permanent record of your session.

The Session commands used to start and stop an outfile, OUTFILE and
NOOUTFILE, are described in Section 2.1. Here we will show you how to
use the dialog boxes corresponding to these commands.

To start an outfile, choose **File > Other Files > Start Recording Session**. This gives the dialog box shown in Exhibit 2.35. Since the default

Exhibit 2.35 First Dialog Box for Start Recording Session

Exhibit 2.36 Second Dialog Box for Recording a Session

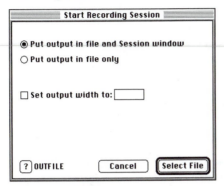

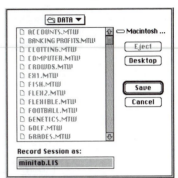

version of this dialog box is what we want, choose Select File. This gives the dialog box shown in Exhibit 2.36. As usual, the folder that is shown is the last one you used. If it is not what you want, move to the appropriate drive and folder. Files that contain Minitab output have the extension .LIS. Type a name for your file and click Save.

From this point on, all commands and output that appear in the Session window will also be put into this file. (High-resolution graphs, however, will not. You must store these separately.) When you want to stop recording your session, select **Files > Other Files**. You will then see the entry **Stop Recording Session**. Choose this entry.

If, at a later time, you want to record more of your session, select **Files > Other Files > Start Recording Session**. If you choose the same file name as before, your new output will be appended to the old. If you type a new file name, the new output will go into the new file. Once you have closed an outfile with **Stop Recording Session**, or have left Minitab, you can edit that outfile in an editor or word processor.

If you want a copy of your data set on paper, start an outfile if you don't already have one, then use Minitab's PRINT command to display the data on the screen (you can use the corresponding dialog box by selecting **Edit > Display Data**). The data will be put it into the outfile.

Graphs

There are two types of graphs. Character graphs appear in the Session window. High-resolution graphs appear in their own windows. For most analyses

Exhibit 2.37 Session Window with Character Histogram

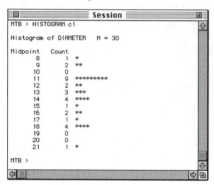

Exhibit 2.38 High-Resolution Histogram

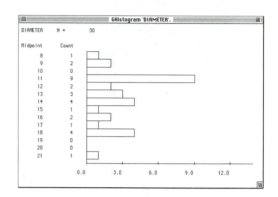

Exhibit 2.39 Dialog Box for a Histogram

character graphs are good enough, although they do not look as pretty as the high-resolution graphs and lack a few of their options.

Let's look at Minitab's histogram. Retrieve the trees data. To get a character histogram of C1, type HISTOGRAM C1 in the Session window. Output is in Exhibit 2.37. If you want a high-resolution histogram, type GHISTOGRAM C1. (Session commands for high-resolution graphs all start with G.) The graph appears in its own window, shown in Exhibit 2.38.

Exhibit 2.39 shows the dialog box from **Graph > Histogram**. The "High resolution" box is checked. By default, you will get a high-resolution graph. If you want a character graph, uncheck the box.

If you want to save a high-resolution graph in a file, choose **File > Save Window As** while the graph is the active window. If you want to print the graph, choose **File > Print Window** while the graph is the active window.

2.4   Release 9 Windows[1]

Starting Minitab

Start your computer. Find Minitab's blue icon—usually it is in the Minitab program group—and double-click it. You are now in Minitab. Click in the "About Minitab" box to get rid of it, or just wait a few moments.

Using Minitab's Windows

Minitab has six types of windows, and they can all be open at the same time. The Data window displays your worksheet. You can enter and edit data here. The Session window displays output. You can also type commands here. The History window contains a record of previously executed commands. The Info window summarizes the data in the current worksheet. A graph window is created each time you create a graph. All of these windows are listed in the Window pull-down menu, and can be made active from there. The Help window gives you Help information on Minitab, and is available in the Help pull-down menu.

You can use Release 9 as if it were a command-only version of Minitab. Make the Session window active (see below) and type Minitab's commands.

Exhibit 2.40 shows two windows and two icons at the bottom. The icons are for two windows that are not open: the Info window and the History window. Double-click an icon and its window will open. You can "iconize" an opened window by clicking the down arrow on the right side of the title bar.

Session Window

The Session window[2] is where you can type Minitab commands and where Minitab displays output. To make this window active, select it from the Window menu, or press Ctrl+M, or click it if it is visible on your screen. If you want to type a command, type it after the MTB > prompt and press Enter; the command will be executed.

In Exhibit 2.41, we use the READ command to enter seven rows of data into C1–C3, name the columns, and DESCRIBE C1 and C2. The Session

---

[1]Release 10 will be available soon after this book is published. We indicate some of the important changes in footnotes.
[2]In Release 10, the Session window has titles, typed commands can be disabled and hidden if you wish, and the window can be edited. Make the Session window active and the Editor menu will be for session editing.

Exhibit 2.40 Overview of Minitab's Windows and Icons

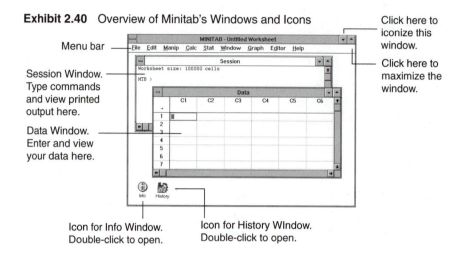

Click here to
iconize this
window.

Click here to
maximize the
window.

Menu bar

Session Window.
Type commands
and view printed
output here.

Data Window.
Enter and view
your data here.

Icon for Info Window.
Double-click to open.

Icon for History Window.
Double-click to open.

Exhibit 2.41 Session Window with READ,
NAME, and DESCRIBE Commands

```
MTB > read c1-c3
DATA>     41    55    1
DATA>     50    41    1
DATA>     48    53    2
DATA>     46    38    2
DATA>     31    48    2
DATA>     29    41    1
DATA>     31    40    2
DATA> end
      7 rows read.
MTB > name c1='Score' c2='Score2' c3='Method'
MTB > describe c1 c2

              N     MEAN   MEDIAN   TRMEAN   STDEV   SEMEAN
Score         7    39.43    41.00    39.43    8.96     3.39
Score2        7    45.14    41.00    45.14    6.82     2.58

            MIN      MAX      Q1       Q3
Score     29.00    50.00    31.00    48.00
Score2    38.00    55.00    40.00    53.00

MTB >
```

window scrolls as more output goes into it. You can scroll up and down to
see various parts of your output.

The Session window is limited in size. The number of screens you can
scroll back through depends on how much is on each line. Usually it is
between 20 and 60 screens. If you want to save more of your output, you
can create an outfile. (See "Saving and Printing the Session Window," later
in this section.)

You can use features from the Edit menu to edit in the Session window.
For example, you can copy anything from the Session window into another

Exhibit 2.42   File Menu with Submenu

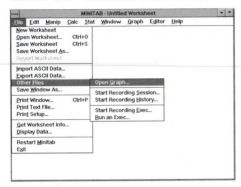

application. You can also copy Minitab commands, subcommands, and data from another part of the Session window or from an editor such as Notepad, and paste them at the current prompt. When you press Enter, the material will be executed. You can edit the material before you press Enter; just make sure that you use Ctrl + Enter if you want to insert a line. You cannot edit any of Minitab's output in the Session window.

Menus

The menu bar[3] across the top of Exhibit 2.40 lists Minitab's main menus (File, Edit, Manip, Calc, Stat, Window, Graph, Editor, and Help). You can execute nearly all of Minitab's commands through these menus. Minitab's menus work the same as menus in any other Windows application. To open a menu, click its name. A pull-down list appears. Exhibit 2.42 shows the File menu and a submenu.

Let's look at the items in the File menu. Most end with an ellipsis (...). If you choose one of these, a dialog box will appear. One item, Other Files, ends in ▶. If you click this, a second list, called a submenu, appears. In Exhibit 2.42, we clicked Other Files to get the submenu. Five items do not end in an ellipsis or in ▶. If you choose one of these, Minitab will execute the corresponding command immediately. For example, if you choose Exit, Minitab will exit (the corresponding Session command is STOP). A few menu entries have keyboard shortcuts, shown to the right of the entry. For example, Open Worksheet is followed by Ctrl+O. Any time you type Ctrl in combination with the letter O, you will get the Open Worksheet dialog box directly.

---

[3]In Release 10, menu items are in a slightly different order, and some submenus contain additional entries.

Dialog Boxes

Suppose you want to use the menus to execute Minitab's command DE-
SCRIBE. First open the Stat menu, then open the Basic Statistics submenu,
then select the Descriptive Statistics entry. This gives the dialog box shown
in Exhibit 2.43.

Note: In most of this book we will abbreviate a menu/submenu selection in the following style: **Stat > Basic Statistics > Descriptive
Statistics**.

Let's look at the parts of this dialog box. The box on the left lists all the
worksheet columns you currently have. It is called a list box. The button
under this box, Select, is used to select variables. The box on the right is a
text box, where you enter the variables (columns) you want to use in DE-
SCRIBE. The entry "By variable" is an option. It corresponds to the BY
subcommand of DESCRIBE. The last line has three buttons. The first is ?.
Click this, and you will get help on this dialog box. The word next to ? is
the name of the corresponding Minitab Session command. This is just for
your information; it does not do anything. Next is the OK button. Click this
and the command is executed. Last is the Cancel button. Click this and the
command is canceled.

Notice that the Select button is "grayed out." That is because you cannot use it at this point. An item can be "grayed out" for any number of
reasons. If it is, you cannot use it until you make some change in the dialog
box. In this case, we need to highlight some variables.

Exhibit 2.43   Dialog Box for DESCRIBE

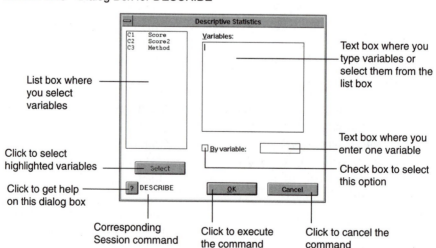

Selecting Variables. Suppose you want to enter Score1 and Score2 into the Variables box. You have two choices.

Typing. First select the Variables box by clicking in it. Type the variables you want, using either column numbers, such as C1 C2, or names, such as Score1 Score2. If you type a name, you do not have to enclose it in quotes (unless the name contains blanks or special characters, or you are using the dialog box for the LET command). You can use a dash to indicate a list of consecutive columns. If you make a typing mistake, edit your text in the usual way.

Selecting from the list box. First click in the Variables box to select it. Then click the variables you want in the list box to highlight them. If you change your mind, click a variable again to unhighlight it. Finally, click Select to transfer all the selected variables into the Variables box. You can also select variables one at a time. If you double-click a variable, it will be written in the Variables box. Continue double-clicking until you have all the variables you want.

   When you are finished, click OK. Minitab writes the corresponding Session command into the Session window, just as if you had typed it there yourself. It then does the calculations and displays output.

Check Boxes. These are small boxes used for checking certain options. For example, the Descriptive Statistics dialog box contains a check box, "By variable", corresponding to the BY subcommand. Some check boxes, such as this one, have a text box that goes with them.

   Let's use the DESCRIBE command with a BY variable. Open the Descriptive Statistics dialog box and put Score1 and Score2 into the Variables box. Next, click the check box before "By variable" to select the entry. To uncheck a box, click it again. Now click in the box to the right. This is another text box. You can type the column number or name of the variable you want. You can also select it from the list box. Select C3. Finally, click OK to have Minitab do the calculations.

Radio Buttons. Exhibit 2.44 shows a dialog box that calculates one statistic (just one number) for one column. We got to this dialog box by starting with the Calc menu, then selecting the entry for Column Statistics. In abbreviated form, this is **Calc > Column Statistics**.

   Look at the items listed under Statistic. Each is preceded by a circle. These are radio buttons. You select one and only one of them. The currently selected button, Sum, has a dot inside it. Suppose you want the minimum value. Click its radio button and it is selected.

Exhibit 2.44   Example with Radio Buttons

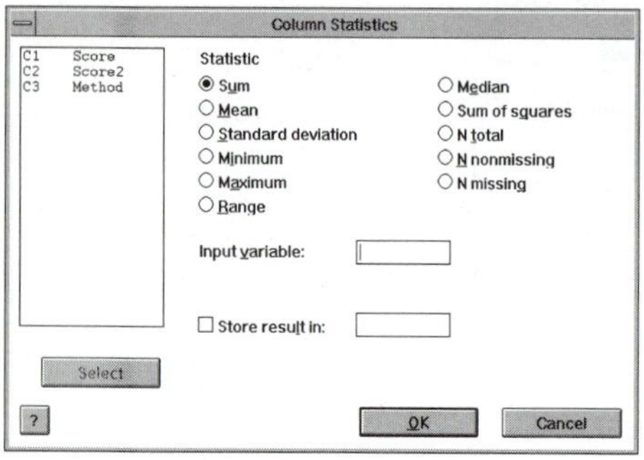

Exhibit 2.45   Data Window

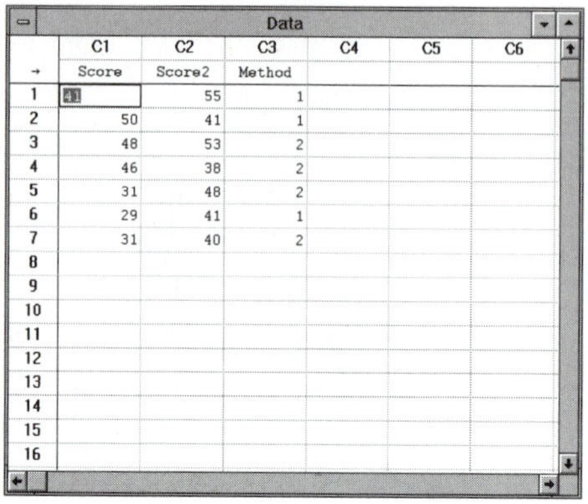

Data Window

The Data window shows the columns in your worksheet. Here you can enter, edit, and view your data. To make this window active, select it from the Window menu, or press Ctrl+D, or click it if it's visible.

Exhibit 2.45 shows the Data window, containing the numbers that are currently in the worksheet. Recall that these were entered using the READ command in Exhibit 2.41. You could also have entered them directly into the Data window.

The current cell is highlighted. You can move about in the window as you do in other Windows applications, using the scroll bars, the arrow keys, and the PgUp and PgDn keys. In addition, Ctrl plus an arrow key moves one screen in the direction of the arrow, Ctrl+Home moves to the first row of the first column of the data set, and Ctrl+End moves to the last row of the last column. Click in a cell to make it the current cell.

Entering and Editing Data. Click in the cell in row 8 of C1, the beginning of the first empty row. Type the number 45. The number appears in the cell. If you press Tab or Enter, the number will be put into that cell and the cursor will move to the right. If you press any arrow key, the number will also be put into the cell and the cursor will move in the direction of the arrow you pressed. If you press Esc, the cell will be returned to its original value. Press Tab.

Now you are in C2. Type 49 and press Tab. Finally, enter 1 in C3. You have now reached the end of row 8. Press Ctrl+Tab to go to the beginning of the next row. Now enter the numbers 27, 42, and 1. Note that pressing Ctrl+Enter puts the number in the current cell and moves to the top of the

Exhibit 2.46 Data Window with New Data

Data

	C1	C2	C3	C4	C5	C6
→	Score	Score2	Method			
1	41	55	1			
2	50	41	1			
3	48	53	2			
4	46	38	2			
5	31	48	2			
6	29	41	1			
7	31	40	2			
8	45	49	1			
9	27	42	1			
10						
11						
12						
13						
14						
15						
16						

Exhibit 2.47 Edit and Editor Menus When Data Window Is Active

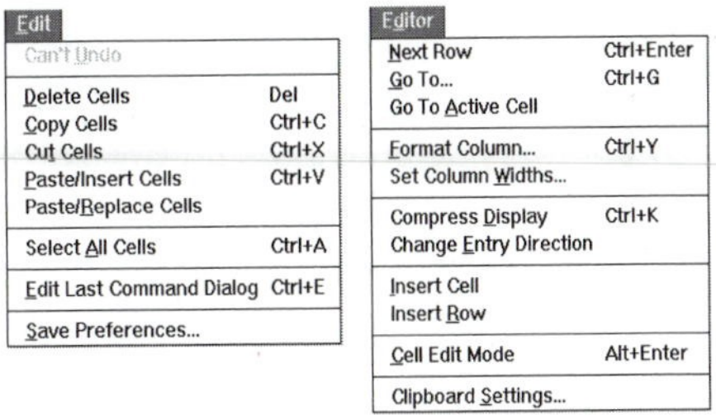

next column. Exhibit 2.46 shows the Data window after the new data have been entered.

If at any time you realize the number in a cell is incorrect, move to that cell and type the correct number.

Column Names. You can enter or change the name of a column. Click in the cell just below the column number and type the name for the column.

The Edit and Editor Menus. Two menus, Edit and Editor, have entries that affect the Data window.[4] The Edit menu is similar to Edit menus found in most Windows applications. Editor contains functions that are special to Minitab. These are both shown in Exhibit 2.47.

Suppose you want to delete the next-to-last row of data (row 8). Click the row number on the left and the entire row is highlighted. Select **Edit > Delete Cells**, or just press [Del], and the row is deleted. You can highlight a block of rows and delete them all together. Click in the first row, hold down the mouse button, drag to the last row, and release the button. Then select **Edit > Delete Cells**.

Suppose you want to insert a row between rows 7 and 8. Click any cell in row 8, then choose **Editor > Insert Row**. The data set is opened up and a row of stars is filled in for row 8. You can now type the data you want in that row.

You can also delete just one cell. For example, highlight the cell in row 8 of C2, then choose **Edit > Delete Cell**. The cell is deleted and now C2 is shorter than the other columns. You can also insert one cell in one column.

[4]In Release 10, the items in the Editor window change depending on what type of window is active: Data, Session, or Graph.

Highlight the cell you want to insert and choose **Editor > Insert Cell**; the column opens up and the cell contains a star. Now type your data.

You can use Copy, Cut, and Paste from the Edit menu as you can in other Windows applications. Thus you can highlight a block of rows and columns in the Data window, copy or cut them, then paste them somewhere else in the Data window, or in another application such as Notepad. If you want to paste data from another application into Minitab's Data window, the numbers must be separated by tabs.

Notice the little arrow in the upper-left part of the Data window. Selecting **Editor > Change Entry Direction** changes the way this arrow points (you can also click the arrow with a mouse). The arrow's direction controls the following keys:

Key	*Action When Arrow Points Right*	*Action When Arrow Points Down*
Home	Moves to beginning of current row.	Moves to top of current column.
End	Moves to end of current row.	Moves to bottom of current column.
Enter	Enters number and moves right.	Enters number and moves down.
Ctrl + Enter	Enters number, moves to beginning of next row.	Enters number, moves to beginning of next column.

Help on Minitab

Help is available from the Help menu, shown in Exhibit 2.48. Minitab uses the standard Windows Help system. You can learn about this by selecting the entry **Help > How to Use Help**.

You can enter the Help system through any of the entries under the Help menu (except About Minitab), or by clicking the ? button in a dialog box. Once you are in the Help system, you have full access to all Help topics. How you invoke Help just determines where you start. Minitab's Help contains answers to most of the questions you may have about using Minitab under Windows. It provides general introductory information as well as detailed explanations on how to use all dialog boxes.

Exhibit 2.48   Minitab's Help Menu

Help
Contents
Getting Started...
How do I...
Search for Help on...
How to Use Help
About Minitab

In this chapter, we give an overview of the user interface and other special features of Minitab under Windows. You can learn other things by reading through the Help information included with the software.

Saved Worksheets

Once you have entered data into Minitab, you will usually want to store it in a computer file so that you have it available for use in another session on another day. The most convenient type of file for use with Minitab is a Minitab saved worksheet. This file contains all columns, stored constants, and column names you have used.

The Session commands used with saved worksheets, SAVE and RETRIEVE, are described in Section 2.1. Here we will show you how to use the dialog boxes for SAVE and RETRIEVE.

Saving Worksheets. Suppose you want to save the data currently in the worksheet on a floppy disk in drive B. Select **File > Save Worksheet As**.

The dialog box, shown in Exhibit 2.49, gives you a choice of three formats.[5] The default is Minitab worksheets saved in DOS format. These have the extension .mtw. The second is portable format, used when you want to move a Minitab worksheet from one type of computer to another, for example from a DOS machine to a Macintosh. These have the extension .mtp. The third is Lotus format, used to move data between Minitab and Lotus or Symphony. These have the extensions .wk1, .wks, .wr1, and .wrk.

To select the default file format, a Minitab worksheet, just click Select File. This gives the dialog box shown in Exhibit 2.50. This is a standard Windows dialog box for saving files. In the Drives box, select b: and the

Exhibit 2.49 First Dialog Box for Save Worksheet As

Exhibit 2.50 Second Dialog Box for Save Worksheet As

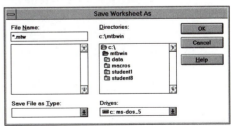

---

[5]In Release 10, the dialog box shown in Exhibit 2.49 is not used. The type of file is selected in the dialog box shown in Exhibit 2.50, in an item called List Files of Type. In addition, there are some new file types and new options.

Exhibit 2.51 Dialog Box to Save
Worksheet as File ex1.mtw

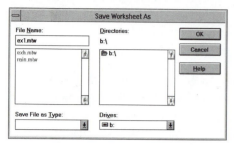

Exhibit 2.52 Dialog Box for Open
Worksheet

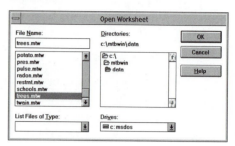

contents of the floppy disk in drive B will be displayed. This floppy disk
happens to contain two Minitab worksheets. In the File Name box, type **ex1**
as the name for your file. Exhibit 2.51 shows the dialog box after the file
name has been typed. Click OK and your file is saved as ex1.mtw on the
floppy disk. If you do not add an extension, Minitab adds the extension
.mtw.

You can also save a file by choosing **File > Save Worksheet**. In that
case, you will not be given a dialog box. Your worksheet will be saved in
the same file from which you opened it.

Retrieving Worksheets. Minitab comes with many data sets stored as
saved worksheets. Let's look at the one called "trees". Select **File > Open
Worksheet**. The dialog box that appears gives you a choice of three formats.[6] What you want is the default, a Minitab worksheet. Click Select File
and you will get a dialog box that looks similar to the one in Exhibit 2.51.
In general, dialog boxes for files show the directory that you last used.
Since you just stored a file on a floppy disk that is still in drive B, the
contents of this disk are shown.

The file we want is in c:\minitab\data. In the Drives box, select c: to get
back to the Minitab directory on the C drive. Highlight "data" and click OK
to get to the data subdirectory. This gives the dialog box in Exhibit 2.52. It
contains a list of files. Scroll down until you highlight "trees.mtw" and then
click OK. The data set is now in Minitab. You can view it in the Data window if you like.

Note: Anything you have in the worksheet when you retrieve a data set
is erased before the saved worksheet is brought in.

---

[6]In Release 10, this type of file is selected in the item List Files of Type, as it is for Save.

Exhibit 2.53 First Dialog Box for
Exporting ASCII Data

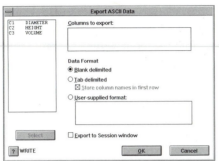

Exhibit 2.54 First Dialog Box for
Importing Data to an ASCII File

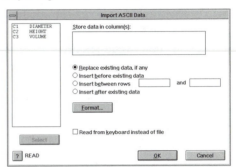

ASCII Data Files

You can transfer data between Minitab and other applications by using a data, or ASCII, file. Almost every application—every editor, spreadsheet, and word processor—can input and output data in ASCII format.

The session commands used with ASCII files—READ, SET, INSERT, and WRITE—are described in Section 2.1. Here we will show you how to use the dialog boxes corresponding to these commands.

Exporting ASCII Files. You can save the columns of your current work-sheet in an ASCII file. Select **File > Export ASCII Data** to get the dialog box in Exhibit 2.53.[7] Enter the columns to be exported. Only columns of data can be stored in an ASCII file. Stored constants and column names will be lost. Click OK and you will get a file dialog box, similar to the one in Exhibit 2.51. Select the directory for your file, then type a name in the File Name box and click OK. Minitab automatically adds the file extension .dat if you do not specify an extension.

When Minitab stores the specified columns, it adjusts the format to make the data as compact as possible. If all columns do not fit on one line, Minitab puts the continuation symbol, &, at the end of the line and continues the data on the next line. If the columns are of different lengths, Minitab makes them all the same length by adding the missing-values code (*) to the shorter columns.

Importing ASCII Files. To import an ASCII data file into Minitab, first select **File > Import ASCII Data**. This gives the dialog box shown in

---

[7]In Release 10, Export ASCII Data and Import ASCII Data are under File > Other Files.

Exhibit 2.54. This box corresponds to two Minitab Session commands, READ and INSERT. The first radio button is for READ. READ replaces any existing data. The other three buttons are for the different cases of IN-SERT. The first puts the new data at the tops of the existing columns. The second choice requires two consecutive row numbers: the new data are put between these rows. The last choice puts the new data at the bottom of the worksheet.

Next, select columns where you will store the data. For example, suppose you have a file called mydata.dat that contains 10 lines of data, each with 4 numbers. Suppose you want to put the data into C1–C4, replacing anything that is in these columns. Select the first radio button, enter C1–C4 in the box titled "Store data in column(s)", and click OK. This gives the usual box for opening files. Data files have the extension .dat, so only these are displayed. Now go to the subdirectory that contains the file mydata.dat, highlight the file name, and click OK.

The dialog box in Exhibit 2.54 contains a Format button. Most ASCII files you use will contain rows and columns of numbers, with at least one space or comma between numbers. In these cases, you do not need this option. However, if the file contains alpha data, if there are no spaces or commas between numbers, if blanks are used for missing values, if two or more lines of data are to go into one row of Minitab's worksheet, or if the file has any other unusual properties, then you will need to use the Format option to describe the data format. Formats are discussed in Section 17.5.

Saving and Printing the Session Window

You can save everything that appears in the Session window in a file, called an outfile. Then, after you have left Minitab, you can print the file or import it into a word processor to create a report.

Minitab's outfile option is basically a switch. When you turn it on, everything in the Session window is put into the outfile. When you turn it off, Minitab stops storing the Session window. When you turn it on again, Minitab stores what appears in the Session window again.

It is a good practice to begin every Minitab session by starting an outfile, so that you will have a permanent record of your session.

The Session commands used to start and stop an outfile, OUTFILE and NOOUTFILE, are described in Section 2.1. Here we will show you how to use the dialog boxes corresponding to these commands.

To start an outfile, choose **File > Other Files > Start Recording Session**. This gives the dialog box shown in Exhibit 2.55. Since the default version of this dialog box is what we want, choose Select File. This gives

Exhibit 2.55 First Dialog Box for Recording a Session

Exhibit 2.56 Second Dialog Box for Recording a Session

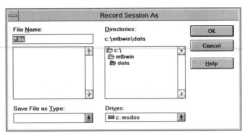

the dialog box shown in Exhibit 2.56. As usual, the directory that is shown is the last directory you used. If it is not what you want, move to the appropriate drive and directory. Files that contain Minitab output have the extension .lis, so only these files are displayed. Type a name in the File Name box and click OK.

From this point on, all commands and output that appear in the Session window will also be put into this file. (High-resolution graphs, however, will not. You must store these separately.) When you want to stop recording your session, select **Files > Other Files**. You will now see the entry **Stop Recording Session**. Choose this entry.

If, at a later time, you want to record more of your session, select **Files > Other Files > Start Recording Session**. If you choose the same file name as before, your new output will be appended to the old. If you type a new file name, the new output will go into the new file.

If you want a copy of your data on paper, you have two choices. Start an outfile if you don't already have one, then use Minitab's PRINT command to display the data on the screen (you can use the corresponding dialog box by selecting **Edit > Display Data**). The data will be put into the outfile. You can also print the data from the Data window directly. Make the Data window active and choose **File > Print Window**.

Graphs

There are two types of graphs. Character graphs appear in the Session window. High-resolution graphs appear in separate windows. For most analyses character graphs are good enough, although they do not look as pretty as the high-resolution graphs and lack their sophisticated options. Character graphs are listed in the menu under **Graphs > Character Graphs**.

Exhibit 2.57 Dialog Box for a Character Histogram

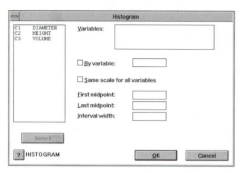

Exhibit 2.58 Session Window with Character Histogram

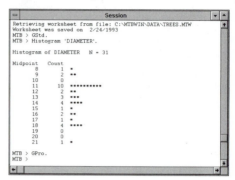

Let's look at Minitab's histogram. Retrieve trees.mtw. Suppose you want a character histogram of C1, DIAMETER. Select **Graph > Character Graphs > Histogram** to get the dialog box shown in Exhibit 2.57. Then enter C1 in the Variables box and click OK. Output is shown in Exhibit 2.58.

Notice the commands GStd and GPro that Minitab typed before and after the histogram. Minitab really has two separate graphics systems. GStd says you want character graphs; GPro says you want high-resolution graphs. If you never type a graph command, but always use dialog boxes, Minitab will always choose the appropriate type of graph.

If you want to type graph commands, you must first tell Minitab which type you want. This is easy. If you want character graphs, just type GSTD. From that point on, Minitab will interpret any graph command you type as a character graph. If you type GPRO, from that point on Minitab will interpret any graph command you type as a high-resolution graph. Almost all the graphs in the examples in this book are character graphs.

If you select **Graph > Histogram**, you will get the dialog box shown in Exhibit 2.59. It has many options. The high-resolution graphs in the Windows version of Minitab are quite sophisticated.[8] However, if you just want the default histogram, all you need to do is enter C1 in the first row of the box labeled Graph variables, and click OK. The graph appears in its own window, as shown in Exhibit 2.60. If you want to get a copy of this graph on paper, first make sure it is the active window, then select **File > Print Window**. You will get a dialog box for printing. Just click OK.

---

[8]Release 10 contains several new graphs and a Graph Editor (essentially a drawing package) that is under the Editor menu when a graph window is active.

Exhibit 2.59 Dialog Box for
High-Resolution Histogram

Exhibit 2.60 High-Resolution Histogram

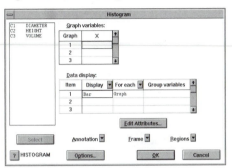

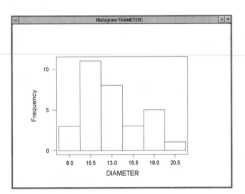

2.5   Release 9 on VAX/VMS and UNIX Workstations

Release 9 on VAX/VMS and UNIX workstations has the same statistical
functionality as Release 9 Windows. It does not, however, have a menu
interface or a data editor. Read Section 2.1 to learn about the commands
you will need.

There are two versions of Release 9 for VAX/VMS and UNIX: the
Standard Version and the Enhanced Version. The main difference is graphics. The Enhanced Version has a sophisticated high-resolution graphics system, called Professional Graphics, that the Standard Version does not have.
Look at the top line on your screen when you start Minitab, or type HELP,
to see which version you have.

Standard Version

The graphs in the Standard Version are the same graphs that are available
in Release 7. Read the material on graphs in Section 2.1 to learn about them.

Enhanced Version

The Enhanced Version essentially has two separate graphics systems, character graphs and Professional Graphics. The commands GSTD and GPRO
allow you to switch between them. If you want character graphs, type
GSTD. From that point on, Minitab will interpret any graph command you
type as a character graph. If you type GPRO, from that point on Minitab
will interpret any graph command you type as a Professional Graph. For

most analyses, character graphs are good enough, and we will use these for most of the examples in this book.

   Let's look at Minitab's histogram. Retrieve the Trees data set. The following commands produce two histograms. The first is a character graph; the second is a high-resolution graph.

```
GSTD
HISTOGRAM C1    # character histogram
GPRO
HISTOGRAM C1    # high-resolution histogram
```

Exhibits 2.58 and 2.60 show these histograms as produced on Release 9 Windows. Graphs produced on VAX/VMS and UNIX workstations are essentially the same.

Exercises

2-1   Students in a small class were given three exams, with the following results:

ID Number	Exam 1	Exam 2	Exam 3
4234	92	82	96
6457	84	84	80
5534	75	79	83
6213	98	60	72
9766	62	55	40
4538	79	72	81
4235	81	70	78

   (a)   Enter these data into Minitab. Use appropriate column names. If your version has a data editor, use the data editor. If not, use READ or SET.
   (b)   Calculate the average exam score for each student.
   (c)   Print a table that contains the ID number and average exam score for each student.
   (d)   Which student had the highest average? The lowest?
   (e)   Put your table on paper.

2-2   (a)   Retrieve the Minitab data set TREES.
   (b)   How tall is the tallest tree? The shortest tree? Use Minitab to help answer these questions.
   (c)   Create three new columns that contain the data for the trees that are at least 75 feet tall.
   (d)   How many trees are at least 75 feet tall? What is the average diameter of these trees? The average volume?

2-3 (a) Retrieve the Minitab data set TREES.
     (b) How many trees have a diameter under 10 feet?
     (c) Delete all trees with a diameter under 10 feet.
     (d) The last tree has a diameter of 20.6 feet. Change this diameter to 21.6 feet.
     (e) Add the following two rows of data to the end of the worksheet:

Diameter	Height	Volume
22.1	89	66.5
22.5	86	62.0

     (f) Display a character histogram of DIAMETER. If your version of Minitab has high-resolution graphics, display a high-resolution histogram as well.

PART TWO
Basic Methods

3

One Variable:
Displays and
Summaries

Much can be learned from data by looking at appropriate plots and tables. Sometimes such displays are all we need to answer our questions. In other cases they will help guide us to appropriate follow-up procedures. In fact, one great advantage of computers is their ability to produce a variety of data displays quickly and easily. In this chapter we introduce some of the most useful displays and some simple summary measures, such as the mean and the median.

We begin with a description of three basic types of data, because the type of analysis you use depends on the type of data you have.

(The section at the end of this chapter shows how to use high-resolution graphs in Releases 8 and 9.)

Note for Release 9 Users: If you are using Release 9 of Minitab (for Windows, VAX, or UNIX workstations) and want to type commands for the character graphs described in this chapter, remember to first type the command GSTD. From then on, you will get the character graphs. To return to typing high-resolution graphs, type GPRO.

3.1   Three Basic Types of Data

Not all numbers are created equal. Categorical data act merely as names; they tell us nothing about order or size. Ordinal data tell us about order but not about size. Interval data give information about size as well as about order.

Categorical Data

Simple examples of categorical variables are sex, which has two values (male and female) and state in the United States, which has fifty values (Alabama, Alaska, . . . , Wyoming). When such data are stored in the computer, they are often converted to numbers. This is usually just for the convenience of the computer. Sex might be coded 1 = male and 2 = female, or as 1 = female and 2 = male, or even as 1410 = male and 2063 = female. State might be coded in alphabetical order, going from 1 = Alabama to 50 = Wyoming.

One problem with computers is that they will do whatever you ask— even if it's nonsense. For example, a computer will gladly average categorical data, even though that average probably has no meaning. Suppose, for example, you have a data set of 30 men (coded 1) and 70 women (coded 2). A computer will calculate the average sex as 1.7. With computers, as with other tools, it is up to you to see that they are used properly. One of the goals of this book is to help you do that.

Categorical data are also called nominal or classification data.

Ordinal Data

One example of ordinal data is army rank: private, corporal, sergeant, lieutenant, major, colonel, and general. We know that a general is one rank higher than a colonel and a corporal is one rank higher than a private. But is the distance from private to corporal the same as the distance from colonel to general? Does distance between army ranks really have any meaning? Probably not.

Perhaps the most common occurrence of ordinal data is in surveys and questionnaires. For example:

"The President is doing a good job." Check one:

Strongly Disagree	Disagree	Indifferent	Agree	Strongly Agree

When entered into the computer, ordinal data are often converted to numbers—for example, 1 = strongly disagree, 2 = disagree, . . . , 5 = strongly agree. Here we would know that a 4 was more favorable toward the President than a 3, but we would not have any clear idea how much more favorable.

Interval Data

These are usually based on measurements such as length, weight, or time. On an interval scale, 4 is halfway between 3 and 5. For example, the difference between a 4-centimeter rod and a 3-centimeter rod is the same as the difference between a 5-centimeter rod and a 4-centimeter rod. Unlike categorical and ordinal data, interval data occur naturally as numbers.

Exercises

3-1 For each of the following, indicate whether the data are best considered as nominal, ordinal, or interval, and justify your choice.
   - (a) The response of a patient to treatment: none, some improvement, complete recovery.
   - (b) The style of a house: split-level, one-story, two-story, other.
   - (c) Income in dollars.
   - (d) Temperature of a liquid.
   - (e) Area of a parcel of land.
   - (f) Highest political office held by a candidate.
   - (g) Grade of meat: prime, choice, good, or utility.
   - (h) Political party.

3-2 Appendix A contains a collection of data sets. Look through these data sets and select at least two examples of each of the three basic data types.

3.2 Histograms

We will get a histogram of the variable PULSE1 from the Pulse data set described in Appendix A. First we need to enter this data set into Minitab. If you're using Release 8 or Release 9, Windows, use the Open Worksheet menu item; otherwise, type the command RETRIEVE. In Exhibit 3.1, we used the Minitab command HISTOGRAM to get a (character-based) histogram. Minitab grouped the pulse rates into 11 intervals, each of width 5. The first interval has a midpoint of 50, goes from 47.5 to 52.5, and contains one observation. The second interval has a midpoint of 55, goes from 52.5 to 57.5, and contains two observations.

A histogram gives a graphical summary of the data: In Exhibit 3.1, we can see that the lowest pulse rate is about 50, the highest is about 100, and the most popular interval is the one at 70—one-quarter of the pulse rates are in this interval.

When we talk about the scale of a histogram, we mean the intervals:

Exhibit 3.1   Histogram of Pulse Rates

```
MTB > histogram 'pulse1'

Histogram of PULSE1    N = 92

Midpoint    Count
       50       1    *
       55       2    **
       60      17    *****************
       65       9    *********
       70      23    ***********************
       75      10    **********
       80      11    ***********
       85       6    ******
       90       9    *********
       95       3    ***
      100       1    *
```

how many, how wide, and where they start. Minitab automatically chooses a scale. If you want a different scale, you can specify it with subcommands. For example, suppose you wanted the intervals to be 40 to 50, 50 to 60, 60 to 70, . . . , 100 to 110. In this example the width, or increment, of each interval is 10 and the starting midpoint is 45. The following instructions can be used:

```
HISTOGRAM 'PULSE1';
   INCREMENT = 10;
   START = 45.
```

Any pulse rate that falls on a boundary between two intervals is put in the higher interval. For example, a pulse of 60 would be put in the interval from 60 to 70.

The histogram is designed primarily for interval data, although it can be used with ordinal and even categorical data as well.

Most of the commands in this chapter have a subcommand, BY, that allows you to produce separate displays for different groups. In the Pulse data set, there are 92 people: 57 men and 35 women (we used the command TALLY, described later in this chapter, to determine these numbers). To get two histograms, one for the pulse rate of the men and the other for the pulse rate of the women, you can use

```
HISTOGRAM  'PULSE1';
   BY 'SEX'.
```

HISTOGRAM C...C

INCREMENT = **K**
START with midpoint at **K** [end with midpoint at **K**]
SAME scale for all columns
BY C

Prints a histogram for each column. Observations on the boundary between two intervals are put in the interval with higher values.

You may specify your own scale with the subcommands. INCREMENT specifies the distance between midpoints or, equivalently, the width of each interval. START specifies the midpoint for the first and, optionally, the last interval. Any observations beyond these intervals are omitted from the display.

If you use SAME, the same scale is used for all the columns listed on HISTOGRAM.

If you use BY, a separate histogram is created for each value in C. All histograms are put on the same scale. The column C must contain integers from -10000 to $+10000$.

Release 8: **Graph > Histogram**
Windows: **Graph > Character Graphs > Histogram**

Exercises

3-3   Consider the histogram in Exhibit 3.1.
   (a)   How many pulse rates fell in the interval with midpoint 65?
   (b)   Are there more pulse rates above 75 or below it? Can you tell from this histogram?
   (c)   Were there any outliers (that is, any pulse rates that were much smaller or much larger than the others) in this data set?

3-4   (a)   Make a histogram of the following numbers:

        36, 43, 82, 84, 81, 84, 45, 60, 64, 71, 81, 78, 79, 43, 79

(b)   Make a histogram for numbers that are 10 times the numbers in (a). (The LET command may be useful here.) Compare this display to the one in part (a). Is the overall impression about the same?

(c)   Repeat (b) with numbers 5 times those in part (a). Compare this display to those in parts (a) and (b).

(d)   Make a histogram of the numbers in part (a) using the subcommand START 37.5. The intervals will be five units wide, as they are in part (a), but will have different midpoints. The first interval will contain values from 35 through 39, the second interval will contain values from 40 through 44, the next 45 through 49, and so on. Compare the overall shape of this display to the one in part (a).

3-5   Use the Pulse data set (described in Appendix A). Do histograms of both first and second pulse rates for all the students. Use the SAME subcommand to put these two histograms on the same scale. Interpret the output.

3.3   Dotplots

Exhibit 3.2 produces a dotplot of the pulse data we used in Exhibit 3.1. This display is very similar to a histogram with many small intervals. In Exhibit 3.2, there are over 50 spaces. Each space represents 1 beat per second. The numbers along the axis refer to the middle of each space. Thus the space labeled 70 goes from 69.5 to 70.5. An observation that falls on the boundary between two spaces goes in the lower interval. Thus, 70.5 would be put in the interval labeled 70. (This convention is the opposite of the one we used for HISTOGRAM. However, it seems to be the more natural one for DOT-PLOT.)

Exhibit 3.2   Dotplot of Pulse Rates

```
MTB > dotplot 'pulse1'

                                   .
                          .        :
                          :        :
                          :   . : : :  . . .
                      . : : : : : : : : :  . .  :     . :
             .        :      : : . : : : : : : : : : : . . : : : . . :        .
          ---+---------+---------+---------+---------+---------+---
PULSE1
```

Exhibit 3.3   Dotplots of Pulse Rates, Done Separately for Men and Women

```
MTB > dotplot 'pulse1';
SUBC>    by 'sex'.


                                          :
   SEX                         :       : :      .
   1                     . : .  . : : : : .
              .      :    : : : : : : : : : : .  . . : .   . : :
              ---+---------+---------+---------+---------+---------+---
   PULSE1
   SEX                           .        .          :
   2                      . ..: . : :    :    : : : : : ...: .   . :   .
              ---+---------+---------+---------+---------+---------+---
```

A histogram groups the data into just a few intervals. For example, HISTOGRAM used 11 intervals for PULSE1. A dotplot, on the other hand, groups the data as little as possible. Ideally, if we had wider paper or a printer with higher resolution, we would not group the data at all. Histograms tend to be more useful with large data sets; dotplots, with small data sets. Histograms show the shape of a sample; dotplots do not.

Dotplots are useful if you want to compare two or more sets of data. Exhibit 3.3 shows two dotplots, one for the men and one for the women. They are on the same scale, so you can compare these two groups. The two groups do overlap quite a bit, as you would expect. However, the lowest pulse rates are for men and the highest are for women.

Exercise

3-6   (a)   The Lake data (described in Appendix A) provide measurements for 71 lakes in northern Wisconsin. Display a dotplot of the areas of these lakes. What are the striking features of this plot?

(b)   How many lakes are under 2,000 acres? How many are under 1,000 acres?

(c)   Remove the lake in row 55. Display a dotplot of the areas of the remaining lakes. How does this plot compare to the one in part (a)?

DOTPLOT C...C

INCREMENT = K
START at **K** [end at **K**]
SAME scale for all columns
BY C

Makes a dotplot for each column. Observations on a boundary are put in the lower (smaller values) interval.

You may specify your own scale on DOTPLOT using the subcommands. INCREMENT specifies the distance between tick marks (the + signs) on the axis. Since there are 10 spaces between tick marks, the width of each space will be K /10. START specifies the first, and optionally the last, tick mark on the axis. Any points outside are omitted from the display.

If you use SAME, the same scale is used for all the columns listed on DOT-PLOT.

If you use BY, a separate dotplot is done for each value in C. All dotplots are put on the same scale. The column C must contain integers from −10000 to +10000.

Release 8: **Graph > Dotplot**
Windows: **Graph > Character Graphs > Dotplot**

3.4   Stem-and-Leaf Displays

A stem-and-leaf display is similar to a histogram, but uses the actual data to create the display. The display is a relatively new technique that was introduced by statistician John Tukey in the late 1960s. It is designed primarily for interval data, although it can be used with any set of numbers.

Suppose we make a stem-and-leaf display of the pulse data used in Exhibit 3.1. Exhibit 3.4 shows the display after we have entered the first four pulse rates: 64, 58, 62, and 66. The digits to the left of the vertical line are called the stems. The digits to the right are called the leaves. To create the display, we split each pulse rate into two parts: the tens digit became the stem, and the ones digit became the leaf. For example, 64 was split into 6 = stem and 4 = leaf, 58 was split into 5 = stem and 8 = leaf, and so on. The stems for the entire data set were listed to the left of the vertical line. Each leaf was put on the same line as its stem. At this point, the line with stem = 5 has just one leaf, an 8. This represents the pulse rate 58. The line with stem

Exhibit 3.4   Stem-and-Leaf Display of First Four Pulse Rates

4	
5	8
6	426
7	
8	
9	
10	

Exhibit 3.5   Stem-and-Leaf Display of 92 Pulse Rates

4	8 .
5	84848
6	42648280268628220288842088814062688
7	46020408826442060608024282686
8	402802844840276
9	026600240
10	0

= 6 has three leaves, 4, 2, and 6. These represent the three pulse rates 64, 62, and 66.

Exhibit 3.5 shows the stem-and-leaf display for all 92 people. Reading from the top of the display, we see that the pulse rates are 48, 58, 54, 58, 54, 58, 64, 62, 66, . . . , 94, 90, 100. This display contains the same information as the original list of numbers, but presents it in a more compact and usable form. The numbers are closer to being in order. We can easily see the range of the data (from a low of 48 to a high of 100) and the most popular categories (the 60s, followed by the 70s, 80s, and 90s). The general shape of the picture is nearly symmetric. There are no gaps (stems with no observations) and no outliers (observations that are much smaller or much larger than the bulk of the data).

Looking more closely, we see that all but two of the numbers are even. Why? A reasonable conjecture, and a correct one, is that pulses were counted for 30 seconds and then doubled to get beats per minute. But what about the other two, the 61 and 87? These two people may have multiplied incorrectly, or counted a half beat, or written down the wrong number. (For example, the 87 could be a transposition error; perhaps 78 was correct.)

Exhibit 3.6 shows the display produced by Minitab's STEM-AND-LEAF command. This differs from our hand-drawn display in several

Exhibit 3.6 Stem-and-Leaf Display Produced by Minitab

```
MTB > stem 'pulse1'

Stem-and-leaf of PULSE1     N  = 92
Leaf Unit = 1.0

      1      4 8
      3      5 44
      6      5 888
     24      6 000012222222224444
     40      6 6666688888888888
    (17)     7 00000022222244444
     35      7 6666688888
     25      8 0002224444
     15      8 67888
     10      9 0000224
      3      9 66
      1     10 0
```

ways. First, an extra column, called depths, was added to the left of the display. Further, the message Leaf Unit = 1.0 was added. (We will discuss both of these additions under Further Details.) Second, each stem is listed on two lines, with leaf digits 0, 1, 2, 3, and 4 on the first line and 5, 6, 7, 8, and 9 on the second. This gives a display that is more spread out than our hand-drawn one.

Minitab also ordered the leaves on each line, making it easier to see what values we have. For example, now it is clear that the most common pulse rate is 68 beats per minute.

Further Details

The depth of a line indicates how many leaves lie on that line or "beyond." For example, the 6 on the third line from the top tells us there are six leaves on that line and above it. The 10 on the third line from the bottom tells us there are ten leaves on that line and below it. The line with the parentheses will contain the middle observation if the total number of observations, N, is odd. It contains the middle two observations if N is even, as it is here. The parentheses enclose a count of the number of leaves on this line. Note that if N is even and if the two middle observations fall on different lines, no parentheses are used in the depth column.

In this example, Minitab listed each stem on two lines. In some cases, Minitab will use five lines for each stem. The number of lines per stem is always one, two, or five, and is determined by the range of the data and the number of values present.

In our example, all but one of the pulse rates contained two digits, so it was easy to split each number into a stem and a leaf. When numbers contain more than two digits, the STEM-AND-LEAF command drops digits that don't fit. For example, the number 927 might be split as stem = 9, leaf = 2, and 7 dropped.

Decimal points are not used in a stem-and-leaf display. Therefore, the numbers 260, 26, 2.6, and .26 would all be split into stem = 2 and leaf = 6. The heading Leaf Unit tells you where the decimal point belongs: For the number 260, Leaf Unit = 10; for 26, Leaf Unit = 1; for 2.6, Leaf Unit = .1; and for .26, Leaf Unit = .01.

STEM-AND-LEAF has a subcommand, INCREMENT, that allows you to control the scale of a stem-and-leaf display. For example, suppose we wanted Minitab to produce the display shown in Exhibit 3.5. There the first stem contains all the numbers in the 40s and the second stem contains all the numbers in the 50s. Therefore, the distance, or increment, from one stem to the next is 10. To specify this scale, use

```
STEM-AND-LEAF 'PULSE1';
  INCREMENT = 10.
```

STEM-AND-LEAF C...C

INCREMENT = K
BY C

Prints a stem-and-leaf display for each column.

INCREMENT specifies the distance from one stem to the next. The increment must be 1, 2, or 5, with perhaps some leading or trailing zeroes. Thus, examples of allowable increments are 1, 2, 5, 10, 20, 50, 100, 200, 500, 1, 0.2, 0.5, 0.01, 0.02, 0.05.

If you use BY, a separate dotplot is done for each value in C. All dotplots are put on the same scale. The column C must contain integers from −10000 to +10000.

Release 8: **Graph > Stem-and-Leaf**
Windows: **Graph > Character Graphs > Stem-and-Leaf**

Exercises

3-7   Consider the stem-and-leaf display in Exhibit 3.6.
    (a)   What were the lowest and highest pulse rates?
    (b)   How many people had a pulse rate of 58?
    (c)   How many had a pulse rate in the eighties?
    (d)   How many had a pulse rate of 64 or lower?
    (e)   Could you answer any of the questions in parts (a)–(d) using only the histogram in Exhibit 3.1? Could you use the histogram to get approximate answers? For each question in parts (a)–(d), give the best answer or range of possibilities you can using only the histogram in Exhibit 3.1.

3-8   (a)   Do a stem-and-leaf display of the following numbers by hand:

        36, 43, 82, 84, 81, 84, 45, 60, 64, 71, 81, 78, 79, 43, 79

    (b)   Use Minitab to do a stem-and-leaf display of the numbers in part (a). Compare it to your hand-drawn display.
    (c)   Multiply the numbers in part (a) by 10. Then use Minitab to get a stem-and-leaf display. Explain the differences between the displays in (b) and (c).
    (d)   Multiply the numbers in part (a) by 2. Use Minitab to get a display of these numbers. How does this display compare to the one in part (b)?
    (e)   Multiply the numbers in part (a) by 5 and get a stem-and-leaf display. Compare this display to the one in part (a).

3-9   (a)   Use Minitab to get a stem-and-leaf display of the variable HEIGHT from the PULSE data set.
    (b)   Convert HEIGHT from inches to centimeters (to do this, multiply by 2.54) and then get a stem-and-leaf display. Comment on the "unusual" appearance of this display. Compare it to the display in part (a), where height was measured in inches.

3-10 (a)   Make a stem-and-leaf display of the variable WEIGHT from the PULSE study. Do you see any special pattern in the leaf digits of the display?
    (b)   What increment did Minitab choose for this display? What is the next smaller increment? Use this value with the INCREMENT subcommand to make a display. Compare this display to the one in part (a). Now use the next larger increment to make a display. Compare this display to the one in part (a).

3.5 One-Number Statistics

We often want to summarize an important feature of a set of data by using just one number. For example, we might use the mean to indicate the center or typical level of the data. We could use the range, the largest value minus the smallest value, to indicate how spread out the data are.

In this section we first discuss DESCRIBE, a command that prints a table of summary numbers. Then we show how these summaries can be computed individually, first for columns of data, then across rows.

The DESCRIBE Command

The Pulse data set (described in Appendix A) contains the weights of 92 people. Exhibit 3.7 provides some summaries of these numbers. The stem-

Exhibit 3.7   A Summary of the Weights in the Pulse Data Set

```
MTB > stem-and-leaf 'weight'

Stem-and-leaf of WEIGHT    N  = 92
Leaf Unit = 1.0

    1      9 5
    4     10 288
   13     11 002556688
   24     12 00012355555
   37     13 0000013555688
  (11)    14 00002555558
   44     15 0000000000355555555557
   22     16 000045
   16     17 000055
   10     18 0005
    6     19 00005
    1     20
    1     21 5

MTB > describe 'weight'

                N      MEAN    MEDIAN    TRMEAN     STDEV    SEMEAN
WEIGHT         92    145.15    145.00    144.52     23.74      2.48

               MIN       MAX        Q1        Q3
WEIGHT       95.00    215.00    125.00    156.50
```

and-leaf display shows several interesting things. For example, most people reported their weight to the nearest 5 pounds, especially the heavier people, and the heaviest person weighed 215 pounds.

The DESCRIBE command printed the following statistics:

N = 92. This tells us there were 92 people in the study who reported their weights.

MEAN = 145.15. This is the average of all 92 weights. The mean, often written as μ, is the most commonly used measure of the center of a batch of numbers.

MEDIAN = 145.00. To find the median, first order the numbers. If N, the number of values, is odd, the median is the middle value. If N is even, the median is the average of the two middle values. Here $N = 92$, so the median is the average of the 46th and 47th values. These are both 145, so their average is 145. The median is another value used to indicate where the center of the data is.

TRMEAN = 144.52. This gives a 5% trimmed mean. First the data are sorted. Then the smallest 5% and the largest 5% of the values are trimmed; the remaining 90% are averaged. Here $N = 92$, and 5% of 92 is 4.6. This is rounded to 5. Thus the five smallest values (95, 102, 108, 108, 110) and the five largest values (190, 190, 190, 195, 215) are trimmed. The remaining 82 values are averaged to give the trimmed mean.

STDEV = 23.74. This is the standard deviation. It is the most commonly used measure of how spread out the data are. The general formula is

$$\text{STDEV} = \sqrt{\frac{(1-4)^2 + (3-4)^2 + (6-4)^2 + (4-4)^2 + (6-4)^2}{5-1}} = \sqrt{4.5} = 2.12$$

SEMEAN = 2.47. This is the standard error of the mean. The formula is STDEV/ $\sqrt{N}$. For the Pulse data,

$$\text{SEMEAN} = \text{STD}/\sqrt{N} = 23.7/\sqrt{92} = 2.5$$

MIN = 95.00.   The minimum, or smallest, value.

MAX = 215.00.   The maximum, or largest, value.

Q1 = 125.00.   The first, or lower, quartile.

Q3 = 156.50.   The third, or upper, quartile.

The median is the second quartile, Q2. The three numbers Q1, Q2, and Q3 split the data into four essentially equal parts. The concept of a quartile is very simple. However, when we try to give a formal definition, we must handle details such as how to divide ten observations into four equal parts.

DESCRIBE C...C

BY C

Prints the following statistics for each column:

N   Number of nonmissing values in the column

NMISS   Number of missing values (omitted if there are no missings)

MEAN   Mean or average

MEDIAN   Median or middle value

TRMEAN   5% trimmed mean

STDEV   Standard deviation

SEMEAN   Standard error of the mean

MAX   Maximum value

MIN   Minimum value

Q3   Third quartile

Q1   First quartile

If you use BY, statistics are calculated separately for each value in C. The column C must contain integers from −10000 to +10000.

Stat > Basic Statistics > Descriptive Statistics

There are several ways to do this, and you will find that different books define quartiles differently. All, however, give answers that are very close.[1] NMISS. This is the number of values recorded as "missing." Here no weights were missing, so the NMISS line was not printed by DESCRIBE.

---

[1]Here is the definition Minitab uses: First order the observations from smallest to largest. Then Q1 is at position $(N + 1)/4$ and Q3 is at position $3(N + 1)/4$. If the position is not an integer, interpolation is used. For example, suppose $N = 10$. Then $(10 + 1)/4 = 2.75$ and Q1 is between the second and third observations (call them x_2 and x_3) and it is three fourths of the way up. Thus, $Q1 = x_2 + .75 (x_3 − 2)$. For Q3, $3(10 + 1)/4 = 8.25$. Thus, $Q3 = x_8 + .25(x_9 − x_8)$, where x_9 and x_8 are the eighth and ninth observations. In the weight example, $N = 92$, so $(N + 1)/4 = 23.25$ and Q1 is between the 23rd and 24th observations. These are both 125, so $Q1 = 125$. Q3 is between the 69th and 70th observations and is equal to 156.5.

One-Number Statistics for Columns

DESCRIBE prints a collection of summary statistics. Minitab also has commands that calculate and store each statistic separately.

Three of these statistics are not printed by DESCRIBE: SUM, SSQ, and COUNT. The command SUM just adds all the values in the column. SSQ is the sum of the squares of the values. For example, for the data 1, 3, 5, and 4, SSQ = $1^2 + 3^2 + 5^2 + 4^2 = 51$. The command COUNT gives the total number of entries in a column. Thus, COUNT = N + NMISS.

Column Statistics with LET. All the one-number statistics for columns can also be used in a LET statement. Here are some examples:

```
LET K1 = MEAN(C1)
LET C2 = C1 - MEAN(C1)
LET K2 = SUM(C1)/N(C1)
LET K3 = MEDIAN('HEIGHT')
```

Notice that you must enclose the column in parentheses when you use column statistics such as MEAN, SUM, or MEDIAN in a LET command.

By using LET, you easily can calculate many statistics that are not built into Minitab. For example, another measure of the center of a set of numbers is the midrange. This is the average of the smallest and largest values. It can be calculated by

```
LET K1 = (MAX(C1) + MIN(C1))/2
```

N	**C** [put in **K**]
NMISS	**C** [put in **K**]
MEAN	**C** [put in **K**]
MEDIAN	**C** [put in **K**]
STDEV	**C** [put in **K**]
MAX	**C** [put in **K**]
MIN	**C** [put in **K**]
SUM	**C** [put in **K**]
SSQ	**C** [put in **K**]
COUNT	**C** [put in **K**]

Each command calculates one number. This statistic is printed; storage is optional.

Release 8: **Calc > Functions and Statistics > Column Statistics**
Windows: **Calc > Column Statistics**

One-Number Statistics for Rows

The statistics we have just described for columns are also available for rows. The command names are the same except that an R (for row) has been added. Here are two examples:

```
RSUM C1-C3   C4
RMAX C1-C3   C5
```

C1	C2	C3		C4	C5
1	7	3		11	7
4	2	3		9	4
1	3	2		6	3
3	5	5		13	5

Exercises

3-11   Consider the output in Exhibit 3.7.
   (a)   How many people are exactly 150 pounds? Over 150 pounds?
   (b)   The mean weight was 145.15. How many people weigh less than the mean?

RN	C...C put in C
RNMISS	C...C put in C
RMEAN	C...C put in C
RMEDIAN	C...C put in C
RSTDEV	C...C put in C
RMAX	C...C put in C
RMIN	C...C put in C
RSUM	C...C put in C
RSSQ	C...C put in C
RCOUNT	C...C put in C

These commands compute summaries across rows rather than down columns. The answers are always stored in a column.

Release 8: **Calc > Functions and Statistics > Row Statistics**
Windows: **Calc > Row Statistics**

(c)   The range of a set of numbers is defined as (maximum value) – (minimum value). Find the range for the weight data.

(d)   There is a connection between the range and the standard deviation of a data set. In many data sets, the range is approximately four times the standard deviation. Is this true for the weight data?

(e)   Calculate the two values (MEAN minus STDEV) and (MEAN plus STDEV). In many data sets, approximately two-thirds of the observations fall between these two values. Is this true for the weight data?

3-12 Consider the following 11 numbers: 5, 3, 3, 8, 9, 6, 9, 9, 10, 5, 10.

(a)   By hand, calculate each of the 10 statistics printed by DESCRIBE.

(b)   Use DESCRIBE to check your answers.

3-13 Suppose Minitab did not have the command STDEV. Show how to use LET and the formula on page 94 to calculate the standard deviation of the 11 observations in Exercise 3-12.

3-14 The median is said to be "resistant" to the effects of a few outlying points in a data set. That is, the median will not be very different even if there are a few unusually large values or abnormally small values in the data set. However, the mean, is not resistant to outliers.

(a)   Enter the observations in Exercise 3-12 into C1. Use DESCRIBE to find the mean and median.

(b)   Use the command LET C1(1) = 25 to change the 5 in row 1 to a 25. Again, use DESCRIBE to find the mean and median. How have they changed?

(c)   Use the command LET C1(1) = 100 to change the first observation to a 100. Find the mean and median. How have they changed?

3-15 (a)   Use Minitab to compute the standard deviation of the numbers

6, 8, 4, 10, 12, 3, 4, 10

(b)   Add 29 to each number. Now compute the standard deviation. How does your answer compare with that in (a)?

(c)   Multiply the data in (a) by 16 and compute the standard deviation. How does your answer compare with that in (a)? If you are not sure, divide the standard deviation in (c) by the one in (a).

Exhibit 3.8   Boxplot of the Weights from the Pulse Data Set

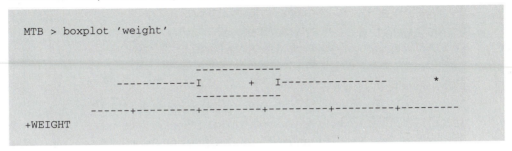

```
MTB > boxplot 'weight'
```

3.6   Boxplots

Boxplots, also called box-and-whisker plots, display the main features of a batch of data. They are especially useful for comparing two or more groups of data, as we will do in later chapters.

Exhibit 3.8 is a boxplot of the weights of the 92 people in the Pulse data set. The plus symbol (+) is the median of the data. The left "I" is the lower hinge, H_L, essentially the lower quartile; the right "I" is the upper hinge, H_U, essentially the upper quartile. A box is drawn from the lower to the upper hinge and represents the middle half of the data. Whiskers are the lines that extend from the ends of the box to the adjacent values. The adjacent values are the lowest and highest observations that are still within the following limits:

Lower limit: $H_L - 1.5(H_U - H_L)$
Upper limit: $H_L + 1.5(H_U - H_L)$

Observations outside these limits are plotted with asterisks. These points are far from the mass of the data, and are often called outliers. There is one outlier in Exhibit 3.8, a weight of over 200 lb.

Exercises

3-16   Use BOXPLOT to summarize each variable in the Pulse data set. When is BOXPLOT most useful? What does the display show you in each case?

3-17   Use BOXPLOT to study two variables, AREA and DEPTH, in the Lake data. What do the plots tell you about the data?

3-18   Use BOXPLOT with its subcommand BY to compare PULSE1 for those who smoke and for those who do not.

BOXPLOT C

> **INCREMENT** = K
> **START** with midpoint at **K** [end with midpoint at **K**]
> **BY C**

Prints a boxplot for the column.

You may specify your own scale with the subcommands. INCREMENT specifies the distance between tick marks (the + signs) on the axis. START specifies the first and, optionally, the last tick mark on the axis. Any observations beyond these intervals are omitted from the display.

If you use BY, a separate dotplot is done for each value in C. All dotplots are put on the same scale. The column C must contain integers from −10000 to +10000.

Release 8: **Graph > Boxplot**
Windows: **Graph > Character Graphs > Boxplot**

3.7   Summarizing Categorical and Ordinal Data

The preceding sections described techniques that are designed primarily for interval data. This section introduces TALLY, a command that is designed especially for categorical and ordinal data. It gives four statistics: count, cumulative count, percent, and cumulative percent.

The Pulse data set contains several categorical variables (what are they?). Exhibit 3.9 is a TALLY of sex and activity level. Sex was coded as 1 = male and 2 = female. Activity was recorded as 1 = low, 2 = moderate, and 3 = high. The first table tells us there are 57 men and 35 women, for a total of N = 92. The second table has a curious feature: One person had an activity level of 0. (Does this mean no activity?) We should find the correct value or, if we cannot, change the 0 to * (Minitab's missing-data code) before doing further analyses.

In Exhibit 3.10, we changed the 0 activity level to * (missing), then did a table with all four statistics calculated by TALLY. The first column tells us that N = 91 and * = 1; that is, there are 91 nonmissing observations and 1 missing. CUMCNT gives the cumulative counts. Thus, there were 9 people with low activity level, 9 + 61 = 70 with low to moderate activity level, and 9 + 61 + 21 = 91 with low to high activity level (that's everyone). PERCENT is (COUNT/N) × 100. Thus, (9/91) × 100 or 9.89% of the

Exhibit 3.9 TALLY of Sex and Activity

```
MTB > tally 'sex' 'activity'

    SEX   COUNT    ACTIVITY   COUNT
     1      57        0         1
     2      35        1         9
    N=      92        2        61
                      3        21
                     N=        92
```

Exhibit 3.10 TALLY with Corrected Data and All Statistics

```
MTB > let 'activity(54) = '*'
MTB > tally 'activity';
SUBC> all.

ACTIVITY   COUNT CUMCNT PERCENT   CUMPCT
    1        9      9     9.89     9.89
    2       61     70    67.03    76.92
    3       21     91    23.08   100.00
   N=       91
   *=        1
```

students, had a low activity level. Similarly, $(61/91) \times 100$, or 67.03%, had a moderate level. CUMPCT gives the cumulative percent. This is (CUMCNT/N) $\times$ 100. You can use subcommands to ask for each of these statistics individually or use the subcommand ALL to get all four, as we did in Exhibit 3.10.

Exercises

3-19   The Furnace data set (described in Appendix A) contains several categori-
       cal variables. These include the type of furnace in a house and the shape of
       the chimney. Use TALLY to help you answer the following questions.
       (a)   What was the most popular chimney type? The least popular?
       (b)   What percent of the houses had a forced-water furnace?
       (c)   Why does chimney shape have N = 89 and chimney type have N =
             90?
       (d)   What percent of the houses had a round chimney?

TALLY C...C

    COUNTS
    PERCENTS
    CUMCNTS   Cumulative counts
    CUMPCTS   Cumulative percents
    ALL   Same as using all four preceding subcommands

TALLY prints a separate frequency table for each column listed. The columns must contain integers from -10000 to $+10000$. Any one or more of the subcommands may be used. They specify what to print in the table. If no subcommands are given, just COUNTS are printed.

Stat > Tables > Tally

3-20 (a)   Two more variables in the Pulse data set are best suited for TALLY. Which are they?

      (b)   Make a tally of each. Look carefully at the results. Are there any things that seem strange?

3-21 Use TALLY to help you answer the following questions concerning the Cartoon data set (described in Appendix A).

      (a)   How many people got a perfect score on the immediate cartoon test? On the delayed cartoon test?

      (b)   How many people failed to take the delayed cartoon test?

      (c)   What percentage of people got a score of 5 or less on the immediate cartoon test? Of 7 or higher?

3.8 Optional Material on High-Resolution Graphs

In this section, we show you how to create high-resolution versions of HISTOGRAM and BOXPLOT. DOTPLOT and STEM-AND-LEAF do not have high-resolution versions. This material is optional; you can just use the character versions to do your data analysis if you prefer.

There are four parts in this section. Read the part that applies to the version of Minitab that you are using.

Exhibit 3.11   High-Resolution Histogram, Using Release 7

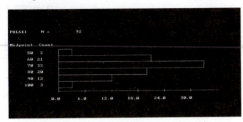

Exhibit 3.12   High-Resolution Boxplots, Using Release 7

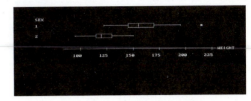

Release 7

There is a high-resolution version of HISTOGRAM and BOXPLOT. These have the same options as the character versions. The only difference is that the command names start with the letter G.

HISTOGRAM. The following command creates a high-resolution histogram for the variable PULSE1 in the Pulse data set. We use an increment, or bin width, of 10. Output is shown in Exhibit 3.11.

```
GHISTOGRAM   'PULSE1';
   INCREMENT 10.
```

BOXPLOT. We will get high-resolution boxplots for the weights in the Pulse data set, separately for the men and women. The following command produced the boxplot shown in Exhibit 3.12.

```
GBOXPLOT 'WEIGHT';
   BY 'SEX'.
```

Release 8, DOS and Macintosh

There are high-resolution versions of HISTOGRAM and BOXPLOT. These have the same syntax and subcommands as the character versions. The only difference is that the command names start with the letter G.

    Note for DOS users: The exhibits in this section and in Section 4.5 are for the Macintosh. DOS versions are essentially the same, with three exceptions: (1) DOS dialog boxes for graphs have an option to store the graph in a file. Macintosh uses a menu entry to store graphs. (2) The OK and Cancel buttons are in the opposite order in DOS. (3) Dialog boxes and output are drawn a little differently.

Exhibit 3.13   Dialog Box for Histogram

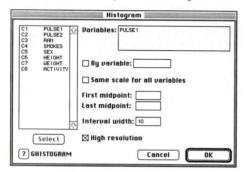

Exhibit 13.14   High-Resolution Histogram

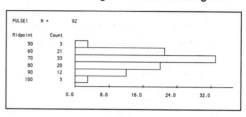

Exhibit 3.15   Dialog Box for Boxplot, Using Release 8

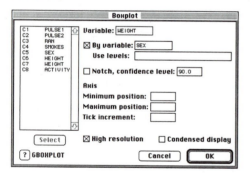

Exhibit 3.16   High-Resolution Boxplots, Using Release 8

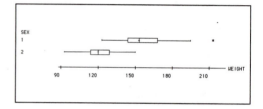

HISTOGRAM. We will get a high-resolution histogram for the variable PULSE1 in the Pulse data set, using an increment, or bin width, of 10. The command is

```
GHISTOGRAM  'PULSE1';
  INCREMENT 10.
```

The dialog box, already filled in, is shown in Exhibit 3.13. Exhibit 3.14 shows the output.

BOXPLOT. We will get high-resolution boxplots for the weights in the Pulse data set, separately for the men and women. The command is

```
GBOXPLOT 'WEIGHT';
  BY 'SEX'.
```

Exhibit 3.17 Dialog Box for Histogram
Using Release 9, Windows

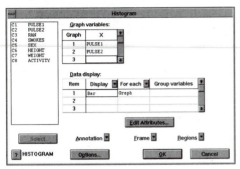

Exhibit 3.18 Dialog Box
for Multiple Graphs Option

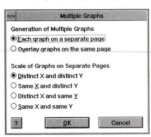

The dialog box, already filled in, is shown in Exhibit 3.15. There are three options we have not discussed: Use levels, Notch, and Condensed display. These options are also available in the character version of BOXPLOT. Read Help or a manual if you want to find out about them. Exhibit 3.16 shows the output from this dialog box.

Release 9, Windows

Graphs in Release 9 have many options; when you look at the graph dialog boxes, you will see them all. If you do not understand an option, you can usually just leave it at the default setting. You can use the Help facility to learn more, and you can just experiment to see what happens. If you change the defaults and want to get back to the original version of a dialog box, just press the ⬚ F3 ⬚ key when the first dialog box for the command is active. Everything will be set back to the defaults.

We will explain a few of the more common graph options in this section and later in this book.

HISTOGRAM. In Chapter 2, we showed an example of the dialog box and output for a high-resolution histogram. Exhibit 3.17 shows a dialog box for a high-resolution histogram, using the Pulse data set. Here we entered two variables. This will give us two graphs, each in its own window. You can enter many variables in one dialog box if you wish.

Each of the three choices, Annotation, Frame, and Regions, contains a list of options. Click the down arrow next to each one to see these. These options and the dialog boxes that go with them are essentially the same for all high-resolution graphs in Minitab.

Exhibit 3.19 Dialog Box for Histogram
Showing the Frame Options

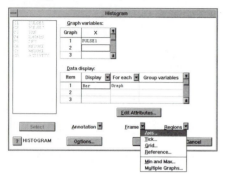

Exhibit 3.20 Dialog Box for Axis Options
Under Frame

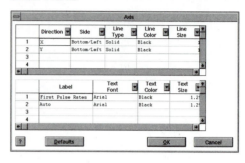

Suppose you want all the graphs in one dialog box on the same scale.
Click the down arrow next to Frame, then click the Multiple Graphs option
to get the dialog box shown in Exhibit 3.18. In the first set of options, we
want the default, "Each graph on a separate page". The second set of options gives four choices for common scales. We want the last, "Same X and
same Y". Click this, then click OK. Finally, click OK in the main dialog box
to get two histograms.

Let's look at some of the other graph options. Select Graph > Histogram from the menu. Notice that the dialog box is filled in with the last
settings you entered. This is a nice feature when you want to do the same
graph again but with some minor changes. This time, however, we want to
start clean, so press F3 to set everything at its default. Now enter the variable PULSE1 in the Graph variables box.

Click the down arrow next to Frame. A list of six options appears, as
shown in Exhibit 3.19. Click Axis to get a dialog box like the one in Exhibit
3.20. The column headed Label allows you to type your own axis labels.
Exhibit 3.20 shows the label we typed for the X, or horizontal, axis. You
can also specify font, color, text size, and many other things for this label.
We left all these at their default settings. Type in the label as we did and
click OK. This label will be used on all histograms specified in the first
dialog box.

Note: If you want to remove a label, just type a blank character in the
column headed Label.

Now click the down arrow next to Annotation, then click the Title option to get a dialog box like the one shown in Exhibit 3.21. You can type

Exhibit 3.21 Dialog Box for Title Option Under Annotation

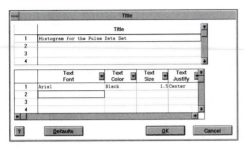

Exhibit 3.22 Dialog Box for Histogram Options

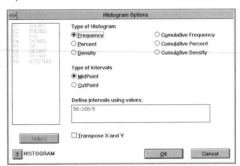

several titles and specify the font, color, text size, and many other things for each. Exhibit 3.21 shows the title we typed. We left everything else at the default settings. Type this title and click OK to get back to the main dialog box, shown in Exhibit 3.13.

Next, click the Options... button to get a dialog box like the one in Exhibit 3.22. The first set of options lets you change the type of histogram. We will stay with Frequency. The second set lets you specify the scale. You can choose either midpoints or cutpoints. In addition, you can give the points you want. We specified midpoints, going from 50 to 100 in steps of 5. The use of colon and slash is the same syntax that is described for the command SET, in Section 1.4. Fill in this dialog box as we did and click OK. Note that in some cases you may need to specify ticks as well as midpoints to get the scale you want. Ticks are described with the boxplot example, below.

Finally, click OK in the main dialog box to get the histogram shown in Exhibit 3.23.

The high-resolution histogram lacks one feature that the character histogram has: there is no BY option. If you want to compare several groups, use boxplots, dotplots, or character histograms, or put the groups into separate columns.

BOXPLOT. We will get high-resolution boxplots for weight in the Pulse data set, separately for the men and women. The dialog box for BOXPLOT, already filled in, is shown in Exhibit 3.24. There are two columns in the Graph variables box. The first is for the measurement variable, PULSE1. The second is optional, and is for a BY variable if you want one. Here we

Exhibit 3.23 Histogram with Axis Label and Title Specified

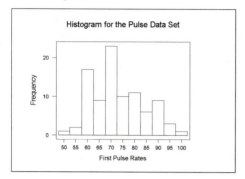

Exhibit 3.24 Dialog box for BOXPLOT

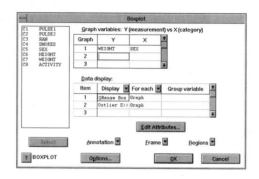

Exhibit 3.25 Dialog Box for Min and Mac Option Under Frame

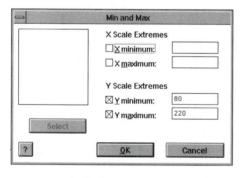

Exhibit 3.26 Dialog Box for Tick Option Under Frame

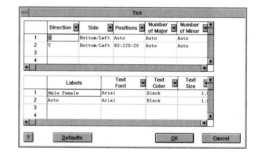

used SEX. This is all you need to specify. If you click OK, you will get a boxplot for weight by sex. However, let's use some options to make the graph look more attractive.

We'll use two options under Frame to specify our own scale for the vertical axis, or *y*-axis. First click "Min and Max" to get a dialog box like the one shown in Exhibit 3.25. Fill in the box as we did, setting the minimum to 80 and the maximum to 220, and click OK.

Next, click the Tick option under Frame to get a dialog box like the one shown in Exhibit 3.26. We made two changes here. First we set the tick positions on the *y*-axis to start at 80 and go to 220 in steps of 20. Then we

Exhibit 3.27   Boxplots with Title, Axis,
Labels, Tick Labels, and Scale Specified

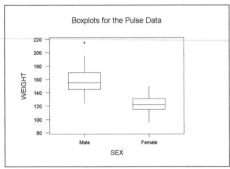

specified labels for the two ticks on the *x*-axis; the first is Male and the
second is Female. Fill in the dialog box as we did, then click OK.

We also used the Title option under Annotation to specify a title, and
the Axis option under Frame to specify axis labels. The graph is shown in
Exhibit 3.27.

Release 9 Commands

The graphics capability in the Standard Version of Release 9 is the same as
that in Release 7. See the material on Release 7 earlier in this section

The high-resolution graphics capability in Release 9 is very sophisticated. Each graph command has many subcommands. The subcommands
even have subcommands of their own. For example, the subcommand
TICK allows you to specify the positions for tick marks. TICK has a
subcommand, LABEL, that allows you to specify a label for each tick mark.
AXIS also has its own subcommand, LABEL. We will see more examples
in Chapter 4.

In this book, we will explain a few of the more common graph options.
Consult the Minitab manual if you want to learn more.

We will give some examples to illustrate these commands and subcommands, using the Pulse data set. Remember, if you want to type high-
resolution commands, make sure the GPRO option is in effect.

To do a histogram of PULSE1, use

```
HISTOGRAM PULSE1
```

HISTOGRAM C...C

MIDPOINTS K...K

Displays one histogram for each column.

MIDPOINTS allows you to specify the midpoint of each bar in the histogram. You may use a colon and a slash, as they are used in SET, to specify midpoints. Any midpoints that are outside the minimum and maximum values for the data (as determined by Minitab if you take the default, or as determined by you if you use MINIMUM and MAXIMUM), are not shown.

BOXPLOT C[*C] ... C[*C]

Displays one graph for each column or pair of columns listed.

There are two forms, C and C*C. The first form displays one boxplot for all the data in C. The second form displays a separate boxplot for each value in the second column. All are put on one graph.

To do histograms for PULSE1 and PULSE2, using the same scales for both the *x*-axis and the *y*-axis on each graph, use

```
HISTOGRAM PULSE1 PULSE2;
   SAME 1 2.
```

Next we do a histogram for PULSE1, give it a title, and specify a label for the *x*-axis (AXIS 1). You must enclose the text for titles and labels in single quotes. (*Note:* If you want to remove an axis label, just type a blank character, enclosed in quotes, for that label.) We also specify midpoints for the bars. These range from 50 to 100 in steps of 5. Output is shown in Exhibit 3.23, in the section for Release 9, Windows, above.

```
HISTOGRAM PULSE1;
   TITLE 'Histogram for the Pulse Data Set';
   AXIS 1;
     LABEL 'First Pulse Rates';
   MIDPOINTS 50:100/5.
```

The following subcommands allow you to customize various features of a graph. They are available for the commands HISTOGRAM and BOXPLOT, discussed in this chapter, and for the commands PLOT and TSPLOT, discussed in Chapter 4.

TITLE 'text'	# adds a title to the graph
FOOTNOTE 'text'	# adds a footnote to the graph
AXIS K	# use K = 1 for the x-axis and K = 2 for the y-axis
LABEL 'text'	# specifies a label for axis K
TICK K [K...K]	# specifies the positions for tick marks on an axis
LABEL 'text' ... 'text'	# specifies a label for each tick mark on axis K
MINIMUM K K	# specifies the minimum data value for an axis
MAXIMUM K K	# specifies the maximum data value for an axis
SAME K [K]	# requests the same scale(s) for all graphs

The first K in AXIS, TICK, MINIMUM, and MAXIMUM specifies the axis. Use K = 1 for the *x*-axis and K = 2 for the *y*-axis. The two Ks in SAME also specify the axis or axes you want.

MINIMUM and MAXIMUM specify the minimum and maximum data values for an axis. Any data points beyond these values are clipped (not displayed on the graph).

TICK lists the positions for the (major) tick marks on an axis. You may use a colon and a slash, as in SET, to specify this list. Any tick marks on TICK that are beyond the minimum and maximum data values (as determined by Minitab if you take the default, or as determined by you if you use MINI-MUM and MAXIMUM) are not shown on the graph. Change MINIMUM and MAXIMUM if you want them displayed.

SAME is used when several graphs are given in the main command. SAME 1 makes the *x*-scale the same for all graphs; SAME 2 makes the *y*-scale the same for all graphs; SAME 1 2 makes both the *x*-scale and the *y*-scale the same for all graphs.

Finally, we create two boxplots, one for the weights of men and one for the weights of women. Axis 1 is for the variable SEX. This is a categorical variable with two values, 1 for men and 2 for women. Therefore, the x-axis has just two tick marks. The default tick labels are 1 and 2. We use the LABEL subcommand of TICK to change these to the words Male and Female. We also change the scale on the y-axis. We enlarge it slightly from the default (use BOXPLOT yourself to see Minitab's default) and specify tick marks of 80, 100, 120, . . . , 220. Output is shown in Exhibit 3.27, in the section for Release 9, Windows, above.

```
BOXPLOT WEIGHT*SEX;
  TITLE 'Boxplots for the Pulse Data';
  TICK 1;
    LABELS 'Male' 'Female';
  MINIMUM 2 80;
  MAXIMUM 2 220;
  TICK 2 80:220/20.
```

4

Plotting Data

The displays we have used so far have involved only one variable at a time. Often we are interested in the relationships between two or more variables, such as the relationship between height and weight, between smoking and lung cancer, or between temperature and the yield of a chemical process. Plots allow us to investigate several variables at a time.

The section at the end of this chapter shows how to use high-resolution graphs in Releases 8 and 9.

Note for Release 9 Users: If you are using Release 9 (Windows, or the Enhanced Version on VAX or UNIX workstations) and want to type commands for the character graphs in this chapter, remember to first type the command GSTD. From then on, you will get character graphs. To return to typing high-resolution graphs, type GPRO.

4.1 Scatterplots

Furnace Example

If both variables are interval or ordinal, the most useful display is the familiar scatterplot. Table 4.1 presents data from a study of the effectiveness of an energy-saving furnace modification. The data are for one house over an 11-week period. The first column gives the week. The second column gives the average amount of natural gas used per hour (per square foot of house area). The third column gives the average daily temperature during the week. The fourth column, Furnace Modification, will be explained in Section 4.2.

Exhibit 4.1 shows how we produced a scatterplot of the amount of gas used versus the outside temperature. Notice that the amount of natural gas

Table 4.1 Furnace Data

Week	Gas Used (BTU/Hour)	Average Temperature (°F)	Furnace Modification
1	2.10	45	Out
2	2.55	39	In
3	2.77	42	Out
4	2.40	37	In
5	3.31	33	Out
6	2.97	34	In
7	3.32	30	Out
8	3.29	30	In
9	3.62	20	Out
10	3.67	15	In
11	4.07	19	Out

Exhibit 4.1 Plot of Furnace Data in Table 4.1

```
MTB > set c1
DATA>    2.10  2.55  2.77  2.40  3.31  2.97  3.32  3.29  3.62  3.67
4.07
MTB > end
MTB > set c2
DATA>    45  39  42  37  33  34  30  30  20  15  19
DATA> end
MTB > name c1 'Gas'  c2 'Temp'
MTB > plot 'Gas' 'Temp'

      4.20+
         -              *
 Gas     -
         -
         -    *         *
      3.50+
         -                        2      *
         -
         -
         -                            *
      2.80+                                              *
         -
         -                                        *
         -                                     *
         -
      2.10+                                        *          *
         -
         --------+---------+---------+---------+---------+---------
 Temp
             18.0      24.0      30.0      36.0      42.0
```

consumed by this house decreased more or less linearly as the average temperature varied from 15° F to 45° F. The 2 on the plot means that two points fell there—that is, that there were two weeks when the average temperature was about 30° F and the gas consumption was about 3.3. Reviewing the data, we see that these were weeks 7 and 8.

Cartoon Example

The Cartoon data (in Appendix A) were collected in an experiment to assess the effectiveness of different sorts of visual presentations for teaching. In one part of the experiment, participants were given a lecture in which cartoons were used to illustrate the material. Immediately after the lecture, a short quiz was given. In addition, each participant was given an OTIS test, a test that attempts to measure general intellectual ability. First we entered the CARTOON data set into Minitab (with Release 8 or Release 9 Windows, use the Open dialog box; otherwise, use Minitab's RETRIEVE command). Then we used the PLOT command to plot each participant's cartoon score, CARTOON1, versus his/her OTIS score. Exhibit 4.2 gives the output.

Exhibit 4.2 Plot of Immediate Cartoon Scores Against OTIS Scores

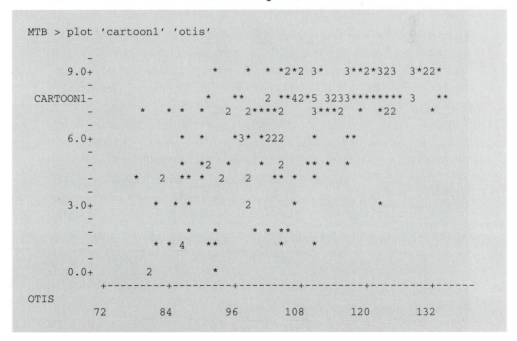

Several interesting facts can be observed. People with higher OTIS scores tend to get higher cartoon scores, as you might expect. Note also that the highest cartoon score possible is 9, so that a person with very high ability cannot really demonstrate the full extent of his/her knowledge. The best a person can get is a 9. In this plot, the scores of people who had OTIS scores below 100 show no evidence that they were held down to the

---

PLOT C C

XINCREMENT = K
XSTART at **K** [end at **K**]
YINCREMENT = K
YSTART at **K** [end at **K**]

TITLE   'text'
FOOTNOTE 'text'
XLABEL   'text'
YLABEL   'text'

SYMBOL   'symbol'

PLOT gives a scatterplot of the data. The first column is put on the vertical axis and the second column on the horizontal axis. Ordinarily each point is plotted with the symbol *. When several points falls on the same plotting position, a count is given. When the count is over 9, the symbol + is used.

You may specify your own scales with subcommands. XINCREMENT is the distance between tick marks (the + symbols) on the x-axis. XSTART specifies the first and, optionally, the last point plotted on the x-axis. Any points outside are omitted from the plot. YINCREMENT and YSTART are for the y-axis.

You may add titles, footnotes, and axis labels with subcommands. In all cases, the text must be enclosed in single quotes. Note, if your text has a single quote in it, replace this with two single quote marks.

TITLE may be used up to three times, to add up to three titles. FOOTNOTE can be used up to two times. XLABEL changes the default label on the x-axis to the text you specify. YLABEL does the same for the y-axis label.

By default, Minitab plots points with the symbol *. To plot points with a symbol of your choosing, specify one symbol, enclosed in single quotes, in the subcommand SYMBOL.

Release 8: **Graph > Scatter Plot**
Windows: **Graph > Character Graphs > Scatter Plot**

maximum score. But the scores of people with OTIS scores over 100 do show the effect of this limitation. This effect is sometimes called truncation; it results in some loss of information. If the experiment were to be done over again, it might be better to develop a slightly longer or harder test so that everyone would have a chance to demonstrate his/her full ability.

Exercises

4-1   The following questions refer to the plot in Exhibit 4.2.
       (a)   How many people got a perfect score on the cartoon test?
       (b)   How many people with an OTIS score of 90 or less got a perfect score on the cartoon test?
       (c)   Approximately what is the lowest OTIS score? The highest?

4-2   (a)   Exhibit 4.1 shows a plot of gas consumption versus temperature. Suppose we want some idea of how much gas would be consumed at a temperature of 25°F. Use this plot to make an estimate. Explain how you arrived at your estimate.
       (b)   Repeat part (a) for 10°F. How much faith do you have in your estimate?

4-3   Table 4.1 shows data on gas consumption over an 11-week period. Plot weekly "average temperature" versus week. How did temperature change during this period?

4-4   In this problem we will examine the Trees data (described in Appendix A) with an eye to developing a way to predict volume from measurements that are easier to make.
       (a)   Plot volume versus diameter.
       (b)   Plot volume versus height. Compare this plot to the one in part (a). Which seems to be a better predictor of volume: height or diameter? Which do you think would be easier to measure in a forest?

4-5   The table below gives the marriage rate and divorce rate per 1,000 from 1920 to 1990. Let's see how these rates have changed over the years.
       (a)   Plot marriage rate versus year. What patterns do you see? You might put your plot on paper and connect the dots. This often helps to reveal patterns.
       (b)   Plot divorce rate versus year. What patterns do you see? Again, you might put your plot on paper and connect the dots to help you see patterns.

Year	Marriage	Divorce
1920	12.0	1.6
1925	10.3	1.5
1930	9.2	1.6
1935	10.4	1.7
1940	12.1	2.0
1945	12.2	3.5
1950	11.1	2.6
1955	9.3	2.3
1960	8.5	2.2
1965	9.3	2.5
1970	10.6	3.5
1975	10.0	4.8
1980	10.6	5.2
1985	10.1	5.0
1990	9.8	4.7

4-6   The following table contains track records as of September 5, 1992.

Distance (meters)	Women's Track		Men's Track	
	Minutes	Seconds	Minutes	Seconds
100	0	10.49	0	9.86
200	0	21.34	0	19.72
400	0	47.60	0	43.29
800	1	53.28	1	41.73
1500	3	52.47	3	28.86
2000	5	28.69	4	50.81
3000	8	22.62	7	28.96
5000	14	37.33	12	58.39
10000	30	13.74	27	8.23

(a)   First we will look at women's track. Calculate the speed at which each race was run in miles per hour. Note that there are 1609.4 meters in a mile. What is the fastest speed? The slowest?

(b)   Plot the record speed versus distance in meters for women's track. What is the pattern?

(c)   Repeat parts (a) and (b) for men's track.

(d)   How do men's and women's track compare?

4.2   Plots with Groups

A scatterplot shows the relationship between two variables. We can add information about a third variable by using different symbols for different points. For an example, let's take a closer look at the furnace modification data in Table 4.1. The last column in the table indicates whether or not the energy-saving modification was installed in the furnace that week. This column is an example of a categorical variable. For Minitab, we usually code categorical data using numbers.

Exhibit 4.3 continues the session begun in Exhibit 4.1. First, we added a third variable to specify the modification, using 1 for In and 2 for Out. Next, we used Minitab's LPLOT command to plot these data with labels. Each observation with a 1 in C3 was plotted with the first letter of the alphabet, an A. Similarly, each observation with a 2 in C3 was plotted with a B. Two points fell on one spot and were plotted with a 2. To determine

Exhibit 4.3   Plot of the Furnace Data with Symbols

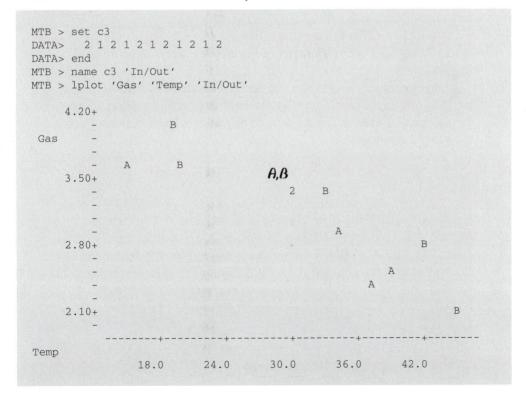

LPLOT C C, using labels as coded in **C**

XINCREMENT = K
XSTART at **K** [go to **K**]
YINCREMENT = K
YSTART at **K** [go to **K**]

TITLE   'text'
FOOTNOTE 'text'
XLABEL   'text'
YLABEL   'text'

LPLOT (the L is for labels or letters) plots data with labels given by

```
. . . -2 -1 0 1 2 3 . . . 24 25 26 27 28 . . .
. . .  X  Y Z A B C . . .  X  Y  Z  A  B . . .
```

The subcommands are the same as in PLOT.

Release 8: **Graph > Scatter Plot,** choose option for tags
Windows: **Graph > Character Plots > Scatter Plot,** choose option for labels

---

their symbols, we checked the data listing in Table 4.1. These points are for weeks 7 and 8, so one is an A and one is a B. We wrote these by hand on the plot.

Looking closely at this plot, we see some indication that the furnace modification might produce a slight savings in gas consumption. Overall, for a given temperature, the As seem to be just a bit lower than the Bs, but the effect is not dramatic.

Exercises

4-7   Exhibit 4.3 used A and B for plotting symbols. A better choice would be I for "In" and O for "Out." The number 9 is LPLOT's code for I and 15 is the code for O. Redo the LPLOT using these symbols. Probably the easiest way to change the codes is to retype the data. Notice that this minor change in plotting symbols makes the plot easier to interpret and easier to explain to others.

4-8   (a)   Use the Pulse data to plot weight versus height.
       (b)   Use LPLOT to make the same plot, but use one symbol for men and another for women. Interpret the plot.

4.3   Plots with Several Variables

Table 4.2 gives the winning times for three Olympic races. There were no games in 1916 because of World War I and none in 1940 and 1944 because of World War II. A quick look tells us that the winners have gotten faster over the years. Suppose we want to compare the "progress" made in the different races. We could plot these winning times, but because the races are of different lengths, probably the average speeds are easier to compare. So let's plot speed versus year for each of the three races. Recall that speed = distance/time. We'll put all three plots on the same set of axes, to make comparisons easier.

Once the data have been entered into C1–C4, we can use the following commands:

Table 4.2   Winning Times in Men's Olympic
Track Races

Year	100-Meter Race	200-Meter Race	400-Meter Race
1900	10.8	22.2	49.4
1904	10.8	21.6	49.2
1908	11.0	22.4	50.0
1912	10.8	21.7	48.2
1920	10.8	22.0	49.6
1924	10.6	21.6	47.6
1928	10.8	21.8	47.8
1932	10.3	21.2	46.2
1936	10.3	20.7	46.5
1948	10.3	21.1	46.2
1952	10.4	20.7	45.9
1956	10.5	20.6	46.7
1960	10.2	20.5	44.9
1964	10.0	20.3	45.1
1968	9.9	19.8	43.8
1972	10.1	20.0	44.7
1976	10.1	20.2	44.3
1980	10.3	20.2	44.6
1984	10.0	19.8	44.3
1988	9.9	19.8	43.9
1992	10.0	20.0	43.5

Exhibit 4.4   Winning Speeds in Men's Olympic Track Races

```
                              Men's Olympic Track
W          -
i          -                                                    2        B 2
n                                                          A     2 2 B A    2
n       9.80+                                              2 B
n          -                           A 2      A 2 B             A
i          -                                    B   A
n          -                     A   B
g          -        A 2    2   A B 2                                     C
        9.10+              A     B                    C              C
s          -        B   B                          C      C C C C
p          -                                         C
e          -             C         C C
e          -                   C         C
d       8.40+              C C
s          -         C
           -      C C     C
           -         C
         ------+---------+---------+---------+---------+---------+
             1900      1920      1940      1960      1980      2000
         A = 100m vs. Year        C = 400m vs. Year
         B = 200m vs. Year
```

```
NAME C1 = 'Year' C12 = '100M' C13 = '200M' C14 = '400M'
LET C12 = 100/C2
LET C13 = 200/C3
LET C14 = 400/C4
MPLOT C12 C1, C13 C1, C14 C1;
   TITLE 'Men''s Olympic Track Data';
   YLABEL 'Winning Speeds'.
```

The resulting plot is shown in Exhibit 4.4. The letter A identifies the winning speeds for the 100-meter races, B identifies those for the 200, and C those for the 400. Occasionally, two speeds fall on the same position and the number 2 is plotted. By looking at this plot, we can get a very good idea of what happened. On the average, the speeds of all three races increased over the years. Notice that the 100-meter and 200-meter races are run at about the same speed. Some years the 100-meter was faster; some years the 200-meter was faster. The 400-meter race, however, has always been run at a much slower pace. We can also spot several very "good" years, when progress was made in all three races—1924, 1932, 1960, and 1968. We can also see that essentially no progress was made in the three Olympics following World War II. Creative plotting can reveal a lot that is not apparent in a simple table of numbers.

MPLOT C vs C   C vs C ... C vs C

    XINCREMENT = **K**
    XSTART at **K** [go to **K**]
    YINCREMENT = **K**
    YSTART at **K** [go to **K**]

    TITLE   'text'
    FOOTNOTE 'text'
    XLABEL   'text'
    YLABEL   'text'

MPLOT (the M is for "multiple plot") puts several plots all on the same axes. The first pair of columns are plotted with the symbol A, the second pair with the symbol B, and so on. If several points fall on the same spot, a count is given.

The subcommands are the same as in PLOT (p. 116).

Release 8: **Graph > Multiple Scatter Plot**
Windows: **Graph > Character Plots > Multiple Scatter Plot**

Exercises

4-9   The following questions refer to the Track data in Table 4.2 and Exhibit 4.4.

    (a)   Were any records set in these three races in the 1992 Olympics? That is, were any of these three races run in the least time ever?

    (b)   In what year was the record set for the 100-meter race? For the 200-meter race? For the 400-meter race?

    (c)   The best time for the 100-meter race was 9.9 seconds. How fast is this in miles per hour? (*Note:* There are 1,609 meters in a mile.)

    (d)   Consider the 100-meter race. In which years was a new Olympic record set?

4-10   Listed below are the monthly average temperatures for five U.S. cities. The data are stored in the file CITIES. Get an MPLOT of temperature versus month for the five cities. (If your output is on paper, label the cities by hand and connect the points for each city.) Write a paragraph or two comparing the temperature patterns in these five cities.

Month	Atlanta	Bismarck	New York	San Diego	Phoenix
1	42	8	32	56	51
2	45	14	33	60	58
3	51	25	41	58	57
4	61	43	52	62	67
5	69	54	62	63	81
6	76	64	72	68	88
7	78	71	77	69	94
8	78	69	75	71	93
9	72	58	68	69	85
10	62	47	58	67	74
11	51	29	47	61	61
12	44	16	35	58	55

4-11   During the winter months, you have probably noticed that on days when the
wind is blowing it seems particularly cold (at least in colder climates). The
explanation lies in the fact that wind creates a slight lowering of atmospheric pressure on exposed flesh, thereby enhancing evaporation, which
has a cooling effect. Thus the effective temperature (how cold it seems to
you) depends on two factors: thermometer temperature and wind speed.
The following table was prepared by the U.S. Army. It gives the effective
temperature corresponding to a given thermometer temperature and wind
speed.

Wind Speed (mph)	Thermometer Temperature (°F)			
	20	10	0	− 10
10	4	− 9	− 21	− 33
15	− 5	− 18	− 36	− 45
20	− 10	− 25	− 39	− 53

(a)   There is an error in this table. Can you find it? Plotting the data will
help. Use MPLOT to plot the Wind Chill Index versus wind speed for
the four temperatures. Do you see any points that do not seem to follow the overall pattern in the plot?

(b)   Try to correct the point that is in error. What value would you substitute? Give reasons for your choice. Use both the data table and the plot
to help.

4-12   How do automobile accident rates vary with the age and sex of the driver?
The table below gives the numbers of accidents per 100 million miles of
exposure for drivers of private vehicles.

Age of Driver	Accidents That Involved a Casualty		Accidents That Did Not Involve a Casualty	
	Male	Female	Male	Female
Under 20	1436	718	2794	1510
20–24	782	374	1939	851
25–29	327	219	949	641
30–39	232	150	721	422
40–49	181	228	574	602
50–59	157	225	422	618
60 and over	158	225	436	443

(a) Read the four columns of accident rates into C1–C4 of the Minitab worksheet. To enter the "Age of Driver" into the worksheet, we will choose one representative age for each category in the table. For example, 18 for "Under 20", 22 for "20–24", and so on. Now use SET to put the following age categories into C5: 18, 22, 27, 35, 45, 55, and 65.

(b) Plot the casualty accident rate for males versus age and the casualty accident rate for females versus age, using MPLOT. Describe in a sentence or two how these accident rates depend on the age and sex of the driver.

(c) Repeat (b) for the noncasualty accidents. How well do the conclusions in (b) and (c) agree?

(d) Compute overall accident rates for male and female drivers by adding up the casualty and noncasualty rates separately for each sex. For example, use LET C11 = C1 + C3 and LET C12 = C2 + C4. Repeat (b) for these overall rates.

(e) For each age and sex group, compute the proportion of accidents that involve a casualty. Repeat part (b) for these proportions.

(f) Suppose we want to look at the total casualty accident rate (that is, the number of casualty accidents per 100 million miles of exposure) for each age category. Would (C1 + C2) provide us with appropriate data? Can we get the appropriate figures from the data we have? Explain.

4-13 The commands LPLOT and MPLOT both put data for several groups or variables on the same plot. Essentially the only difference between them is how the data are organized in Minitab's worksheet. Any plot that can be done by LPLOT also can be done by MPLOT and vice versa, if you reorganize the data appropriately. How could you do the plot in Exhibit 4.3 using MPLOT?

4.4   Time Series Plots

A sequence of observations taken over time is called a time series. Examples include the milk output of a cow recorded each week, the average temperature in Chicago recorded each day, the consumer price index recorded each month, and the number of homicides in the United States recorded each year. Minitab has a command, TSPLOT, which is designed to plot time-series data.

Table 4.3 gives the number of people employed in the food and kindred-products industry in the state of Wisconsin from January 1970 to December 1974. The data for this series, along with those for two other series, are described in Appendix A. Exhibit 4.5 shows a plot. TSPLOT plotted the first observation with the number 1, the second observation with the number 2, . . . , the ninth observation with the number 9, and the tenth observation with the number 0. Then it started over again. The eleventh observation was plotted with a 1, the twelfth with a 2, and so on.

There are two clear patterns in the plot: a very strong repeating cycle and a slight upward trend in the second half of the series. The cycle comes from the seasonal employment in the food industry—employment is naturally quite high in the summer and low in the winter. The cycle is repeated every 12 months. Many time series contain repeating patterns. For example, if you record temperature every hour for 10 days, you will probably see a repeating cycle every 24 hours. Temperatures generally tend to be high in the daytime and low in the night.

TSPLOT will plot data using special symbols to indicate a cycle. All you need to do is specify the length of the cycle, often called the period of the cycle. Exhibit 4.6 presents an example using the Food data. In this plot, January is plotted with a 1, February with a 2, . . . , September with a 9, and October with a 0 (for 10). Since we ran out of numbers at that point, we

Table 4.3 Number of Employees in Wisconsin in the Food Products Industry (in units of 1000 employees)

	Jan	Feb	Mar	Apr	May	Jun	Jul	Aug	Sep	Oct	Nov	Dec
1970	53.5	53.0	53.2	52.5	53.4	56.5	65.3	70.7	66.9	58.2	55.3	53.4
1971	52.1	51.5	51.5	52.4	53.3	55.5	64.2	69.6	69.3	58.5	55.3	53.6
1972	52.3	51.5	51.7	51.5	52.2	57.1	63.6	68.8	68.9	60.1	55.6	53.9
1973	53.3	53.1	53.5	53.5	53.9	57.1	64.7	69.4	70.3	62.6	57.9	55.8
1974	54.8	54.2	54.6	54.3	54.8	58.1	68.1	73.3	75.5	66.4	60.5	57.7

Exhibit 4.5   TSPLOT of the Food Employment Data

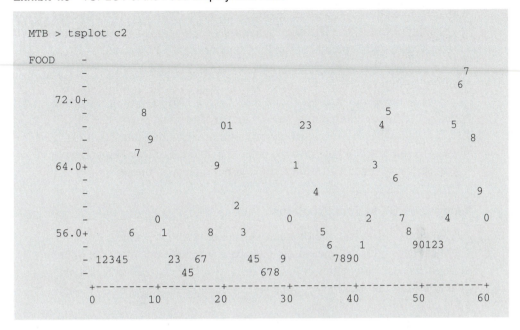

Exhibit 4.6   TSPLOT of the Food Data, with Symbols for the Period

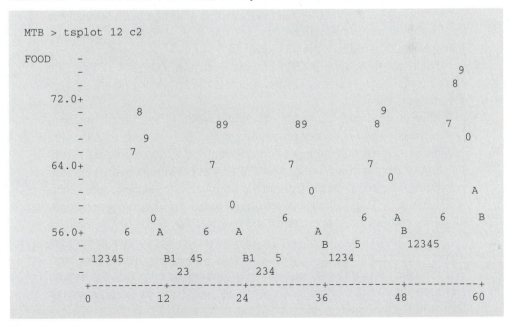

started using letters. Thus November is plotted with an A and December with a B. In Exhibit 4.6 we can see the seasonal pattern more clearly than we could in Exhibit 4.5. The highest symbols each year are 8 and 9. These are for August and September. The lowest symbols are 1, 2, 3, and 4. These are for January, February, March, and April, the months after harvest and before planting.

Note that you can also use PLOT, LPLOT, and MPLOT to display time series data. We use them a number of times in this book. If you use PLOT, LPLOT, and MPLOT, you must have a column with time in it. TSPLOT is most convenient when you have not entered a time variable, when you have a long series, or when you want to look at periodicity in the series.

Using TSPLOT to Find Problems

All data must be collected either across time or across space. For example, chemical and industrial measurements are often made one after the other; agricultural data are often collected on various areas scattered across a farm. Plotting the data in ways that reflect these connections often reveals unusual features or problems.

An Example. Track-etch devices were under consideration as an inexpensive way to measure low-level radiation. An experiment was done to study the usefulness of one such device, called an open cup. A total of 25 open cups were hung along a string that ran the length of a chamber. The chamber was designed to have a uniform level of radiation. The level of radiation was measured by each of the 25 cups.

The data, along with a TSPLOT, are shown in Exhibit 4.7. This plot shows a systematic pattern: Observations at the left of the plot are larger than those at the right.

Why this pattern? These 25 observations are plotted in the order in which the cups had been hung along the string. One conjecture is that the radiation level in the chamber was not constant, but changed gradually from one end of the chamber to the other. Further checking uncovered another possibility. The cups had been put in the chamber in the same order as their serial numbers. Thus a change in sensitivity of the devices during the manufacturing process could account for the observed trend.

As of this writing, the source of the problem is not known, and investigation is continuing. It is important to note that if the cups had not been placed in the chamber in the same order as their serial numbers, but instead had been placed in a random order, we might have been able to pinpoint the source of the problem.

Exhibit 4.7 Time Series Plot of Track-Etch Data

```
MTB > set c1
DATA>  6.6 4.8 5.0 6.3 5.9 6.5 6.1 5.8 6.4 5.3 5.2 5.2 4.8
DATA>  4.9 5.5 5.1 5.0 5.7 4.8 5.3 4.2 5.3 4.5 4.5 4.2
DATA> end
MTB > tsplot c1


          - 1
   C1     -         6  9
          -      4
          -         7
       6.00+
          -         5  8
          -                      8
          -                  5
          -
       5.25+             012       0 2
          -                  6
          -      3          4  7
          -      2          3     9
          -
       4.50+                       34
          -
          -                    1    5
          +---------+---------+---------+
          0        10        20        30
```

Exercises

4-14 In this section, we used TSPLOT to display the employment data for the
      food industry, using the EMPLOY data set (described in Appendix A).
      Make similar displays of the data for the wholesale and retail trade indus-
      tries. Interpret the plot. Did employment tend to increase over this period?
      Is there a seasonal pattern to employment in wholesale and retail trade? Can
      you give any reasons for the patterns in the data?

4-15 Use TSPLOT to display the data for fabricated metals from the EMPLOY
      data set. Describe the pattern of employment in this industry. Did employ-
      ment tend to increase over this period? Is there a seasonal pattern?

4-16 Just by looking at the TSPLOT in Exhibit 4.7, can you figure out the fol-
      lowing?
      (a) Which cup gave the highest reading? Approximately what value did
           the highest reading have?

TSPLOT C

INCREMENT = **K**
START at **K** [go to **K**]
TSTART at **K** [go to **K**]

TSPLOT plots the data in C versus the integers 1, 2, 3, and so on. This type of plot is often used for time series data. If the time series is too long to fit across the page, the plot is automatically broken into several pieces.

TSPLOT period = **K**   **C**

ORIGIN K

Time series data often have an associated period. For example, they may be collected monthly (period = 12) or hourly (period = 24). If you specify a period, plotting symbols that reflect this period are used. You should specify the origin if the first observation in C does not correspond to the first period. For example, if C1 contains monthly data starting in March, use

```
TSPLOT 12 C1;
  ORIGIN 3.
```

The subcommands INCREMENT and START control the y-axis; they are the same as YINCREMENT and YSTART in PLOT. TSTART allows you to plot a subset of your time series. For example, TSTART 15 30 would plot only the 15th to the 30th observation. If you use TSTART and ORIGIN, TSTART refers to the time values in ORIGIN. For example, suppose C1 contains yearly data from 1921 to 1992. Then to plot only the data from 1950 to 1980, use

```
TSPLOT C1;
  ORIGIN 1921;
  TSTART 1950 1980.
```

Release 8: **Graph > Time Series Plot**
Windows: **Graph > Character Plots > Time Series Plot**

(b)   Two cups seem to be tied for lowest. Which are they? Approximately what value did they have?

(c)   How many cups were there in all?

4-17   The data listed below are measurements from a diffusion porometer, an instrument for measuring changes in resistance across a leaf's surface. Any treatment that closes the stomates of a leaf causes the resistance to go up;

when the stomates are open, the resistance goes down. Four treatments were used, with six observations of each treatment:

	Treatment											
	1						*2*					
Resistance	7.6	8.9	8.4	14.5	15.8	12.8	8.2	10.5	6.8	14.3	13.4	13.2
	3						*4*					
Resistance	6.5	9.9	8.9	12.6	13.5	14.6	11.4	8.6	12.2	10.9	9.9	15.3

The observations were taken in the order in which they are listed. Use TSPLOT with period 6 to plot these data in this order. (If your plot is on paper, connect the six points within each treatment.) Write a concise summary of these data, including an opinion as to whether the treatments differ very much. Also discuss any problems you see with the data.

4.5 Optional Material on High-Resolution Graphs

Release 7

Release 7 has high-resolution versions of PLOT, LPLOT, and MPLOT, but not of TSPLOT. The command names for the high-resolution graphs are GPLOT, GLPLOT, and GMPLOT. These have all the subcommands that the character plots have.

GPLOT. The following command creates a high-resolution plot, using the Furnace data. Output is shown in Exhibit 4.8.

```
GPLOT 'Gas' 'Temp'
```

Exhibit 4.8 High-Resolution Plot for the Furnace Data

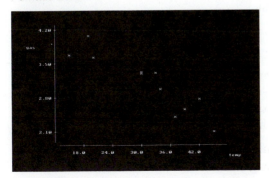

GPLOT, GLPLOT, and GMPLOT have two additional subcommands:

COLOR C
LINE [style = **K** [color = **K**]] connecting pairs in **C C**

COLOR specifies the color of each point. The column C must be the same
length as the data columns and must contain integers. The integer codes that
specify color are given below.

LINE adds a line to the plot. The points in the two columns listed on LINE
are connected in the order in which they are listed. The first column is for
the *y*-coordinates and the second column is for the *x*-coordinates. You can
use up to 3 LINE subcommands in one plot. You can specify the style and
color for a line using the following integer codes:

Styles for Lines		*Colors for Symbols and Lines*	
0 solid	4 dots	0 white	4 blue
1 large dashes	5 medium dash, dot	1 black	5 cyan
2 medium dashes	6 medium dash, small dash	2 red	6 magenta
3 small dashes	7 medium dash, 2 small dashes	3 green	7 yellow

You can draw lines on your plot by using the LINE subcommand. Suppose we draw a straight line through the points in the plot in Exhibit 4.8. In
Chapter 11 we will show you how to determine a good fitting line. For now
we'll just tell you the answer: The line goes from the point (15, 4) to the
point (45, 2.3). We put these numbers into two columns, then list them on
LINE. Output is shown in Exhibit 4.9.

```
NAME C11 'x' C12 'y'
READ C11 C12
   15   4
   45   2.3
END
GPLOT 'Gas' 'Temp';
  LINES 'y' 'x'.
```

You can use the LINE subcommand of GPLOT to draw any line or
curve you want, and you can draw just the curve without any points. For

Exhibit 4.9 Furnace Data with a Line Drawn

Exhibit 4.10 High-Resolution Plot for the Track Data

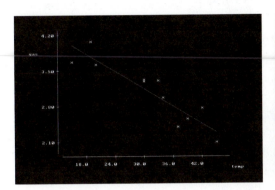

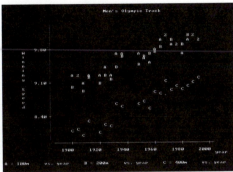

example, to draw a simple quadratic curve that runs from 0 to 10, enter points for the quadratic into two columns, then use GPLOT without any arguments on the main command. Here is an example:

```
NAME C1='x' C2='y'
SET 'x'
  0:10/.1
END
LET 'y' = 'x'**2
GPLOT;
  LINE 'y' 'x'.
```

GLPLOT. Again, we use the Furnace data. The following command produces a high-resolution LPLOT:

```
GLPLOT 'Gas' 'Temp' 'In/Out'
```

GMPLOT. The following command produces a high-resolution version of the plot shown in Exhibit 4.4, using the Olympic track data. Output is in Exhibit 4.10.

```
MPLOT C12*C1 C13*C1 C14*C1;
  TITLE 'Men''s Olympic Track';
  YLABEL 'Winning Speed'.
```

Release 8, DOS and Macintosh

There are high-resolution versions of PLOT, LPLOT, and MPLOT in Release 8, but not of TSPLOT. The command names for the high-resolution graphs are GPLOT, GLPLOT, and GMPLOT. As in the last part of Chapter 3, all exhibits here are for the Macintosh. Plots done in DOS are similar, but may have slightly different scales.

Exhibit 4.11   Dialog Box for PLOT Using the Furnace Data

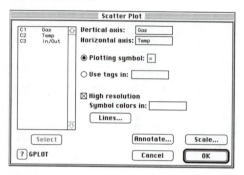

Exhibit 4.12   High-Resolution Plot for the Furnace Data

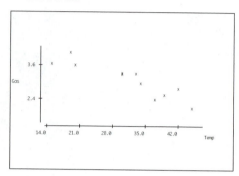

PLOT and LPLOT. Exhibit 4.11 shows the dialog box for PLOT, already filled in using the Furnace data from Table 4.1. Output is shown in Exhibit 4.12. The dialog box for PLOT can also do an LPLOT. Just specify your labels or tags in the entry "Use tags in."

The button for Annotate allows you to specify titles, footnotes, and axis labels. The button for Scale allows you to specify scales. These are the same options provided by subcommands of PLOT.

The high-resolution plot has two options that the character plot does not, symbol colors and lines. To specify symbol colors give a column of integers that is the length of your data. Colors are then assigned to each point as follows:

0 black on DOS, white on Mac   4 blue
1 white on DOS, black on Mac   5 cyan
2 red   6 magenta
3 green   7 yellow

You can draw lines on your plot by using the Lines button. Suppose we draw a straight line through the points in the plot in Exhibit 4.12. In Chapter 11 we'll show you how to determine a good fitting line. For now we'll just tell you the answer: The line goes from the point (15, 4) to the point (45, 2.3). We put these numbers into two columns:

```
NAME C11='x' C12='y'
READ C11 C12
   15   4
   45   2.3
END
```

Exhibit 4.13 Dialog Box for Adding a Line to the Furnace Plot

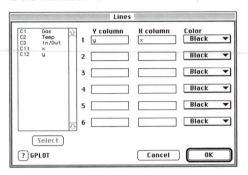

Exhibit 4.14 Furnace Data with a Line Drawn

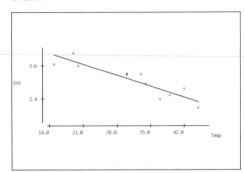

Then we enter these columns in the dialog box for lines as shown in Exhibit 4.13. Output is shown in Exhibit 4.14.

You can define up to three lines for one graph; you can also specify the color and line style for each. Look at the dialog box shown in Exhibit 4.13.

Mouse use. Click a down arrow under Color to see a list of colors. Click on the color you want. Similarly, click a down arrow under Style to see a list of line styles and click on the one you want.

DOS without a mouse. Tab to a box for a color (or line style). Press ⃞F4⃞ to see a list of colors (or line styles). Move to the one you want with the arrow keys and press ⃞F4⃞ to select it.

You can use the Line option of PLOT to draw any line or curve you want, and you can draw just the curve without any points. For example, to draw a simple quadratic curve that runs from 0 to 10, enter the following data:

```
NAME C1='x' C2='y'
SET 'x'
   0:10/.1
END
LET 'y' = 'x'**2
```

Then use the PLOT dialog box. Leave the first dialog box blank, and put x and y into the box for lines.

Exhibit 4.15 Dialog Box for MPLOT
Using the Track Data

Exhibit 4.16 Dialog Box for Annotation,
Using the Track Data

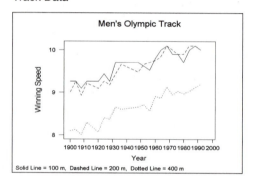

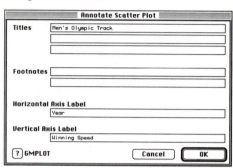

Exhibit 4.17 High-Resolution Plot for the
Track Data

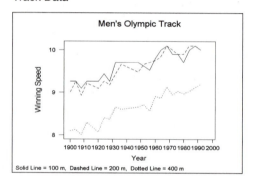

MPLOT. Exhibit 4.15 shows the dialog box for MPLOT using the Track
data from Table 4.2. It is filled in to produce the high-resolution version of
the plot in Exhibit 4.4. In addition, we specified a title and axis labels, as
shown in Exhibit 4.16. Output is shown in Exhibit 4.17.

Release 9, Windows

We will look at two items in the Graph menu, Plot and Times Series Plot.
Plot does all the work that the character commands PLOT, LPLOT, and
MPLOT do.

As we said in Section 3.8, graphs in Release 9 have many options, and
when you look at the graph dialog boxes you will see them all. If you do
not understand an option, you can usually just leave it at the default setting.

Exhibit 4.18   Dialog Box for a Plot Using the Furnace Data

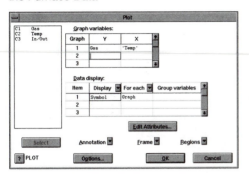

Exhibit 4.19   High-Resolution Plot Using the Furnace Data

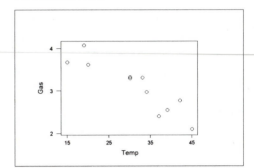

You can use the Help facility to learn more and you can just experiment to see what happens. If you change the defaults and want to get back to the original version of a dialog box, just press ⌐F3⌐ when the first dialog box for the command is active. Everything will be set back to the defaults.

     We will explain a few of the more common graph options in this appendix and later in this book.

PLOT. Exhibit 4.18 shows the dialog box for Graph > Plot, using the Furnace data. We filled in two variables, Gas and Temp, for Graph variables. That's all we need do for a simple plot. Output is shown in Exhibit 4.19.

Drawing Lines on a Plot. You can draw lines on your plot by using the Lines option under Annotation. Suppose we draw a straight line through the points in the plot in Exhibit 4.19. In Chapter 11 we'll show you how to determine a good-fitting line. For now we'll just tell you the answer: The line goes from the point $(15, 4)$ to the point $(45, 2.3)$. We put these numbers into two columns:

```
NAME C11 'x' C12 'y'
READ C11 C12
   15   4
   45   2.3
END
```

Then we enter these columns in the Line dialog box as shown in Exhibit 4.20. Notice that the x variable is listed first, then the y variable. Output is shown in Exhibit 4.21.

Exhibit 4.20 Dialog Box for Adding a Line to the Furnace Data

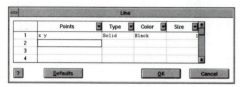

Exhibit 4.21 Furnace Data with a Line Drawn

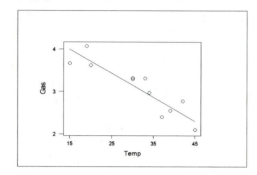

Exhibit 4.22 Dialog Box for a High-Resolution LPLOT

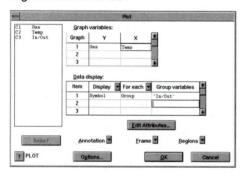

Exhibit 4.23 High-Resolution LPLOT Using the Furnace Data

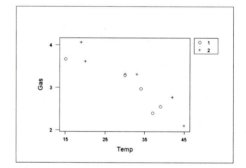

You can add a curve of any type to a graph. Just make sure the columns for *x* and *y* contain enough points so that when Minitab connects the points, the graph is a smooth curve.

LPLOT. The high-resolution version of LPLOT uses Graph > Plot, with a change to the Data display entries. This is shown in Exhibit 4.22. First click the down arrow next to For each to get a list of three choices: Graph, Group, and Point. Click Group to say you want separate symbols for each group. Next, click in the box under Group variables and enter the variable you want to determine the groups. In this case, it is In/Out. Click OK to get the plot shown in Exhibit 4.23.

Exhibit 4.24 Dialog Box Specifying
Several Graphs Using the Track Data

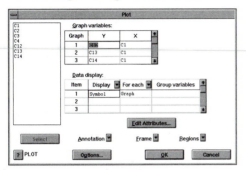

Exhibit 4.25 Dialog Box for Multiple Graphs

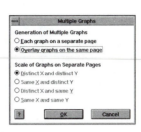

Exhibit 4.26 High-Resolution MPLOT
Using the Track Data

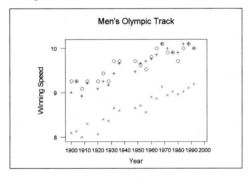

Exhibit 4.27 Dialog Box Specifying Both
Symbols and Lines

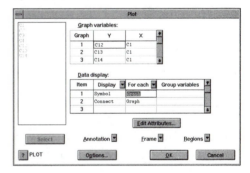

MPLOT. The high-resolution version of MPLOT also uses Graph > Plot. Just list all the plots you want, one by one, in the Graph variables box. Exhibit 4.24 shows the dialog box filled in using the Track data we used in Exhibit 4.4. If we click OK now, we will get three separate graphs, one for each pair of variables, but we want them all as one graph. So click the arrow next to Frame and click Multiple Graphs to get the dialog box shown in Exhibit 4.25. Select Overlay graphs on the same page, then click OK, then click OK in the main dialog box. We also specified a title and a label for the y-axis. Output is shown in Exhibit 4.26.

Now we will use a few other graph options to make a better-looking graph for the Track data. Exhibit 4.27 has Item 2 of the "Data display" table

Exhibit 4.28 Dialog Box for Symbol
Attributes

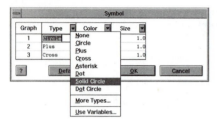

Exhibit 4.29 Improved High-Resolution
MPLOT Using the Track Data

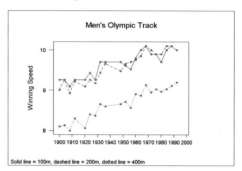

filled in. To do this, click in the box for Item 2 under Display. Then click the down arrow next to Display to get a list of four choices: Area, Connect, Project, and Symbol. Click Connect to say you want to connect the points. Next, click the down arrow next to For each, and click Graph. We have asked for two data display items on our graph: Item 1 says to show symbols for the points in the graph, and Item 2 says to connect the points in the graph, using a different line type for each graph.

The Edit Attributes button allows you to specify the type, color, and size for the symbols and lines in the graph. Click Item 1 to select symbols, then click Edit Attributes to get the dialog box shown in Exhibit 4.28. Click the down arrow next to Type to get the list shown. Click Circle. Repeat this for the three graphs, so that they are all plotted with circles. In the same way, specify circles of size .5 for all three graphs. Click OK. Next, we remove the label from the *x*-axis. In the main dialog box, click Frame, then click Axis. Type a blank character in the label field for the *x*-axis, then click OK. We also added a footnote to the graph to say which line is for which race. Add this footnote (recall that footnotes are under the Annotation button) to get the output shown in Exhibit 4.29.

TSPLOT. We will show just the simplest time series plot, one without a period. Exhibit 4.30 shows the dialog box for TSPLOT, with the variable FOOD from Table 4.3. We also added a title and a label for the *y*-axis. The graph is shown in Exhibit 4.31. In the high-resolution version of TSPLOT, by default the points are all plotted with the same symbol and the points are connected.

Exhibit 4.30 Dialog Box for TSPLOT Using the Wisconsin Employment Data

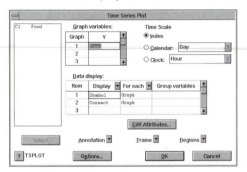

Exhibit 4.31 High-Resolution TSPLOT

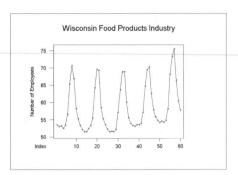

Release 9, VAX/VMS and UNIX Workstations

The graphics capability in the Standard Version of Release 9 is the same as in Release 7. See the material on Release 7, earlier in this section.

The Enhanced Version has many options. We will explain a few of the more common ones. Consult the Minitab manual if you want to learn more.

We will look at two commands, PLOT and TSPLOT. PLOT does all the work that the character commands PLOT, LPLOT, and MPLOT do. PLOT and TSPLOT can use all the subcommands described in Section 3.8 for HISTOGRAM and BOXPLOT. These are TITLE, FOOTNOTE, AXIS and its subcommand LABEL, TICK and its subcommand LABEL, MINIMUM, MAXIMUM, and SAME.

PLOT. The following command creates a high-resolution plot, using the Furnace data. Output is in Exhibit 4.19.

```
PLOT Gas*Temp
```

You can draw lines on your plot by using the LINE subcommand. Suppose we draw a straight line through the points in the plot in Exhibit 4.19. In Chapter 11, we will show you how to determine a good fitting line. For now, we'll just tell you the answer: The line goes from the point (15, 4) to the point (45, 2.3). We put these numbers into two columns, then list them on LINE. Notice that the *x*-variable is listed first, then the *y*-variable. Output is shown in Exhibit 4.21.

The following subcommands are available for PLOT and TSPLOT.

SYMBOL [C]   # specifies symbols for the points

   TYPE   K...K   # specifies the symbol type
   COLOR K...K   # specifies the symbol color
   SIZE   K...K   # specifies the symbol size

CONNECT [C]   # connects points in the graph

   TYPE   K...K   # specifies the line type
   COLOR K...K   # specifies the line color
   SIZE   K...K   # specifies the line size

LINE C C   # adds a line, using (x, y) values in C C

   TYPE   K   # specifies the line type
   COLOR K   # specifies the line color
   SIZE   K   # specifies the line size

OVERLAY   # puts all plots on one graph

SYMBOL says to plot the points with symbols. This is the default. If you use CONNECT, symbols are not automatically drawn. You must use SYMBOL to add symbols if you want them. The column on SYMBOL is for a grouping variable. If you list it, the points in each group are assigned different symbols.

CONNECT connects the points in the plot. Symbols are not shown unless you explicitly use the SYMBOL subcommand. The column on CONNECT is for a grouping variable. If you list it, the points in each group are connected separately, using a different line type for each group.

•

```
NAME C11 'x' C12 'y'
READ C11 C12
   15   4
   45   2.3
END
PLOT Gas*Temp;
  LINES x y.
```

You can add a curve of any type to a graph. Just make sure the columns for x and y include enough points so that when Minitab connects these points, the graph is a smooth curve.

(box continued)

You can specify the type, color, and size of symbols and lines with an integer as shown below. If you use a grouping column on SYMBOL or CONNECT, you can give a different K for each group.

Symbol Types	*Line Types*	*Colors for Symbols and Lines*	
0 none	0 null (invisible)	0 white	5 cyan
1 circle	1 solid	1 black	6 magenta
2 plus	2 dash	2 red	7 yellow
3 cross	3 dot	3 green	
4 asterisk	4 dash-dot	4 blue	
5 dot			

SIZE can be any positive real number. Symbols and connection lines with larger sizes are larger. The default size is 1.

LINE adds a line to the plot. The points in the two columns are connected in the order in which they are listed. The first column is for the x-coordinates; the second column is for the y-coordinates. You can use several LINE subcommands in one plot. Plot scales are not automatically adjusted to accommodate lines that you add, so you may need to enlarge the scales with MINIMUM and MAXIMUM.

OVERLAY puts all the plots listed on the main command on one graph. By default, each graph is assigned a different symbol type and, if you use CONNECT, a different line type.

LPLOT. The high-resolution version of LPLOT is done with PLOT. The following command uses separate symbols based on the variable In/Out. Output is shown in Exhibit 4.23.

```
PLOT Gas*Temp;
  SYMBOLS 'In/Out'.
```

MPLOT. The high-resolution version of MPLOT is done with PLOT. List all the plots you want, one by one, on PLOT. If you do nothing else, each plot will be done separately. We want them all on the same graph. The subcommand OVERLAY requests this.

The following commands produce a high-resolution version of the plot shown in Exhibit 4.4, using the Olympic track data. Output is shown in Exhibit 4.26.

```
PLOT C12*C1 C13*C1 C14*C1;
  OVERLAY;
  TITLE 'Men''s Olympic Track';
  AXIS 2;
    LABEL 'Winning Speed'.
```

Now we will use a few other graph options to make a more attractive graph for the Track data. You can specify the type, color, and size for the symbols and lines in the graph. We use symbol type 1, a circle, and symbol size .5 for all three graphs. We add a footnote to the graph to indicate which line is for which race. We remove the label from the x-axis by specifying a blank label. Output is shown in Exhibit 4.29.

```
PLOT C12*C1 C13*C1 C14*C1;
  OVERLAY;
  SYMBOLS;
    TYPE 1;     # circle
    SIZE .5;    # half the default size
  CONNECT;
  TITLE 'Men''s Olympic Track';
  FOOTNOTE 'Solid line = 100m, dashed line = 200m, dotted line = 400m;
  AXIS 1;
    LABEL ' ';
  AXIS 2;
    LABEL 'Winning Speed'.
```

TSPLOT. We will show just the simplest time series plot, one without a period, using the variable FOOD from Table 4.3. We add a title and a label for the y-axis. The graph is shown in Exhibit 4.31. In the high-resolution version of TSPLOT, by default the points are all plotted with the same symbol and the points are connected.

```
TSPLOT 'FOOD';
  TITLE 'Wisconsin Food Products Industry';
  AXIS 2;
    LABEL 'Number of Employees'.
```

5

Tables

Minitab's TABLE command provides two general types of capabilities. The first, considered in Section 5.1, makes tables of counts and percents. It is a generalization of the TALLY command (see Section 3.7). The second, considered in Section 5.2, provides summary statistics, such as means and standard deviations, for related variables.

Statistical tests of association also can be done with the TABLE command, but the explanation of these tests will be deferred until Chapter 12.

An Example: The Wisconsin Restaurant Survey

Because tables arise naturally in survey work, we will illustrate the TABLE command using data from an actual survey, the 1980 Wisconsin Restaurant Survey, conducted by the University of Wisconsin Small Business Development Center. This survey was done primarily to "allow educators, researchers, and public policy makers to evaluate the status of Wisconsin's restaurant sector and to identify particular problems that it is encountering." A second purpose was to develop data that would "be useful to small business counselors in advising managers as to how to effectively plan and operate their small restaurants."

Nineteen of Wisconsin's counties were selected for study. Lists of restaurants were drawn up from telephone directories and these were sampled in proportion to the population of each county. A sample of 1000 restaurants yielded 279 usable responses.

In a full analysis of these data, we would be concerned about possible biases resulting from the sampling method and from the 28% response rate. Here, however, we are interested only in providing summaries of the returned questionnaires—not in making inferences for all restaurants.

The data base thus consists of 279 cases, one for each restaurant with usable data. In this book, we will use only a few of the many variables in the survey. The data set is described in Appendix A and is saved in the worksheet called RESTRNT. Each of the 279 respondents failed to answer at least one question. Thus there are a number of missing values, each of which is coded as an asterisk (*).

5.1   Tables of Counts and Percents

We begin by describing the basic TABLE command. Then, as needed, we will introduce other features.

As a first example, we will classify the restaurants by type of ownership and size. The variable OWNER takes on three values: 1 = sole proprietorship; 2 = partnership; and 3 = corporation. SIZE also takes on three values: 1 = under 10 employees; 2 = from 10 to 20 employees; and 3 = over 20 employees. Exhibit 5.1 uses the TABLE command to print the appropriate table.

The first column of this table gives counts for restaurants with SIZE 1. There were 83 restaurants with OWNER = 1 and SIZE = 1; that is, there were 83 sole proprietorships with under 10 employees. Similarly, there were 16 partnerships and 40 corporations with under 10 employees. Altogether, there were 139 restaurants with under 10 employees.

The grand total, 261, is the total number of restaurants in this table. This is less than the 279 restaurants that are in the data base. Therefore, only 261

Exhibit 5.1   Number of Restaurants Classified by Ownership and Size

```
MTB > table 'owner' 'size';
SUBC>   counts.

 ROWS: OWNER      COLUMNS: SIZE

                1          2          3        ALL

     1         83         18          2        103
     2         16          6          4         26
     3         40         42         50        132
   ALL        139         66         56        261

     CELL CONTENTS --
                    COUNT
```

restaurants had usable data for both OWNER and SIZE. The other 18 had missing data for either OWNER or SIZE or both.

The last row and last column are called the margins of the table, and the counts there are called marginal statistics. The other nine counts form the main body of the table.

Percents

It is often easier to interpret a table if we convert counts to percents. TABLE has three subcommands to do this: ROWPERCENTS calculates row percents, COLPERCENTS calculates column percents, and TOTPERCENTS calculates total percents. Exhibit 5.2 presents an example using all three types of percents.

Exhibit 5.2   Percents for Restaurants

```
MTB > table 'owner' 'size';
SUBC>   rowpercents;
SUBC>   colpercents;
SUBC>   totpercents.

 ROWS: OWNER       COLUMNS: SIZE

               1          2          3        ALL

    1      80.58      17.48       1.94     100.00
           59.71      27.27       3.57      39.46
           31.80       6.90       0.77      39.46

    2      61.54      23.08      15.38     100.00
           11.51       9.09       7.14       9.96
            6.13       2.30       1.53       9.96

    3      30.30      31.82      37.88     100.00
           28.78      63.64      89.29      50.57
           15.33      16.09      19.16      50.57

  ALL      53.26      25.29      21.46     100.00
          100.00     100.00     100.00     100.00
           53.26      25.29      21.46     100.00

   CELL CONTENTS --
                        % OF ROW
                        % OF COL
                        % OF TBL
```

The first number in each cell is the row percent. To calculate these, Minitab divided the count in each cell by the row total, then multiplied by 100. Refer back to Exhibit 5.1. There are 103 restaurants in row one, sole proprietorships. Of these, 83 have under 10 employees. Therefore, (83/103) x 100 = 80.58% of the sole proprietorships had under 10 employees. This is the row percent shown in the first cell of Exhibit 5.2. Let's look at the rest of this row. Only 1.94% of the sole proprietorships had more than 20 employees. The remaining 17.48% had between 10 and 20 employees. What about the third row, corporations? They are split fairly evenly among the three sizes: 30.30% have under 10 employees, 31.82% have from 10 to 20 employees, and 37.88% have over 20 employees.

The second number in each cell is the column percent. To calculate these, Minitab divided the count of each cell by the column total, then multiplied by 100. Notice that among restaurants with under 10 employees, the majority, 59.71%, are sole proprietorships, whereas among restaurants with over 20 employees, the overwhelming majority, 89.29%, are corporations.

TABLE the data classified by **C...C**

COUNTS
ROWPERCENTS
COLPERCENTS
TOTPERCENTS

Prints one-way, two-way, and multiway tables of counts. The classification variables must contain integer values between −10000 and +10000. If no subcommands are given, then counts are printed by default. You may use up to 10 classification variables in one table.

The subcommands allow you to control what is printed in each cell. COUNT prints a count of the number of observations in each cell. The other subcommands are defined as follows:

$$\text{ROWPERCENTS} = \frac{\text{(number of observations in the cell)}}{\text{(number of observations in the row)}} \times 100$$

$$\text{COLPERCENTS} = \frac{\text{(number of observations in the cell)}}{\text{(number of observations in the column)}} \times 100$$

$$\text{TOTPERCENTS} = \frac{\text{(number of observations in the cell)}}{\text{(number of observations in the table)}} \times 100$$

Stat > Tables > Cross Tabulation

To calculate the total percent in a cell, Minitab divided the count in that cell by the grand total, 261, then multiplied by 100. For example, (83/261) x 100 = 31.80% of the restaurants were sole proprietorships with fewer than 10 employees. Thus, almost one third of the restaurants in the study were small: one owner and under 10 employees.

The last row, the row margin, tells us that over half the restaurants, 53.26%, have under 10 employees. And the last column, the column margin, tells us that just over half the restaurants, 50.57%, are owned by corporations.

Exercises

5-1 The following questions refer to Exhibit 5.1.
   (a) How many restaurants had corporate ownership? How many were partnerships?
   (b) How many were partnerships with 10 to 20 employees? How many were partnerships with over 20 employees?
   (c) How many restaurants had 10 or more employees? Twenty or fewer employees?
   (d) How many had 10 or more employees and were owned by a corporation? How many had 10 or more and were not owned by a corporation?

5-2 Refer to Exhibit 5.2. In each part, indicate what percent of the restaurants had the specified characteristics.
   (a) What percent had fewer than 10 employees? More than 20 employees? Ten or more employees?
   (b) What percent were partnerships with fewer than 10 employees? Corporations with 10 to 20 employees?
   (c) Of the partnerships, what percent had 10 to 20 employees? More than 20 employees?
   (d) Of the restaurants with fewer than 10 employees, what percent were partnerships?
   (e) Of the partnerships, what percent had 10 or more employees?

5-3 (a) Make a table of the Pulse data (described in Appendix A) in which the two classification variables are SEX and RAN. Calculate row and column percentages.
   (b) What percent of the females ran in place? Males? Do either or both of these results seem surprising to you in light of the data description in Appendix A? Discuss.

5-4   Data on all persons committed voluntarily to the acute psychiatric unit of a
health care center in Wisconsin during the first six months of 1981 are given
in the table below and stored in the file HCC. There are three variables: (1)
reason for discharge, with 1 = normal and 2 = other (against medical advice,
court ordered, absent without leave, and so on); (2) month of admission,
with 1 = January, 2 = February, . . . , 6 = June; and (3) length of stay, in
number of days.

Reason	1	1	1	1	1	1	1	1	1	1	1	1	1	1	1	1	1	1	1	1	1	1	1	1	1	1	1	1
Month	1	1	1	1	1	1	1	1	1	2	2	2	2	2	2	2	2	3	3	3	3	3	3	3	3	4	4	4	4
Length	1	2	5	8	8	9	10	13	25	0	1	7	7	1	8	11	11	0	1	2	4	5	13	1	25	2	4	5	12

Reason	1	1	1	1	1	1	1	1	1	1	1	1	1	2	2	2	2	2	2	2	2	2	2	2	2	2	2	2
Month	4	4	4	5	5	5	5	6	6	6	6	6	6	1	2	3	4	4	3	4	6	6	6	6	5	3	4	4	4
Length	25	35	45	1	11	18	19	1	1	3	15	35	75	0	9	1	2	3	1	5	2	5	6	14	1	4	6	19	25

Make a table of the data in which the two classification variables are REA-
SON and MONTH.
(a)   What month had the highest number of discharges? The lowest?
(b)   What percent of all discharges are normal?
(c)   How many discharges are there in the winter months January through
March? In the spring months April through June?
(d)   Compare the length of stay, month by month, for the two types of
discharge. Is one always longer than the other?

5.2   Tables for Related Variables

So far, we have discussed frequency tables, tables whose cells contain
counts and percents. We can also create tables whose cells contain summary
statistics on related variables.

Any number of summary statistics, on any number of variables, can be
put in a table. Exhibit 5.3 has three statistics in each cell: MEDIANS gives
the median sales in each cell, MINIMUMS gives the minimum sales, and
MAXIMUMS gives the maximum. (Notice that medians do not appear in
the margins of the table. Minitab did this to save computing time. Median
is the only statistic that does not appear in the margins.)

Exhibit 5.3   Sales for Restaurants (in thousands of dollars)

```
MTB > table 'owner' 'size';
SUBC>   medians 'sales';
SUBC>   minimums 'sales';
SUBC>   maximums 'sales'.

 ROWS: OWNER       COLUMNS: SIZE

              1         2         3       ALL

   1      98.00    220.00    579.00       --
           0.00     39.00    425.00      0.00
         435.00    507.00    733.00    733.00

   2      76.00    267.50    750.00       --
           2.00    200.00    480.00      2.00
         250.00    550.00    800.00    800.00

   3     133.50    274.00    565.00       --
           2.00    100.00    225.00      2.00
        3450.00    720.00   8064.00   8064.00

 ALL        --        --        --        --
           0.00     39.00    225.00      0.00
        3450.00    720.00   8064.00   8064.00

   CELL CONTENTS --
              SALES:MEDIAN
                    MINIMUM
                    MAXIMUM
```

Exhibit 5.3 merits careful study. First look at the cell with OWNER = 1, SIZE = 1. The minimum sales figure is 0. Thus, one restaurant reported no sales for 1979. Suppose we look at the data set. This sales figure is on line 203. This restaurant also has its market value listed as 0. All other responses, however, seem reasonable. Perhaps the restaurant owner forgot to answer these two questions, or gave unreadable answers, and the person who entered the data into a computer file mistakenly typed a zero instead of an asterisk. Or perhaps the owner answered zero when he meant to answer, "I don't know." Unfortunately, the original questionnaire had been destroyed by the time these problems were discovered, so the best we can do now is replace the two 0s by *, the missing-data code. The appropriate commands are:

```
LET 'SALES'(203) = '*'
LET 'VALUE'(203) = '*'
```

The problem of a zero value for SALES was very easy to find. If we carefully look at the patterns in the table, we can find another unusual value. Within each OWNER category, minimum sales increase as the number of employees increases. The medians follow the same pattern. Now look at the maximums. In the first two OWNER categories, maximum sales increase as we go across the row. In the last OWNER category, however, this pattern does not hold: Corporations with under 10 employees have a much larger maximum than corporations with 10 to 20 employees. In fact, the sales figure of 3450 (that's $3,450,000) is suspiciously large.

Again, we go back to the data. This sales figure is on line 111. Look at the other information for this restaurant. The owner said he was very pessimistic about the future, had made only $10,000 in capital improvements, and valued his business at just $100,000. These responses do not make much sense for a restaurant that had gross sales of $3,450,000. This sales figure is almost certainly in error. Again, we will replace it by *.

The simple table in Exhibit 5.3 brought two errors to our attention. There are probably many more still in the data set. Most data sets initially contain many errors, and one of the first steps in any analysis is to try to find them. The simple displays and summaries of Chapters 3 and 4, and the TABLE command, can help. Use the TALLY command on each categorical variable to discover values out of range. Use DESCRIBE and perhaps STEM-AND-LEAF on all interval variables to find values out of range and unusual patterns. Both DESCRIBE and STEM-AND-LEAF would have uncovered the zero on SALES. Neither, however, would have revealed the restaurant with $3,450,000 on SALES. In itself, $3,450,000 is not unusual, but in the context of OWNER and SIZE it is. This value is called a multivariate outlier; it is unusual in the context of several variables. The plotting commands of Chapter 4 and the TABLE command display several variables at once and can help find multivariate outliers.

Exercises

5-5   Refer to Exhibit 5.3. If the desired results are not available from the exhibit, please say so.

    (a)   What was the median sales volume of partnerships with under 10 employees? Of all partnerships? Of all restaurants?

    (b)   Which type of restaurant had the highest median sales volume? The lowest?

TABLE the data classified by **C...C**

MEANS   for **C...C**
MEDIANS for **C...C**
STDEVS for **C...C**
MINIMUMS for **C...C**
MAXIMUMS for **C...C**
N   for **C...C**
NMISS   for **C...C**

STATS   for **C...C**
DATA   for **C...C**

TABLE prints one-way, two-way, and multiway tables containing statistics for associated variables. These are listed on the subcommands. The classification variables must contain integer values between −10000 and +10000. You may use up to 10 classification variables in one table.

The first seven subcommands calculate one summary statistic for each related variable, for each cell of the table. N gives the number of observations in a cell that have a value for the related variable. NMISS gives the number of observations that have * (missing value) for the related variable. (Thus, COUNT = N + NMISS.)

STATS is the same as typing N, MEAN, and STDEV. DATA prints all values of the related variable, in each cell.

Stat > Tables > Cross Tabulation, use Summaries button for specifying associated variables

---

5-6   Exhibit 5.3 gives data on SALES classified by OWNER and SIZE. Use the TABLE command to determine how many of the restaurants were used in this table. Compare this to the total number of restaurants in Exhibit 5.1. Why is there a difference?

5-7   The following are based on the Pulse data (described in Appendix A).
   (a)   Find the mean of PULSE1 classified by SEX and ACTIVITY. Discuss the output. How do men and women compare on pulse rate?
   (b)   Find the mean of PULSE2 classified by SEX and RAN. Discuss the output.

5-8   TABLE can be used with just one classification variable. Use the Restaurant data to answer the following:

(a)   Find the mean and median number of seats for each TYPEFOOD. Discuss the results.

(b)   Find the mean and median amount of sales per seat for each different TYPEFOOD. Discuss the results.

5-9   Refer to the HCC data in Exercise 5-4.

(a)   Obtain a table of the average length of stay in the acute psychiatric unit by reason and month.

(b)   Does the average length of stay in April 1981 seem to be the same for both types of discharge?

(c)   Repeat parts (a) and (b) for median length of stay.

(d)   Does average length of stay seem to be constant over time for normal discharges? For other discharges? If any pattern is apparent, is it similar for both types of discharge?

5-10   TABLE can be used with three classification variables. Use the Restaurant data to create the table

```
TABLE 'OWNER' 'SIZE' 'TYPEFOOD'
```

One cell contains the count 32. What does this mean? One cell contains the count 42. What does this mean? One cell contains the count 58. What does this mean? One cell contains the count 108. What does this mean?

6

Statistical Distributions

6.1 The Normal Distribution

Probability Density Function

The normal distribution is probably the most important distribution in statistics. Many populations in nature are well approximated by the normal. As an example, consider the data in Table 6.1. Exhibit 6.1 contains a bar chart (this is very similar to a histogram) of the chest girth proportions. The area of each bar is equal to the proportion of men with the corresponding chest girth. For example, the bar at 32 has a base of 1 and a height of .0673. This gives an area of $1 \times .0673 = .0673$, which is the proportion of men with a chest girth of 32 inches.

In Exhibit 6.2 we drew a curve that approximates the bar chart of Exhibit 6.1. The technical term for this curve is the probability density function (pdf) of the normal distribution. The equation for this curve is given in most statistics textbooks and on p. 179 of this handbook. This formula contains two parameters, the mean, μ, and the standard deviation, σ. The center of the normal curve is at μ and the spread is specified by σ. Large values of σ give curves that are spread out—that is, short and fat. Small values of σ give curves that are tall and thin. The curve in Exhibit 6.2 is the pdf of a normal distribution with $\mu = 35$ and $\sigma = 2$.

Note: High-resolution graphs in this chapter were created using Release 9 of Minitab.

Table 6.1 Chest Girth for 1516 Soldiers of the 1835 Army of the Potomac

Chest Girth (in inches)	Number of Soldiers with This Girth	Proportion with This Girth	Cumulative Proportion with This Girth
28	2	.0013	.0013
29	4	.0026	.0040
30	17	.0112	.0152
31	55	.0363	.0515
32	102	.0673	.1187
33	180	.1187	.2375
34	242	.1596	.3971
35	310	.2045	.6016
36	251	.1656	.7672
37	181	.1194	.8865
38	103	.0679	.9545
39	42	.0277	.9822
40	19	.0125	.9947
41	6	.0040	.9987
42	2	.0013	1.0000

Exhibit 6.1 Bar Chart for Soldier Data

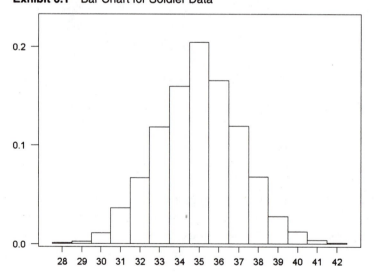

Exhibit 6.2   Bar Chart and Approximating Normal Distribution for Soldier Data

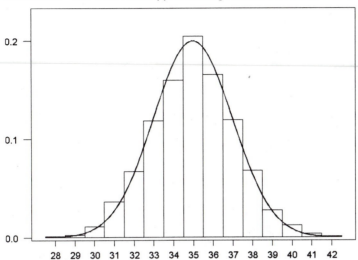

The output in Exhibit 6.3 helped us sketch this curve. Because the chest girths went from 28 to 42, we calculated the normal pdf for values from 28 to 42. The normal pdf is essentially 0 for values below 28 and above 42 (you might check this out yourself using the PDF command). We needed to decide how many values between 28 and 42 to use; we needed enough to be able to draw a smooth curve. Using intervals of .5 seemed about right, so we used SET to put these values into C1. PDF calculated the height of the normal curve at each value and stored these heights in C2. The subcommand NORMAL specifies which parameters to use: $\mu = 35$ and $\sigma = 2$. We then printed the values. We plotted these points on the bar chart and drew a smooth curve through them.

Cumulative Distribution Functions

A cumulative distribution function (cdf) gives the cumulative probability associated with a pdf. In particular, a cdf gives the area under the pdf, up to the value you specify. The cdf is analogous to the cumulative proportions in Table 6.1. Most statistics textbooks have tables that give a range of

Exhibit 6.3   PDF for a Normal Distribution with $\mu = 35$ and $\sigma = 2$

```
MTB > set c1
DATA>    28:42/.5
DATA> end
MTB > pdf c1 c2;
SUBC>    normal 35 2.
MTB > print c1 c2

ROW     C1           C2
  1    28.0     0.000436
  2    28.5     0.001015
  3    29.0     0.002216
  4    29.5     0.004547
  5    30.0     0.008764
  6    30.5     0.015870
  7    31.0     0.026995
  8    31.5     0.043139
  9    32.0     0.064759
 10    32.5     0.091325
 11    33.0     0.120985
 12    33.5     0.150569
 13    34.0     0.176033
 14    34.5     0.193334
 15    35.0     0.199471
 16    35.5     0.193334
 17    36.0     0.176033
 18    36.5     0.150569
 19    37.0     0.120985
 20    37.5     0.091325
 21    38.0     0.064759
 22    38.5     0.043139
 23    39.0     0.026995
 24    39.5     0.015870
 25    40.0     0.008764
 26    40.5     0.004547
 27    41.0     0.002216
 28    41.5     0.001015
 29    42.0     0.000436
```

values for the cdf of a standard normal distribution. With Minitab, you can calculate the cdf for any values you want, and you can calculate it directly for a normal distribution with any mean and standard deviation.

Exhibit 6.4 gives some examples of how Minitab's CDF command can be used to compute cumulative probabilities. In panel (a) we want the area under the normal curve up to the value 33. The CDF command calculates

Exhibit 6.4 Probabilities for a Normal Distribution with $\mu = 35$ and $\sigma = 2$

(a)   **(b)**   **(c)**

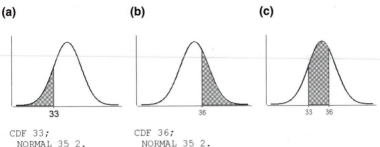

```
CDF 33;             CDF 36;
  NORMAL 35 2.        NORMAL 35 2.
```

this directly as .1587. In panel (b) we want the area above 36. The CDF command only gives areas below a value. Therefore, we used CDF to find the area below 36. This is .6915. The area under the entire normal curve is 1. Thus, $1 - .6915 = .3085$ is the area above 36. In panel (c) we want the area between 33 and 36. This is given by (area below 36) – (area below 33) $= .6915 - .1587 = .5328$.

Exhibit 6.5 gives an example to show how well the normal curve approximates the soldier data. We want the proportion of soldiers with a chest girth of 33. The observed proportion is .1187 and is the area of the bar at 33. To approximate it with the normal curve, we use the region under the curve that approximates the bar at 33. This is the region between 32.5 and

---

PDF for values in **E** [put results into **E**]

NORMAL mu = **K**, sigma = **K**

PDF calculates a probability density function. The subcommand NORMAL says you want the pdf for a normal distribution with the specified mean and standard deviation. If you do not use the subcommand, PDF calculates values for the "standard" normal distribution, with $\mu = 0$ and $\sigma = 1$. If you do not store the results, they are printed. If you do store them, they will not be printed.

Calc > Probability Distributions > Normal, choose option for Probability density

Exhibit 6.5   Using a Normal Curve to Approximate the Soldier Data

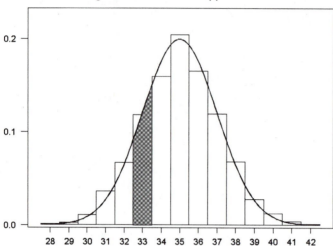

33.5, and is shaded. Note that we cannot just use the value of the PDF at 33. To calculate the area of this bar, we use the CDF command and subtract.

Exhibit 6.6 approximates the proportion of soldiers with chest girths from 31 to 33 inches. The observed proportion is the sum of the areas of the bars at 31, 32, and 33. This sum is .2223. The normal approximation is .2144, which is very close.

CDF for values in **E** [put results into **E**]

    NORMAL mu = **K** sigma = **K**

CDF calculates a cumulative distribution function. The subcommand NORMAL says you want the cdf for a normal distribution with the specified mean and standard deviation. If you do not use the subcommand, CDF calculates values for the "standard" normal distribution, with $\mu = 0$ and $\sigma = 1$. Answers are printed if they are not stored.

Calc > Probability Distributions > Normal, choose option for Cumulative probabilities

---

Exhibit 6.6   Using a Normal Curve to Approximate the Soldier Data

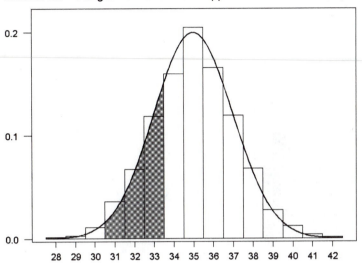

Inverse of the CDF

The command CDF calculates the area associated with a value. The command INVCDF does the opposite: it calculates the value associated with an area. Exhibit 6.7 gives an example using a normal curve with $\mu = 35$ and $\sigma = 2$. We want to find the value x that has an area of .25 below it. The command

```
INVCDF .25;
  NORMAL 35, 2.
```

prints the answer 33.65. Thus, one quarter of the area under this normal curve lies below 33.65.

Exhibit 6.7   Normal Curve with $\mu = 35$ and $\sigma = 2$

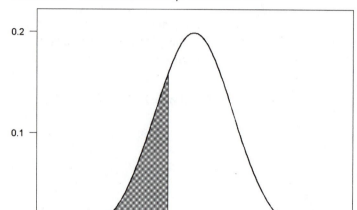

INVCDF for values in **E** [put results into **E**]

NORMAL mu = **K** sigma = **K**

INVCDF calculates an inverse cumulative distribution function. The subcommand NORMAL says you want the values for a normal distribution with the specified mean and standard deviation. If you do not use any subcommand, CDF calculates values for the "standard" normal distribution, with $\mu = 0$ and $\sigma = 1$. Answers are printed if they are not stored.

Note that the values given to INVCDF are probabilities and thus must be between 0 and 1.

Calc > Probability Distributions > Normal, choose option for Inverse cumulative probabilities

Exercises

6-1   In Table 6.1, how many soldiers had chest girths of 37 inches? What pro-
       portion of soldiers had chest girths of 37 inches? What proportion had chest
       girths of 37 inches or less? Of 37 inches or more? Which chest size was the
       most common?

6-2   For a normal distribution with $\mu = 35$ and $\sigma = 2$, use the information in
       Exhibit 6.4 to find:
       (a) The area above 33.
       (b) The area below 36.
       (c) The area between 33 and 35.
       (d) The area between 35 and 36.

6-3   Suppose a normal distribution has $\mu = 100$ and $\sigma = 10$. Use the CDF com-
       mand to find each of the following. In each case, make a sketch like those
       in Exhibit 6.4.
       (a) The area below 90.
       (b) The area above 110.
       (c) The area below 80.
       (d) The area above 120.
       (e) The area between 90 and 110.
       (f) The area between 80 and 120.
       (g) The area between 80 and 110.

6-4   In this exercise you will compute and plot the pdf and the cdf for a normal
       distribution with $\mu = 100$ and $\sigma = 10$. The normal pdf is essentially zero for
       values smaller than $\mu - 3\sigma$ and for values larger than $\mu + 3\sigma$. Therefore, we
       will use the values 70, 72, 74, ..., 130. We can use the commands

```
SET C1
  70:130/2
END
PDF C1 C2;
  NORMAL 100 10.
CDF C1 C3;
  NORMAL 100 10.
PLOT C2 C1
PLOT C3 C1
```

       Run this program, sketch the two curves, and comment on their shape.

6-5   (a) Make a histogram for the SAT math scores for the first set of data in
            the Grades data set (described in Appendix A). Do you think these

scores might have been a random sample from a normal distribution? Why or why not?

(b)   Make a histogram for the SAT math scores for the second set of data in the Grades data set. Compare both histograms. How much do they vary in shape? Do they both look bell-shaped? Do they both look as if they are random samples from a normal population? Do they look as if they both came from the same population?

6.2   The Binomial Distribution

Suppose you bought four light bulbs. The manufacturers claim that 85% of their bulbs will last at least 700 hours. If the manufacturer is right, what are the chances that all four of your bulbs will last at least 700 hours? That three will last 700 hours, but one will fail before that?

Consider another situation. You've just entered a class in ancient Chinese literature. You haven't even learned the language yet, but they've given you a pop quiz. You'll have to guess on every question. It's a multiple-choice test; each of the 20 questions has 3 possible answers. To pass, you must get at least 12 correct. What are the chances that you'll pass?

Questions such as these can sometimes be answered with the help of the binomial distribution. However, for the binomial distribution to hold, the trials must be independent. That is, success on one trial must not change the probability that there will be a success on some other trial. Thus, there must be no special reason why successes or failures should tend to occur in streaks. (Is this condition likely to be met in the light bulb example? In the Chinese literature example?)

Calculating Binomial Probabilities with PDF

A binomial distribution is specified by two parameters: n = the number of trials and p = the probability of success on each trial. The number of successes can be 0, 1, 2, . . . , n. The PDF command will calculate the probability for any possible number of successes. To find out the probability that three of your four light bulbs will be successes (last more than 700 hours) and one will fail, we use

```
PDF 3;
  BINOMIAL n=4, p=.85.
```

Minitab gives the answer .3685.

Exhibit 6.8   Probabilities for a Binomial with $n = 4$, $p = .85$

```
MTB > pdf;
SUBC>   binomial 4  .85.

   BINOMIAL WITH N =    4  P = 0.850000
      K           P(X = K)
      0             0.0005
      1             0.0115
      2             0.0975
      3             0.3685
      4             0.5220
```

PDF for values in **E** [put results into **C**]

BINOMIAL n = **K** p = **K**

Calculates probabilities for the binomial distribution with the specified parameters n and p. If you do not store the results, they will be printed. If you use no arguments on the PDF line, a table of all probabilities will be printed. For example,

```
PDF;
  BINOMIAL n=10, p=.4.
```

prints a table with probabilities for 0, 1, 2, . . . , 10 successes. If any probabilities at the beginning or end of the list are very small (less than .00005), they are not printed.

Calc > Probability Distributions > Binomial, choose option for Probability density

---

If we omit the number of successes from the main command line, PDF will print out the probabilities for all values 0, 1, 2, . . . , n. Exhibit 6.8 contains an example.

Cumulative Distribution Function

The CDF command calculates cumulative probabilities. A cumulative probability is the probability that your result will be less than or equal to a

---

CDF for values in **E** [put results into **E**]

BINOMIAL n = **K**   p = **K**

Calculates the cumulative distribution function (cumulative probabilities) for the binomial. If you do not store the results, they will be printed. If you use no arguments on the CDF line, a table of all cumulative probabilities will be printed. Values that are essentially 0 (less than .00005) and values that are essentially 1 (over .99995) are omitted from the table.

Calc > Probability Distributions > Binomial, choose option for Cumulative probabilities

---

particular value. As an example, suppose we calculate the probability you will fail the test in ancient Chinese literature. Here $n = 20$ and $p = .3333$. You will fail the test if you get less than or equal to 11 questions correct. (You will pass if you get 12 or more right.) The command

```
CDF 11;
  BINOMIAL 20, .3333.
```

calculates this probability to be .9870. This is the probability you will fail. The probability you will pass is therefore $1 - .9870 = .0130$. You thus have a very small chance of passing, as you might expect.

Inverse of the CDF

The command CDF calculates the cumulative probability associated with a value. The command INVCDF does the opposite: it calculates the values associated with a cumulative probability. As an example, Exhibit 6.9 contains the CDF for a binomial with $n = 5$ and $p = .4$. The cumulative probability for 3 is .913. Therefore,

```
INVCDF .913;
  BINOMIAL 5, .4.
```

would give the value 3.

Notice that Exhibit 6.9 contains six probabilities, one corresponding to each possible number of successes. These are the only probabilities for which INVCDF can give an exact answer. Exhibit 6.10 shows what happens when there is no exact answer. The value below and the value above the requested probability are both printed. If you put storage on the CDF command, then the larger value (here it is 3) is stored.

Exhibit 6.9   Cumulative Probabilities for Binomial with $n = 5$, $p = .4$

```
MTB > cdf;
SUBC>    binomial 5 .4.

    BINOMIAL WITH N =    5  P = 0.400000
      K  P(X LESS OR = K)
      0            0.0778
      1            0.3370
      2            0.6826
      3            0.9130
      4            0.9898
      5            1.0000
```

Exhibit 6.10   Example of INVCDF for a Binomial Distribution Where the Probability Requested Cannot Be Exactly Attained

```
MTB > invcdf .8;
SUBC>    binomial 5 .4.
     K  P(X LESS OR = K)        K  P(X LESS OR = K)
     2            0.6826        3            0.9130
```

INVCDF for probabilities in **E** [put results into **E**]

   BINOMIAL n = **K** p = **K**

Calculates the inverse cumulative distribution function. Answers are printed if they are not stored.

If there is no exact answer for a probability you specify, both the value below and value above are printed. If you request storage, the value above is stored.

Calc > Probability Distributions > Binomial, choose option for Inverse cumulative probabilities

Exercises

6-6   A total of 16 mice are sent down a maze, one by one. From previous experience, it is believed that the probability a mouse turns right is .38. Suppose their turning pattern follows a binomial distribution. Use the PDF and CDF commands to answer each of the following.
   (a)   What is the probability that exactly 7 of these 16 mice turn right?
   (b)   That 8 or fewer turn right?
   (c)   That more than 8 turn right?
   (d)   That 8 or more turn right?
   (e)   That from 5 to 7 turn right?

6-7   In this exercise, you'll compute and plot some binomial probabilities in order to get a better idea what they look like.
   (a)   First, let's fix n at 8 and vary p. Use $p = 0.01, 0.1, 0.2, 0.5, 0.8$, and 0.9. For each value of p, compute the binomial probabilities and get a plot. Here are the commands to do the first plot:

```
SET C1
   0:8
END
PDF C1 C2;
   BINOMIAL 8 .5.
PLOT C2 C1
```

How does the shape of the plot change as p increases? Compare the two plots where $p = 0.1$ and $p = 0.9$. Do you see any relationship? What about the two plots where $p = 0.2$ and $p = 0.8$?
   (b)   Next, let's fix p at 0.2 and vary n. Use $n = 2, 5, 10, 20$, and 40. For each value of n, compute the binomial probabilities and get a plot as in part (a). How does the plot change as n increases? How about the spread? The shape? Does the "middle" move when n increases?

6-8   Suppose X is a binomial random variable with $n = 16$ and $p = 0.75$.
   (a)   Write a Minitab program to calculate the mean of X using the formula $\mu = \Sigma x \times \Pr(X = x)$. Does the answer agree with the answer you get when you use the formula $\mu = np$?
   (b)   Write a Minitab program to calculate the variance of X using the formula $\sigma^2 = \Sigma(x - \mu)^2 \times \Pr(X = x)$. Use the value of μ from part (a). Does the answer agree with the formula $\sigma^2 = npq$?

6-9   In the Pulse experiment (described in Appendix A), students were asked to toss a coin. If the coin came up heads, they were asked to run in place. Tails meant they would not run in place. Do you think all students who got heads ran in place? Compare the data with the output from the instruction

```
CDF;
  BINOMIAL 92 .50.
```

Does it seem very likely that only 35 of 92 students would get heads if they all flipped coins? Can you make a conjecture as to what might have happened?

6-10 *Acceptance Sampling.* When a company buys a large lot of materials, it usually does not check every single item to see that they all are satisfactory. Instead, some companies pick a sample of items, then check these; if they do not find many defective items, they go ahead and accept the whole lot. In this problem, we'll look at the kind of risk they run when they do this. Suppose the inspection plan consists of looking at 40 items chosen at random from a large shipment, then accepting the entire shipment if there are 0 or 1 defective items, and rejecting the shipment if there are 2 or more defective items.

(a) If in the entire shipment 25% of the items are defective, what is the probability the shipment will be accepted?

(b) Compute the probability of acceptance if the shipment has 0.1% defective, 0.5% defective, 1%, 2%, 3%, 5%, and 8% defective. Then sketch a plot of the probability of acceptance versus the percent defective.

(c) About what percent defective leads to a 50-50 chance of acceptance?

(d) Repeat part (b) but use a different plan where 100 items are checked and the lot is accepted if 0, 1, or 2 are found defective, and rejected if 3 or more are found defective.

(e) Sketch both graphs from (b) and (d) on the same plot. Discuss the advantages and disadvantages of the two different plans.

6.3   Normal Approximation to the Binomial

For many combinations of n and p, the binomial distribution can be well approximated by a normal distribution that has the same mean and standard deviation—that is, by a normal distribution that has $\mu = np$ and $\sigma = \sqrt{npq}$.

Let's see how the normal approximates a binomial with $p = 1/2$ and $n = 16$. The approximating normal has $\mu = 16(1/2) = 8$ and $\sigma = \sqrt{(16)(1/2)(1/2)} = 2$.

Exhibit 6.11   Binomial and Approximating Normal Curve (A = binomial; B = normal)

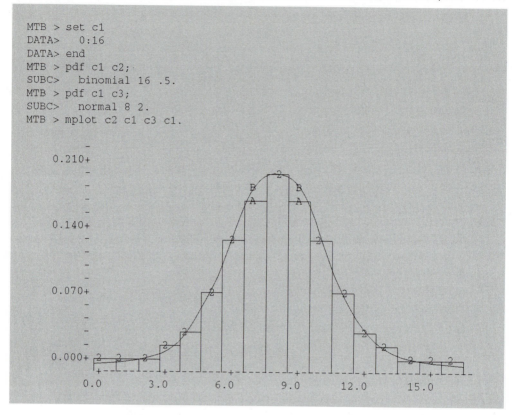

```
MTB > set c1
DATA>    0:16
DATA> end
MTB > pdf c1 c2;
SUBC>    binomial 16 .5.
MTB > pdf c1 c3;
SUBC>    normal 8 2.
MTB > mplot c2 c1 c3 c1.
```

Exhibit 6.11 uses MPLOT to plot the pdf for the binomial and for the normal approximation on the same axes. This will help us see why we can approximate a binomial by a normal and how to do the appropriate calculations. Recall that MPLOT plots the first pair of columns with the letter *A* and the second pair with the letter *B*. When two points overlap, it plots a 2. In this MPLOT, there are many overlaps.

First we converted the binomial probabilities, by hand, to a bar chart. This chart is similar to the one we used for the soldier data in Exhibit 6.2. The height of a bar is the probability the binomial variable is equal to the corresponding value. For example, the height of the bar centered at 5 is the probability the binomial is equal to 5. The base of a bar is 1. Therefore, the area of a bar is equal to its height, and is thus equal to the corresponding probability.

Exhibit 6.12   Binomial and Approximating Normal Probabilities

```
MTB > set c1
DATA>    5 6 7
DATA> end
MTB > pdf c1 c2;
SUBC>    binomial 16 .5.
MTB > print c1 c2

  ROW    C1          C2
    1     5    0.066650
    2     6    0.122192
    3     7    0.174561

MTB > set c11
DATA>    4.5  7.5
DATA> end
MTB > cdf c11 c12;
SUBC>    norm 8 2.
MTB > print c11 c12

  ROW    C11         C12
    1    4.5    0.040059
    2    7.5    0.401294
```

Next we drew the normal curve. To do this, we drew a smooth curve through the values we plotted from the normal pdf.

Exhibit 6.12 does some calculations that will help our explanation. Suppose we want the probability the binomial is from 5 to 7. This probability is the sum of the probabilities at 5, 6, and 7. Thus it is .066650 + .122192 + .174561 = .36340. This is the correct answer. The area of the bar at a given value is equal to the binomial probability of that value. Thus, if we measured the areas of the bars at 5, 6, and 7, we would get .36340. The area under the normal curve that goes from 4.5 to 7.5 approximates the area of the three binomial bars. Exhibit 6.12 gives us the two normal cdf values we need to calculate this area. It is $0.401294 - 0.040059 = 0.361235$, very close to the correct answer.

If we were to use a normal approximation for a binomial with $p = 1/2$ and $n = 30$, the approximation would look even better. In general, for any fixed value of p, the larger n is, the closer the binomial distribution is to a normal. Here we used $p = .5$. In the exercises, we'll look at other values of p.

Exercises

6-11 (a) Make plots as in Exhibit 6.11, but use $p = .4$ instead of $p = .5$. Use $n = 4$ and 16.

(b) Repeat part (a) using $p = .2$.

(c) What can you say about the normal approximation to the binomial? For what values of n and p does it seem to work best?

6-12 Suppose X has a binomial distribution with $p = .8$ and $n = 25$. Use Minitab to calculate each of the probabilities below exactly. Also compute the normal approximation to these probabilities. Remember to go .5 below and above when computing the normal approximation (this is often called using the continuity correction). Compare the binomial results with the normal approximations.

(a) Prob $(X = 21)$.

(b) Prob $(X \leq 21)$.

(c) Prob $(X \geq 24)$.

(d) Prob $(21 \leq X \leq 24)$.

6.4   The Poisson Distribution

A famous statistician once called the Poisson distribution "the distribution to read newspapers by." He said this because so many times we read facts such as "Crime Increases 25% in Year" or "543 Fatalities Expected This Weekend." Such figures frequently can be assessed with the Poisson distribution.

The Poisson distribution arises when we count the number of occurrences of an event that happens relatively infrequently, given the number of times it could happen. For example, the number of automobile accidents that will occur in a given county next weekend could be very large. Any motorist in the area could have an accident. But the chance that any given motorist will have an accident is very small, so the actual number of accidents probably will not be too large.

Calculating Poisson Probabilities with PDF

A Poisson distribution is specified by just one parameter, the mean, μ. For example, if we know that there are, on the average, 6 accidents per weekend, then we can calculate the probability there will be, say, 10 accidents next weekend, or 20 accidents, or no accidents. The PDF command will

Exhibit 6.13   Example of Probabilities for a Poisson Distribution

```
MTB > set c1
DATA>    10 20 0 5
DATA> end
MTB > pdf c1;
SUBC>    poisson 6.
        K              P(X = K)
     10.00              0.0413
     20.00              0.0000
      0.00              0.0025
      5.00              0.1606
```

PDF for values in E [put results into E]

POISSON   mean = **K**

Calculates probabilities for the Poisson distribution with the specified mean. If you do not store the results, they are printed. The mean, μ, must not exceed 100.

If no arguments are given on the PDF line, a table is printed that gives all pdf values greater than .0005.

Calc > Probability Distributions > Poisson, choose option for Probability density

---

calculate probabilities for any values you specify. Exhibit 6.13 gives an example.

Cumulative Distribution Function

Often we are not interested in individual probabilities but in cumulative probabilities, such as the probability of ten or fewer accidents. We then use the CDF command. Exhibit 6.14 gives an example.

From this we see that if the average number of accidents per weekend is six, then the probability of no accidents next weekend is 0.0025, the probability of six or fewer is 0.6063, and the probability of ten or more accidents is $1 - 0.9161 = 0.0839$. What is the probability of having 20 or

Exhibit 6.14   Cumulative Probabilities for a Poisson

```
MTB > cdf;
SUBC>   poisson 6.

   POISSON WITH MEAN =    6.000
       K   P(X LESS OR = K)
       0         0.0025
       1         0.0174
       2         0.0620
       3         0.1512
       4         0.2851
       5         0.4457
       6         0.6063
       7         0.7440
       8         0.8472
       9         0.9161
      10         0.9574
      11         0.9799
      12         0.9912
      13         0.9964
      14         0.9986
      15         0.9995
      16         0.9998
      17         0.9999
      18         1.0000
```

CDF for values in E [put results into E]

POISSON mean = K

Calculates the cumulative distribution function (cumulative probabilities) for the Poisson. If you do not store the results, they will be printed. If you use no arguments on the CDF line, a table of cumulative probabilities will be printed. Values that are essentially 0 (less than .00005) and values that are essentially 1 (over .99995) are omitted from the table. All other probabilities are printed.

Calc > Probability Distributions > Poisson, choose option for Cumulative probabilities

fewer accidents? This value is not in the table. Minitab prints the cumulative probabilities only as far as 18. From there on, the cumulative probabilities are all essentially 1.

Suppose the police decide to crack down on speeders, and that the following weekend there are only three accidents. We can imagine a headline: "Police Crackdown Leads to 50% Reduction in Accidents." What do you think? Suppose we find the probability of three or fewer accidents for $\mu = 6$. It is .1512. That is, there is a 15% chance of a 50% or better reduction in accidents even if the police crackdown has no real effect whatsoever.

Inverse of the CDF

Suppose, in another community, we know that the average number of accidents is 50 per weekend. We want to get some idea of how many accidents we can expect to occur in a given weekend. Exhibit 6.15 gives an example.

As with the binomial distribution, the Poisson does not have an exact answer for every probability we might specify. One way we could summarize this output is as follows: Since $P(54$ or fewer accidents$) = 0.7423$ and $P(44$ or fewer accidents$) = 0.2210$, then $P($from 45 to 54 accidents$) = 0.7423 - 0.2210 = 0.5213$. Thus, over 50% of the time there will be from 45 to 54 accidents. Similarly, $P($from 33 to 69 accidents$) = P(69$ or fewer$) - P(32$ or fewer$) = 0.9957 - 0.0044 = 0.9917$. Thus, over 99% of the time there will be from 33 to 69 accidents. Rarely (about .5% of the time) will there be fewer than 33 accidents and rarely (about .5% of the time) will there be more than 69.

Exhibit 6.15   INVCDF for Poisson with $\mu = 50$

```
MTB > set c1
DATA>    .25    .75   .005   .995
DATA> end
MTB > invcdf c1;
SUBC>    poisson 50.

        K   P(X LESS OR = K)        K   P(X LESS OR = K)
        44              0.2210       45              0.2669
        54              0.7423       55              0.7845
        32              0.0044       33              0.0070
        68              0.9938       69              0.9957
```

INVCDF for probabilities in **E** [put results into **E**]

POISSON mean = **K**

Computes the inverse cumulative distribution function. Answers are printed if they are not stored. If there is no exact answer for a probability you specify, both the value below and the value above are printed. If you request storage, the value above is stored.

Calc > Probability Distributions > Poisson, choose option for Inverse cumulative probabilities

---

Exercises

6-13   A typist makes an average of only one error every two pages, or .5 errors per page. Suppose these mistakes follow a Poisson distribution. Use the PDF and CDF commands to answer each of the following.
   (a)   What is the probability the typist will make no errors on the next page?
   (b)   One error?
   (c)   Fewer than two errors?
   (d)   One or more errors?
   (e)   Sketch a histogram of this distribution (the pdf).
   (f)   Comment on reasons why the Poisson distribution might or might not be a good approximation for the number of errors made by a typist.

6-14   In high-energy physics, the rate at which some particles are emitted has a Poisson distribution. Suppose, on the average, 15 particles are emitted per second.
   (a)   What is the probability that exactly 15 will be emitted in the next second?
   (b)   That 15 or fewer will be emitted?
   (c)   That 15 or more will be emitted?
   (d)   Find a number x such that approximately 95% of the time fewer than x particles will be emitted.
   (e)   Sketch a histogram of this distribution (the pdf).

6-15   In the fall of 1971, testimony was presented before the Atomic Energy Commission that the nuclear reactor used in teaching at Penn State University was causing increased infant mortality in the surrounding town of State

College. This reactor had been installed in 1965. The following data for State College were presented as part of the testimony. In addition, data for Lebanon, Pennsylvania, a city of similar size and rural character, were presented.

	State College			Lebanon		
Year	Live Births	Infant Deaths	Infant Deaths per 1000 Births	Live Births	Infant Deaths	Infant Deaths per 1000 Births
1962	369	4	10.8	666	16	24.0
1963	403	4	9.9	464	15	23.2
1964	365	5	13.7	668	9	13.5
1965	365	6	16.4	582	10	17.2
1966	327	4	12.2	538	8	14.9
1967	385	6	15.6	501	8	16.0
1968	405	10	24.7	439	5	11.4
1969	441	6	13.6	434	3	6.9
1970	452	8	17.7	500	8	16.0

One of the authors (Brian Joiner) was asked to examine the evidence in detail and find out whether there was cause for concern. A wide variety of statistical procedures were used, but one important discussion centered on whether there had been an abnormal peak in infant mortality in State College in 1968.

(a) Suppose we assume that infant deaths follow a Poisson distribution. Over the nine-year period presented in the testimony, there were 53 deaths in State College. On the average, this is $53/9 = 5.9$ per year. Simulate nine observations from a Poisson distribution with $\mu = 5.9$. Repeat this simulation 10 times. Are peaks such as the one in State College in 1968 unusual?

(b) One of the strongest critics of Penn State's reactor drew the following conclusion from the data:

"Following the end of atmospheric testing by the US, USSR, and Britain in 1962, infant mortality declined steadily for Lebanon, while it rose sharply for State College. Using 1962 as a reference equal to 100, State College rose to $(24.7/10.8) \times 100 = 229$ by 1968 while Lebanon declined to $(11.4/24.0) \times 100 = 48$. Furthermore, not only is there an anomalous rise above the 1962 levels in State College after 1963, but there are two clear peaks of infant mortality rates in 1965 and 1968. Especially the high peak in 1968 has no parallel in Lebanon."

Do you think the data support the writer's conclusions that Penn State's reactor has led to a significant rise in infant mortality? What criticisms of his argument can you make?

6-16 If you have a Poisson distribution with a large mean μ, its probabilities can be closely approximated by a normal distribution with the same mean μ and with a standard deviation equal to σ. For each of the following conditions, compute the Poisson and normal pdfs and plot them on the same graph. Comment on the quality of the approximation.

(a)   Use $\mu = 25$. The following program can be used:

```
SET C1
  0:4 0
END
PDF C1 C2;
  POISSON  25.
PDF C1 C3;
  NORMAL 25 5.
MPLOT C2 C1   C3 C1
```

(b)   Repeat (a) for $\mu = 4$.
(c)   Repeat (a) for $\mu = 1$.

6-17 If you have a binomial distribution with a large value of n and a small value of p, its probabilities can be closely approximated by a Poisson distribution with mean equal to np.

(a)   Use $n = 30$ and $p = .01$ and compute the corresponding binomial and Poisson probabilities (pdfs). Do they seem to be fairly close?
(b)   Plot both pdfs on the same plot, using the MPLOT command. Do the pdfs agree fairly well everywhere?
(c)   Compare the binomial and Poisson cdfs. How well do they agree?
(d)   Repeat (a), but use $n = 30$ and $p = .5$. How good is the approximation now?
(e)   Plot both pdfs from part (d) on the same plot. How well do they agree?
(f)   Compare the binomial and Poisson cdfs. How well do they agree?

6.5   A Summary of Distributions in Minitab

The commands PDF, CDF, and INVCDF (discussed in this chapter), and RANDOM (discussed in Chapter 7), all have the same collection of sub-commands to specify different distributions. In this section, we list these distributions and give the corresponding subcommands.

Continuous Distributions

NORMAL μ = **K** σ = **K**

Normal distribution with mean μ and standard deviation σ. The pdf is

$$f(x) = \frac{1}{\sqrt{2\pi}\,\sigma}\, e^{-(x-\mu)^2/2\sigma^2}, \quad \sigma > 0$$

UNIFORM a = **K** b = **K**

Continuous uniform distribution on a to b. The pdf is

$$f(x) = \frac{1}{b-a}, \quad a < x < b$$

CAUCHY a = **K** b = **K**

Cauchy distribution. The pdf is

$$f(x) = \frac{1}{(\pi b)\{1 + [(x-a)/b]^2\}}, \quad b > 0$$

LAPLACE a = **K** b = **K**

Laplace or double exponential distribution. The pdf is

$$f(x) = \frac{1}{2b}\, e^{-|x-a|/b}, \quad b > 0$$

LOGISTIC a = **K** b = **K**

Logistic distribution. The pdf is

$$f(x) = \frac{e^{-(x-a)/b}}{b[1 + e^{-(x-a)/b}]^2}, \quad b > 0$$

LOGNORMAL μ = **K** σ = **K**

Lognormal distribution. A variable X has a lognormal distribution if $\log(X)$ has a normal distribution with mean μ and standard deviation σ. The pdf of the lognormal is

$$f(x) = \frac{1}{x\sqrt{2\pi}\,\sigma}\, e^{-[(\log x)-\mu]^2/2\sigma^2}, \quad x > 0$$

T $v = $ K

Student's t distribution with n degrees of freedom. The pdf is

$$f(x) = \frac{\Gamma[(v+1)/2]}{\Gamma(v/2)\sqrt{v\,\pi}} \frac{1}{(1+x^2/v)^{(v+1/2)}}, \quad v > 0$$

F $u = $ K $v = $ K

F distribution with u degrees of freedom for the numerator and v degrees of freedom for the denominator.

$$f(x) = \frac{\Gamma[(u+v)/2]}{\Gamma(u/2)\Gamma(v/2)}\left(\frac{u}{v}\right)^{u/2} \frac{x^{(u-2)/2}}{[1+(u/v)x]^{(u+v)/2}}, \quad x > 0, u > 0, v > 0$$

CHISQUARE $v = $ K

χ^2 distribution with v degrees of freedom. The pdf is

$$f(x) = \frac{x^{(v-2)/2}\,e^{-x/2}}{2^{v/2}\,\Gamma(v/2)}, \quad x > 0, v > 0$$

EXPONENTIAL $b = $ K

Exponential distribution. (*Note:* Some books use $1/b$ where we have used b.) The pdf is

$$f(x) = \frac{1}{b}\,e^{-x/b}, \quad x > 0, b > 0$$

GAMMA $a = $ K $b = $ K

Gamma distribution. (*Note:* Some books use $1/b$ where we have used b.) The pdf is

$$f(x) = \frac{x^{a-1}e^{-x/b}}{\Gamma(a)b^a}, \quad x > 0, a > 0, b > 0$$

WEIBULL $a = $ K $b = $ K

Weibull distribution. The pdf is

$$f(x) = \frac{ax^{a-1}e^{-(x/b)^a}}{b^a}, \quad x > 0, a > 0, b > 0$$

BETA $a = $ K $b = $ K

Beta distribution. The pdf is

$$f(x) = \frac{\Gamma(a+b)x^{a-1}(1-x)^{b-1}}{\Gamma(a)\,\Gamma(b)}, \quad 0 \le x \le 1, a > 0, b > 0$$

Discrete Distributions

BERNOULLI $p = $ K

This subcommand is available only for the command RANDOM. It simulates Bernoulli trials with probability p of success on each trial.

BINOMIAL $n = $ K $p = $ K

$$f(x) = \binom{n}{x} p^x (1-p)^{(n-x)}, \quad x = 0, 1, \ldots, n$$

POISSON $\mu = $ K

Poisson distribution. The probability of x is

$$f(x) = \frac{e^{-\mu}\mu^x}{x!}, \quad x = 0, 1, 2, 3, \ldots$$

INTEGER $a = $ K $b = $ K

Discrete uniform distribution on the integers from a to b. Each integer has the same probability.

DISCRETE values in C and probabilities in C

Arbitrary discrete distribution. You put the values and the corresponding probabilities into two columns. For example, the following program will print the value .7. This subcommand is most useful for simulating data with the command RANDOM.

```
READ C1 C2
    1   .1
    2   .2
    3   .1
    4   .3
    5   .3
END
CDF 4;
    DISCRETE C1 C2.
```

7

Statistical Distributions and Simulation

7.1   Using Simulation to Learn About Randomness

Most statistical methods described in this book use data to try to answer questions. The data may come from a specially conducted survey, from a carefully designed experiment, from a large data bank, or just from happenstance. In almost all cases, there is variability in data. This could be due to measurement error. Measuring devices (thermometers, IQ tests, scales, and so on) are never perfect. Variability could be due to sampling error (we take a sample from a large population, study the sample, then generalize to the population). Variability could come from environmental conditions (a laboratory gradually warms up during the day, ambient noise varies from testing room to testing room, and so on). Variability could be inherent in the process. You toss a coin; sometimes it comes up heads and sometimes it comes up tails.

If you have the time, money, and awareness, you can often reduce variability. You can use very high-quality measuring devices, take a very large sample, control environmental conditions, and so on—but not always, and never perfectly. This means you must detect the major features of your data in the presence of variability. An engineer would say you must separate the signal from the noise.

To use statistical methods wisely, you need an appreciation of random variability. Simulation can help. Simulation allows you to study the properties of statistical methods in an environment where variability is introduced in a controlled manner. Simulation is also inexpensive and fast, because data are created by a computer rather than through a real experiment or survey. Simulation does have limits, however. All simulated experiments are artificial and idealized.

Simulation is a very good learning tool. It is also used by researchers to study real problems. Economists have developed sophisticated computer models to simulate ways in which the economy might behave under different conditions. Meteorologists also have sophisticated computer models, which they use to predict weather patterns. Statisticians often use simulation to estimate theoretical quantities that are too difficult to calculate directly.

In this chapter, we will use simulation to study properties of the distributions discussed in Chapter 6. First, however, we will look at a very simple random process: Bernoulli trials.

7.2   Bernoulli Trials

There are many cases in which an outcome can be classified into one of two categories: a coin falls either heads or tails; the next child born in a hospital is either a girl or a boy; the next person to walk into a store either buys something or does not; a tomato seed either germinates or does not. These two categories are often labeled, somewhat arbitrarily, as "success" (coin falls heads, baby is a girl, and so on) and "failure" (coin falls tails, baby is a boy, and so on).

Now suppose we imagine a sequence of, say, 20 trials: We toss a coin 20 times, or we observe the next 20 people who walk into a store. If the chance of a success remains the same from trial to trial, and is independent of the outcomes in all previous trials, we have Bernoulli trials. We use the letter p to represent the probability of a success on a single trial. The probability of a failure is then $(1 - p)$.

One way to learn about Bernoulli trials is to toss a coin 20 times, or to observe the next 20 people who walk into a store. Unfortunately, even studies as simple as these can be time-consuming. So, instead of actually doing a study, we can use the computer to simulate the results. Minitab's RANDOM command will simulate data from many different distributions.

Let's look at the gambling game of roulette. Suppose we spin a roulette wheel that has 38 different stopping points. If it stops on any of 18 specific

Exhibit 7.1   Simulation of 20 Spins of a Roulette Wheel
(1 represents a win; 0 represents a loss)

Simulation for Monday

```
MTB > random 15 c1;
SUBC>    bernoulli .4737.
MTB > print c1
```

C1
```
    0     0     0     0     1     1     1     0     0     0     0     0     0     0     0
```

Simulation for Tuesday

```
MTB > random 15 c2;
SUBC>    bernoulli .4737.
MTB > print c2
```

C2
```
    0     0     0     0     1     0     0     1     0     1     0     0     1     0     1
```

Simulation for Wednesday

```
MTB > random 15 c3;
SUBC>    bernoulli .4737.
MTB > print c3
```

C3
```
    0     1     0     1     0     1     0     1     0     1     1     0     1     1     1
```

points, we win. Otherwise, we lose. Thus, the probability of a success is
18/38 = .4737. Suppose that on Monday evening, we play this game for 15
spins. What is likely to happen? Suppose we play for another 15 spins on
Tuesday evening, and for another 15 spins on Wednesday.

Exhibit 7.1 simulates results for these three evenings of gambling. In
each case, we use RANDOM to simulate 15 spins and store them in a column. The subcommand BERNOULLI says to use Bernoulli trials with $p =$
.4737. We then print the column. Here, a 1 represents a win and a 0 represents a loss. How did we do? On Monday, not too well: only 3 wins and 12
losses. We did better on Tuesday: 5 wins. Wednesday was a good day: 9
wins and only 6 losses. On all three days, the same process generated the
results, but we did not get the same results each time. That is because there
is inherent variability in this process.

RANDOM K observations into each of **C...C**

    BERNOULLI p = **K**

RANDOM simulates data, putting K observations into each column listed. The subcommand says to simulate Bernoulli trials. The value p is the probability of a success on each trial. Each success is assigned the value 1, and each failure is assigned the value 0.

Calc > Random Data > Bernoulli

Random Sequences

It is important to understand that randomness has to do with the process that generates observations, not with the observations themselves. However, even if the data come from a random process, this still does not guarantee that a particular set of outcomes will look random. It is possible to spin the roulette wheel 15 times and win each time, or lose each time, or get a sequence that alternates win, lose, win, lose, win, lose, and so on. Such clearly patterned sequences are possible. They are not very likely, however, if the process that generated the observations is truly random.

    Now suppose we collect some real data. How can we know whether we have a random sequence? In practice, we can never know for certain. We can try to understand how the data arose or how they were collected, and then try to determine whether it is likely that they came from a random process. We can also look at the sequence itself. If we see a clear pattern, we might suspect that the data did not come from a random process.

Exercises

7-1   Suppose a basketball team wins its games, on the average, 55% of the time. Suppose we make the simplifying assumptions that there is a 55% chance of winning each game and that these odds are not influenced by what the team has done so far. Then we can simulate a "season" for the team using Minitab. Suppose a season consists of 30 games. Simulate one season.

    (a)   Did the team have a winning season (win more games than it lost)?

    (b)   How long was the team's longest winning streak? How long was its longest losing streak?

(c)   Simulate four more seasons and answer the questions in (a) and (b) for each season. Roughly how much variation should you expect from season to season in this team's performance, even when its basic winning capability is unchanged?

7-2   Suppose someone gives you a coin that is loaded so that heads comes up more often than tails. You want to know exactly how loaded the coin is. To estimate the probability of a head, p, you could toss the coin, say, 100 times.

(a)   How could you simulate this, assuming that p is really 0.6? How could you estimate p from the data? Will your estimate be exactly equal to p?

(b)   Suppose you asked five friends to estimate the probability of a head by having each toss the coin 100 times. Will their estimates all be equal to p? Will they all get the same estimate? How can you simulate the results for all five friends?

(c)   Perform the simulation for yourself and your five friends. How much do the six estimates vary?

7-3   The World Series in baseball pits the winner of the American League against the winner of the National League. These two teams play until one team wins a total of four games.

(a)   Pretend that the two teams are perfectly matched and that there is a 50-50 chance of each team's winning any given game. Further, suppose the games are independent; that is, the chance of winning a particular game does not depend on whether other games have been won or lost. Simulate 20 World Series by using these instructions:

```
RANDOM 20 C1-C7;
   BERNOULLI 0.5.
PRINT C1-C7
```

Now pretend a 1 means the National League won that game and 0 means the American League won. For each series, figure out which team won the series (that is, won four games first). In how many series did the American League win? The National League? How many series lasted only four games? How many lasted five games? Six games? Seven games?

(b)   Repeat (a), but now pretend the National League had a probability of $p = 0.6$ of winning each game. How do these results compare with those in (a)?

7-4   Consider a simple gambling game. You toss a fair coin repeatedly. For each head, you win \$1; for each tail, you lose \$1. The program below simulates this game for 50 tosses. RANDOM simulates the result of each trial, using

1 for a win and 0 for a loss. LET converts the data to winnings, using +1 for win $1 and −1 for lose $1. PARSUMS computes the cumulative winnings. For example, the number in row 10 is the amount of money you have after 10 tosses. (*Note:* This process is an example of what is called a simple random walk.)

```
RANDOM   50 C1;
  BERNOULLI .5.
LET C2 = 2*C1-1
PARSUMS C2 C3
NAME C3 'Winnings'
TSPLOT C3
```

(a) Run this simulation. How did you do?

(b) How often were you ahead? Were you ahead at the end of 50 tosses?

(c) What was your longest winning streak? Losing streak?

(d) Run the simulation four more times. How do the results compare?

7.3 Simulating Data from a Normal Distribution

Let's see how data can vary when you take a sample from a normal population. We will use a simple example. For the population of males between 18 and 74 years old in the United States, systolic blood pressure is approximately normally distributed with mean 129 millimeters of mercury and standard deviation 19.8 millimeters of mercury.[1]

Suppose we were to take a random sample of 25 men and measure their systolic blood pressure. Would the sample look normal? Would its mean be 129, or at least close? Would you expect any observation to be over 190? (That's more than three standard deviations above the population mean.)

We do not have the time or money to take a random sample of 25 men, so we will use simulation. In Exhibit 7.2, we use RANDOM to draw three samples of $n = 25$ men. The subcommand NORMAL says we want normal data with mean = 129 and standard deviation = 19.8. We use HISTOGRAM and DESCRIBE to summarize the data. Notice that the results for the three samples differ. Each time we take a sample, we get different men, with different blood pressures. As a result, the histograms vary in shape; the sample means and other statistics vary in value. This variation is called sampling error.

---

[1]M. Pagano and K. Gruvreau, *Principles of Biostatistics.* North Scituate, Mass.: Duxbury Press, 1993, p. 163.

Exhibit 7.2   Three Samples of 25 Observations from a Normal Distribution

```
MTB > random 25 c1-c3;
SUBC>   normal 129 19.8.
MTB > histogram c1-c3;
SUBC>   same.

Histogram of C1   N = 25

Midpoint   Count
      80       1   *
      90       0
     100       0
     110       4   ****
     120       4   ****
     130       3   ***
     140       5   *****
     150       1   *
     160       5   *****
     170       2   **


Histogram of C2   N = 25

Midpoint   Count
      80       0
      90       0
     100       2   **
     110       2   **
     120       3   ***
     130       9   *********
     140       5   *****
     150       1   *
     160       3   ***
     170       0

Histogram of C3   N = 25

Midpoint   Count
      80       0
      90       1   *
     100       2   **
     110       5   *****
     120       4   ****
     130       7   *******
     140       2   **
     150       1   *
     160       3   ***
     170       0
```

Exhibit 7.2　*(continued)*

```
MTB > describe c1-c3

                N      MEAN    MEDIAN    TRMEAN     STDEV    SEMEAN
C1             25    134.91    135.95    135.63     21.75      4.35
C2             25    132.01    132.26    131.99     15.59      3.12
C3             25    124.92    125.49    124.81     19.23      3.85


              MIN       MAX        Q1        Q3
C1          84.17    168.97    118.26    156.80
C2         101.12    163.23    121.38    138.50
C3          87.95    164.40    112.43    135.52
```

Let's see how each sample compares to the population. Do the histograms look bell-shaped? Not the first one. It is fairly uniform over the interval from 110 to 170, and has one "unusual" observation around 80. This sample, however, came from a normal population. The other two histograms are a little closer to bell-shaped, and do not have any unusual observations. Now look at the sample means. The first is 134.91, almost 6 above the population mean of 129. The other two, 132.01 and 124.92, are closer.

In Exhibit 7.3, we again draw samples from the population of blood pressures, but now we use a sample size of $n = 100$. When n is larger, a sample will, in general, look more like the population from which it was drawn; these three do. The histograms are all approximately bell-shaped. The sample means, 128.17, 128.18, and 131.79, are all within three units of the population mean. There is still variability from sample to sample, but less than in Exhibit 7.2. If we were to use a still larger sample size, say $n = 500$ or $n = 1000$, then, in general, the histograms would look even more bell-shaped, and the sample mean would be even closer to the population means.

Exhibit 7.3   Three Samples of 100 Observations from a Normal Distribution

```
MTB > random 100 c11-c13;
SUBC>   normal 129 19.8.
MTB > histogram c11-c13;
SUBC>   same.

Histogram of C11   N = 100

Midpoint   Count
      70       1   *
      80       1   *
      90       4   ****
     100       6   ******
     110      14   **************
     120      14   **************
     130      21   *********************
     140      16   ****************
     150      16   ****************
     160       6   ******
     170       1   *
     180       0


Histogram of C12   N = 100

Midpoint   Count
      70       0
      80       1   *
      90       1   *
     100       5   *****
     110      15   ***************
     120      30   ******************************
     130      15   ***************
     140      16   ****************
     150       6   ******
     160       8   ********
     170       3   ***
     180       0
```

Exhibit 7.3 *(continued)*

```
Histogram of C13    N = 100

Midpoint   Count
      70      1   *
      80      0
      90      3   ***
     100      5   *****
     110     10   **********
     120     18   ******************
     130     17   *****************
     140     20   ********************
     150     12   ************
     160     11   ***********
     170      2   **
     180      1   *


MTB > describe c11-c13

              N      MEAN    MEDIAN    TRMEAN     STDEV    SEMEAN
C11         100    128.17    130.13    128.75     19.86      1.99
C12         100    128.18    124.73    128.02     17.60      1.76
C13         100    131.79    133.50    132.30     20.42      2.04


            MIN       MAX        Q1        Q3
C11       72.96    172.49    114.80    141.93
C12       77.68    170.00    116.67    138.99
C13       69.19    178.22    118.47    146.32
```

RANDOM K observations into each of C...C

NORMAL mu = K sigma = K

Puts a random sample of K observations into each column. The subcommand says you want to simulate data from a normal distribution with the specified μ and σ. If no subcommand is given, RANDOM simulates data from a standard normal distribution, with $\mu = 0$ and $\sigma = 1$.

Calc > Random Data > Normal

A Useful Trick for Simulating Data

Suppose we wanted 50 random samples, each containing 10 observations. There are two ways we could do this in Minitab:

```
RANDOM 10 C1-C50
RANDOM 50 C1-C10
```

With the first command, we consider each column to be a separate sample. With the second command, we consider each row to be a separate sample. Sometimes it is better to have samples in rows; sometimes it is better to have them in columns. It all depends on what we want to do with the samples, and which Minitab commands we want to use for further calculations.

As an example, suppose we simulate 100 samples, each of size 25, from a normal distribution with $\mu = 129$ and $\sigma = 19.8$, and then determine how many of these samples have at least one observation smaller than 84. Exhibit 7.4 shows the commands and output. We put the samples into rows. Because of this, we needed just three commands to do the work.

Remember the first sample in Exhibit 7.2. It had one observation of 84.17. Is this observation really so unusual? Exhibit 7.4 simulates 100 samples from the population of blood pressures. We then find the minimum value in each row (each sample), store the minimums in C50, and do a stem-and-leaf display. Notice that in 22 of the samples, there was at least one observation smaller than 84. Thus, 84.17 is not so unusual at all.

Exercises

7-5   In samples from the normal distribution, about 68% of the observations should fall between $\bar{x} - s$ and $\bar{x} + s$. About 95% should fall between $\bar{x} - 2s$ and $\bar{x} + 2s$. Simulate a random sample of size 100 from a normal distribution. What percentages do you find in these two regions (you choose μ and σ)?

7-6   Repeat Exercise 7-5 using real data. For example, use the SAT math scores from sample A of the Grades data (described in Appendix A).

7-7   In this exercise we look at the distribution of a linear combination of normal variables. Simulate 200 observations from a normal distribution with $\mu = 20$ and $\sigma = 5$ into C1. Simulate another 200 with $\mu = 7$ and $\sigma = 1$ into C2. Then compute the following:

Exhibit 7.4   Simulating Data in Rows of Minitab's Worksheet

```
MTB > random 100 c1-c25;
SUBC>    normal 129 19.8.
MTB > rminimum c1-c25   c50
MTB > stem-and-leaf c50

Stem-and-leaf of C50           N  = 100
Leaf Unit = 1.0

    1      5 4
    1      5
    2      6 3
    4      6 79
    7      7 034
   15      7 55567889
   25      8 0012333444
   42      8 55666888888889999
  (18)     9 001122233344444444
   40      9 556666777888888999
   22     10 0000122222333344
    6     10 5568
    2     11 11
```

```
LET  C3=3*C1
LET  C4=2*C2
LET  C5=3*C1+10
LET  C6=2*C2+5
LET  C7=C1+C2
LET  C8=3*C1+2*C2+3
```

Now use DESCRIBE C1–C8 and HISTOGRAM C1–C8. Try to give formulas indicating how the means and standard deviations of the various columns are related. Use DESCRIBE to support your formulas. What sort of distributions do the various columns seem to have?

7-8   In this exercise, we look at a mixture of two normal populations.

   (a)   The heights of women are approximately normal with $\mu = 64$ and $\sigma = 3$. The heights of men are approximately normal with $\mu = 69$ and $\sigma = 3$. Suppose you took a sample of 200 people—100 men and 100 women—and drew a histogram of the 200 heights. Do you think it would look normal? Should it look normal? Use Minitab to simulate this process. Simulate 100 men's heights and put them in C1. Simulate 100 women's heights and put them in C2. Then use the STACK command to put the combined sample into C3 (see Section 17.2 for a discussion):

```
STACK C1 on top of C2, put stacked data into C3
```

Get a histogram of the data. Does it look normal?

(b) Part (a) is an example of a mixture of two populations. Let's try a slightly more extreme example. Take a sample of 200 observations: 100 of them from a normal population with $\mu = 5$ and $\sigma = 1$, and the other 100 from a normal population with $\mu = 10$ and $\sigma = 1$. Get a histogram of the 200 observations. Does this histogram seem to indicate that there might be two populations mixed together in the sample?

7-9 Exercise 7-4 contained an example of a random walk in which each "step" was +1 or −1. Now we will look at an example in which the size of each step has a normal distribution. A process such as this is often used to build models for the way the stock market operates.

```
RANDOM 50 C1;
   NORMAL 0 2.
PARSUMS C1 C3
NAME C3 'Price'
TSPLOT C3
```

Run this simulation five times. Compare the various patterns.

7.4 Simulating Data from the Binomial Distribution

In Chapter 6, we discussed an example concerning light bulbs. Each package contains four bulbs; the manufacturer claims that 85% of its bulbs will last at least 700 hours. Suppose we were to keep track of the next 50 packages sold. For each package, we record the number of bulbs that last at least 700 hours. What might happen? Exhibit 7.5 shows a simulation. We use TSPLOT to display the results over time.

In the first two packages, all four bulbs lasted over 700 hours. In the third package, only three bulbs lasted over 700 hours. The next five packages were good, and so on. Package 15 had only two bulbs that lasted over 700 hours; the same was true of packages 18, 33, and 38. This simulation gives you some idea of how a sequence of 50 binomials with $n = 4$ and $p = .85$ might look. To develop your appreciation of randomness, you might try repeating this simulation several times.

Exhibit 7.5 Simulation Binomial Data

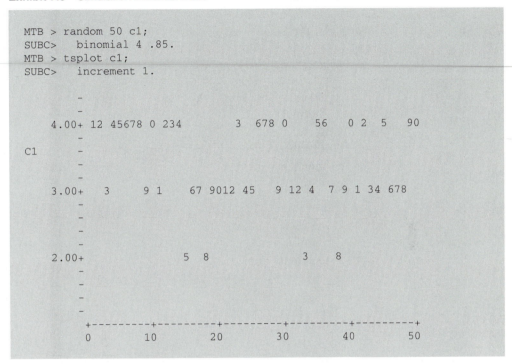

```
MTB > random 50 c1;
SUBC>   binomial 4 .85.
MTB > tsplot c1;
SUBC>   increment 1.
```

Let's use the intuition about randomness that you are developing to look at an example with data. Suppose you actually check the next 50 packages of light bulbs from this manufacturer and get the data shown in Exhibit 7.6. We connected the points in the plot by hand to help you see patterns more clearly. What do you see? Something seems to have changed around package 40. The numbers seem to be smaller, on the average, than they were for the first 39 packages. The mean of a binomial with $n = 4$ and $p = .85$ is $\mu = np = 4 \times .85 = 3.4$. We drew this line on the plot. Notice that the last 10 observations are all below the mean. Do you think these observations came from a binomial with $n = 4$ and $p = .85$? It looks as if something may have changed in the manufacturing process, something that decreased the lifetimes of the bulbs.

Exhibit 7.6   Data for 50 Packages of Light Bulbs

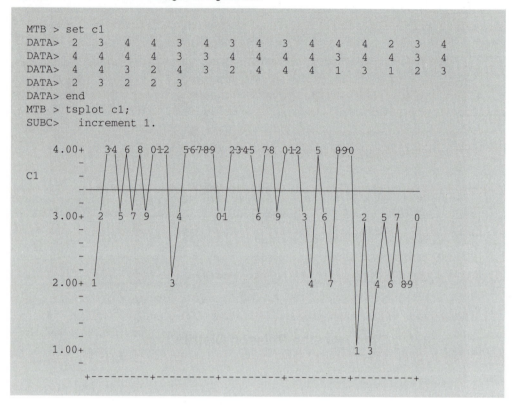

```
MTB > set c1
DATA>  2    3    4    4    3    4    3    4    3    4    4    4    2    3    4
DATA>  4    4    4    4    3    3    4    4    4    4    3    4    4    3    4
DATA>  4    4    3    2    4    3    2    4    4    4    1    3    1    2    3
DATA>  2    3    2    2    3
DATA> end
MTB > tsplot c1;
SUBC>    increment 1.
```

RANDOM K observations into each of C...C

    BINOMIAL n = **K** p = **K**

Simulates random observations from a binomial distribution.

Calc > Random Data > Binomial

Exercises

7-10 (a) Imagine flipping six fair coins and recording the number of heads. Simulate this process 100 times. Use the "Useful Trick" described on p. 192. Get a histogram of your results.

(b) How often did all six coins come up heads? How often would you expect this event to occur in the 100 simulations? (PDF can help you answer this.)

(c) How often did you get more heads than tails? How often would you expect this event to occur in the 100 simulations? (PDF can help you answer this.)

(d) Does your histogram look symmetric (more or less)? Would you expect it to?

(e) Convert your observed frequencies of 0, 1, 2, . . . , 6 heads to relative frequencies. Compare these relative frequencies with the theoretical probabilities from the PDF command.

(f) Repeat (a) through (e), but assume the coin is biased so that the probability of a head is 0.9.

7.5   Simulating Data from the Poisson Distribution

Let's look again at the example for weekend automobile accidents discussed in Chapter 6. Exhibit 7.7 simulates data for one year. We use the POISSON subcommand of RANDOM to simulate 52 observations from a Poisson distribution with mean = 6.

Can you imagine what would appear in the newspaper each Monday? For the first five weeks, you might read: "Weekend accidents a little high, sheriff's program not working yet." The next two Mondays, we might read: "Sheriff's program finally taking effect, accidents are down." Now what would appear the next week? A major headline: "Carnage on the highway, 14 accidents this weekend." We could continue writing the Monday story on weekend accidents for the rest of this year. The assessments and explanations in these stories would all be fiction. Only one thing is really going on: ordinary random variation in a Poisson distribution with mean 6.

Exhibit 7.7   Simulating Poisson Data

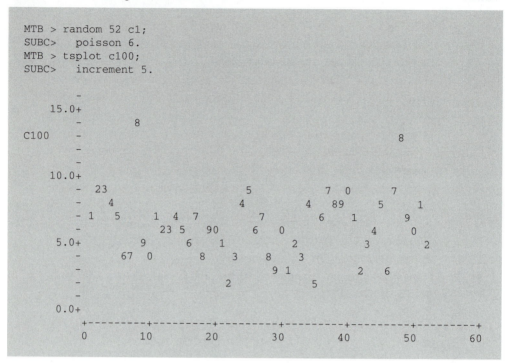

```
MTB > random 52 c1;
SUBC>   poisson 6.
MTB > tsplot c100;
SUBC>   increment 5.
```

RANDOM K observations into each of **C...C**

   POISSON mean = **K**

Simulates random observations from a Poisson distribution.

Calc > Random Data > Poisson

Exercises

7-11 (a) Run the simulation in Exhibit 7.7. Compare your results to those in Exhibit 7.7.

(b) How many times did you get 12 or more accidents? What is the probability of getting 12 or more accidents? How many times would you expect to get 12 or more accidents in one year?

(c) What was the mode—that is, the number that occurred most often? TALLY can help you.

(d) Was there ever a weekend that was accident-free? Would you expect that this might happen within a year?

(e) What was the total number of accidents for the year?

7.6 Sampling Finite Populations

So far in this chapter, we have talked about random samples from theoretical distributions. However, sometimes we are interested in a random sample from an actual finite population.

Suppose we have a very large set of data and want to do some preliminary analyses. If we use only a portion of the data, then our work will be easier, faster, and cheaper. Once we have some idea of the structure of the data, we can analyze the full set. One way to select a portion of a data set is to take a random sample from that data set.

SAMPLE K observations from **C...C** put sample into **C...C**

REPLACE

Takes a random sample of size K from the data in the first group of columns and stores the sample in the second group of columns. If you do not use the subcommand REPLACE, sampling is done without replacement. If you use REPLACE, sampling is done with replacement.

Calc > Random Data > Sample From Columns

Exercises

7-12   The Pulse experiment was run in one class. There were 92 students in the class.

(a)   Use SAMPLE to take a random sample of 20 students from this class.

(b)   What proportion of the students in your sample are women? How does this compare with the full class?

(c)   What is the average height of the students in your sample? How does this compare with the full class?

(d)   Take another sample of 20 students and answer (b) and (c) for this sample.

7.7   Simulation Data from Other Distributions

Minitab can simulate data from all the distributions listed in Section 6.5. RANDOM uses the same subcommand for a given distribution as do PDF, CDF, and INVCDF.

Exercises

7-13   *Discrete Uniform Distribution.* Suppose there are 30 people at a party. Do you think it is very likely that at least two of the 30 people have the same birthday? Try estimating this probability by a simulation. To simplify things, ignore leap years (assume all years have 365 days) and assume all days of the year are equally likely to be birthdays. Use the following commands to simulate 10 sets of 30 birthdays:

```
RANDOM 30 observations into C1-C10;
  INTEGERS 1 to 365.
TALLY C1-C10
```

How many of your 10 sets of 30 people had no matching birthdays?

7-14   *Continuous Uniform Distribution.* The uniform distribution and the *t*-distribution with 2 degrees of freedom are examples of very nonnormal distributions. Their histograms should look quite different from those for normal data sets.

(a)   Simulate 50 observations from a uniform distribution and make a histogram. Repeat five times. Comment on the general shape of the histograms.

(b)   Do (a), but simulate data from a *t*-distribution with 2 degrees of freedom.

7-15   *Chi-Square Distribution.* The mean, standard deviation, and shape of the chi-square distribution change as the number of degrees of freedom changes. For a chi-square distribution with n degrees of freedom, the mean is n and the standard deviation is $\sqrt{2n}$.

    (a)   Simulate 200 observations from a chi-square distribution with 1 degree of freedom into C1. Use DESCRIBE to compute the sample mean and the standard deviation. Are they approximately 1 and $\sqrt{2} = 1.414$, respectively? Now make a histogram and sketch the shape of the distribution. For all histograms in this exercise, use the subcommands START = 0.1 and INCREMENT = 0.2.

    (b)   Repeat (a), but use 2 degrees of freedom.

    (c)   Repeat (a), but use 5 degrees of freedom.

    (d)   Repeat (a), but use 10 degrees of freedom.

7-16   *Chi-Square Distribution.* The chi-square distribution arises in several ways in statistics. Here we will use simulation to illustrate three of those ways.

    (a)   If X is from a standard normal distribution (with mean 0 and standard deviation 1), X^2 has a chi-square distribution with 1 degree of freedom. Simulate 200 standard normals into a column, then square them. Then simulate 200 chi-square observations with 1 degree of freedom. Now make a histogram of the X^2s and a histogram of the chi-squares. The two histograms should be very similar.

    (b)   Another way the chi-square distribution arises is from the variance of samples from a normal distribution. Simulate 200 rows of standard normal observations into C1–C4. Use RSTDEV to compute the standard deviations of the 200 rows. Square the results to get variances. Make a histogram of the variances. Compare this to a histogram of 200 chi-square observations with 3 degrees of freedom.

    (c)   Still another source of the chi-square distribution is as follows: Simulate 200 Poisson observations with $\mu = 5$ into C1–C3. Now compute the 11 chi-square test, which we will discuss in Chapter 11, using the following commands:

```
LET C4 = (C1+C2+C3)/3
LET C6 = (C1-C4)**2+(C2-C4)**2+(C3-C4)**2
LET C7 = C6/C4
```

    Now make a histogram of C7 and compare it to a histogram of a chi-square with 2 degrees of freedom.

7-17   *The t Distribution.* The t distribution arises most naturally as $(x - \mu)/(s/\sqrt{n})$. To illustrate the development of a t distribution, simulate 200 rows of normal data into C1–C3. You choose μ and σ. Use RMEAN and RSTDEV to compute $\bar{x}$ and s for C1–C3. Use LET to compute $t = (\bar{x} - \mu)/(s/\sqrt{3})$.

Make a histogram of t. Compare this histogram to one you obtain by using RANDOM to simulate directly 200 observations for a t distribution with 2 degrees of freedom. Describe the shapes of the two histograms.

7.8   The Base for the Random Number Generator

To simulate random data, Minitab uses a mathematical function that creates a very long list of numbers, which appear to be random. When you use RANDOM, Minitab haphazardly chooses a place to start reading from this list. Each time you use the RANDOM command, Minitab chooses a different place to start, and therefore you get a different set of simulated data.

   Occasionally you may want to generate the same set of "random" data several times. To do this, use the command BASE. BASE has one argument, K, that tells Minitab where to start reading in its list of random numbers. For example, the following commands put the same set of random numbers into C1 and C2, but a different set into C3:

```
BASE 23
RANDOM 10 C1;
   BERNOULLI .2.
BASE 23
RANDOM 10 C2;
   BERNOULLI .2.
RANDOM 10 C3;
   BERNOULLI .2.
```

BASE = K

BASE is used to generate the same set of random data at different times. The value of K tells Minitab where to start reading in its list of random numbers. (*Note:* K determines where in the list to start. It is not the first random number Minitab gives you.) If you type two RANDOM commands in the same program, the second RANDOM will continue reading the list of random numbers where the first RANDOM stopped.

K should be a positive integer. The BASE command applies to all distributions used with RANDOM.

Calc > Set Base

8

One-Sample Confidence Intervals and Tests for Population Means

We often want to know something about a population. The population could be all the 10-year-old girls in Philadelphia and we might want to know the mean blood pressure. The population could be all the light bulbs produced by a plant last month and we might want to know what proportion are defective. In many cases, we cannot afford to study the entire population. So we take a small sample—perhaps 200 girls or 80 light bulbs—and we use information about the sample to estimate what we want to know about the population. If our sample is representative of the population, then our estimate may be quite good.

In the example with 10-year-old girls, we want to know the mean blood pressure of the population. This mean is denoted by the letter μ. We can use the mean of the sample, $\bar{x}$, as a guess or estimate of μ. In the light-bulb example, we want to know the proportion of defective bulbs in the population. This population proportion is usually denoted by p. We can estimate p by the proportion of defectives in the sample of 80 bulbs. This sample proportion is often denoted by $\hat{p}$.

But how do we get a representative sample? One way is to take what is called a simple random sample. A simple random sample has two important properties: Each member of the population has an equal chance of being chosen for the sample, and the observations form a random sequence, as discussed in Chapter 7. The command RANDOM, discussed in Chapter 7, simulates drawing a simple random sample from a population with a specified distribution.

Here is one straightforward way to get a simple random sample of 200 girls: Start by writing the name of each 10-year-old girl in Philadelphia on a slip of paper. We might get a good list from school records. Then put all the slips in a large box. Mix them up and draw one slip. Mix them up again and draw a second slip, and so on until we have 200 slips. We could then contact these girls and measure their blood pressure. In this example, everything seems fairly simple. Unfortunately, in most studies it is very difficult to get a truly random sample. In Section 8.6 we will briefly discuss some ways to spot nonrandomness.

In this chapter we study different methods of learning about the mean of a population. These methods require two things: We must have a random sample and we must have a normal population. The first requirement is very important. The second, however, can be relaxed somewhat. In Section 8.6 we'll briefly discuss the ways a nonnormal population might affect your conclusions and how you can spot nonnormality using a sample of data.

8.1   How Sample Means Vary

Imagine drawing a random sample from a population with a normal distribution and calculating the mean of the sample. Imagine doing this 100 times so that we have 100 values of $\bar{x}$. Exhibit 8.1 shows the results using 100 samples, each with 9 observations. We did two dotplots: one of the data in C1, a sample of 100 individual observations, and one of the 100 sample means. We put both on the same scale so we could compare them more easily. There is one striking difference: The 100 sample means are much less spread out than the 100 individual observations.

Exhibit 8.2 contains output from STEM-AND-LEAF and DESCRIBE for the 100 $\bar{x}$s, so we can examine them more carefully. Most of the sample means are fairly close to the population mean, $\mu = 80$. All of them are between 75 and 84, and half are between 79 and 81. In real life we usually have just one sample, not 100. That one sample would give just one $\bar{x}$, not 100 of them. We could be lucky and get a sample that's near 80—most are close—or we could be unlucky and get that sample that's down at 75.853.

Exhibit 8.1   Samples of 100 Individual Observations and 100 Sample Means with $n = 9$

```
MTB > rand 100 c1-c9;
SUBC>    norm 80 5.
MTB > rmean c1-c9 c20
MTB > name c20 = 'xbars'   c1 = 'x-values'
MTB > dotplot c1 c20;
SUBC>    same.


                                        :  .
                             .    .  :  :     .  .   :
                      .      : .:   :  :.:::::::.:
               . .   ::  .:::::..:: :::::::::::::.::..::  ...: .
             ---+---------+---------+---------+---------+---------+---x-
values


                                        .
                                       ::
                                      .::
                              :  .:::
                              ::  :::::
                             .::  ::::::::.
                             :::.::::::::::
                         .  .:::::::::::::::
             ---+---------+---------+---------+---------+---------+---
xbars
```

Notice that the mean of the 100 $\bar{x}$s is 80.160, which is very close to 80, the population mean. Even if we take samples from nonnormal distributions, the mean of the population of $\bar{x}$s will always be equal to the mean of the population of individual observations. Some of the exercises allow you to explore this fact. In symbols, we write

$$\mu_{\bar{x}} = \mu$$

As we noticed before, the distribution of $\bar{x}$s does differ from that of the individual observations in one important respect—it is less spread out. In fact, for random samples of size n from any population,

$$\sigma_{\bar{x}} = \frac{\sigma}{\sqrt{n}}$$

In our simulation, $\sigma = 5$ and $n = 9$; $\sigma_{\bar{x}} = 1.667$. This is the standard deviation for the population of $\bar{x}$s. The standard deviation of our particular collection of 100 $\bar{x}$s is 1.588, which is fairly close.

Exhibit 8.2   Sample Means from a Normal Population with $n = 9$

```
MTB > stem-and-leaf 'xbars'

Stem-and-leaf of xbars      N  = 100
Leaf Unit = 0.10

     1    75 9
     1    76
     1    76
     4    77 034
     8    77 5568
    16    78 01111224
    27    78 55666667789
    31    79 2334
    39    79 55566788
   (17)   80 00012222233333344
    44    80 556667889
    35    81 000011222222444
    20    81 667778899
    11    82 023
     8    82 56679
     3    83 23
     1    83 6



MTB > describe 'xbars'

                   N      MEAN    MEDIAN    TRMEAN     STDEV    SEMEAN
  xbars          100    80.152    80.295    80.160     1.588     0.159


                 MIN       MAX        Q1        Q3
  xbars       75.853    83.607    78.760    81.215
```

Notice that in Exhibit 8.1, rarely did a sample mean $\bar{x}$ deviate from μ by more than $2\sigma_{\bar{x}} = 3.33$. This fact forms the basis for the confidence intervals described in Section 8.2 and for the tests described in Section 8.4.

Exercises

8-1   (a) Run the program in Exhibit 8.1, but now use a sample of size $n = 2$ instead of 9. Discuss the output.

      (b) Run the same program using $n = 9$ and $n = 25$. Compare the output for $n = 2$, 9, and 25.

8-2   Exhibit 8.1 helped illustrate the fact that the means of random samples from
       normal distributions are normal. An equally important fact is that the means
       of samples of at least moderate size from most nonnormal distributions are
       also approximately normal. This approximation gets better as the sample
       size n gets larger. Simulate 200 observations from a continuous uniform
       distribution into C1–C6. Then compute the means for samples of sizes $n =$
       1, 2, 3, 4, 5, and 6. Commands to do this are given below.

```
RANDOM 200 C1-C6;
  UNIFORM 0 1.
RMEAN C1        C11
RMEAN C1-C2     C12
RMEAN C1-C3     C13
RMEAN C1-C4     C14
RMEAN C1-C5     C15
RMEAN C1-C6     C16
DOTPLOT C11-C16;
  SAME.
```

8-3   In this exercise, we will take a look at how the distribution of the sample
       mean relates to the distribution of the individual observations in samples
       from some nonnormal distributions.

       (a)   Use RANDOM with the subcommand UNIFORM 0, 1 to simulate
              100 observations from a distribution that is uniform on the interval 0
              to 1. Put the observations into C10. Construct displays of these data
              with the following commands:

```
DOTPLOT C10
HISTOGRAM C10
HISTOGRAM C10;
  START 05;
  INCREMENT .1.
```

       (b)   How do these displays convey the feature that leads to the descriptive
              name *uniform distribution*? How do the results of the two HISTO-
              GRAM commands differ? If particular bars of the two histograms
              look different, try to give an explanation. Discuss why we need to be
              careful when choosing cell boundaries in histograms, especially when
              the data are constrained to lie in a particular interval.

       (c)   Again use the uniform distribution on 0 to 1, but now simulate 100
              samples, each of size 5, into C1–C5. Then use the command RMEAN
              to put the means of these 100 samples into C6. How does the distri-
              bution of the 100 means in C6 relate to the distribution of the 100
              individual observations in C10? Use DESCRIBE and DOTPLOT to
              look at the differences.

(d)   Compare the shape of the distribution of sample means in C6 and individual observations in C10. Does the shape of the distribution of the sample means look familiar?

(e)   Theory tells us that the mean of C6 should be approximately equal to the mean of C10. Is it? Theory also tells us that the standard deviation of C6 should be approximately equal to the standard deviation of C10 divided by $\sqrt{5}$. Is it? Why did we divide by $\sqrt{5}$?

8-4   In Exercise 8-3, we looked at data from a uniform distribution. Now we will look at another nonnormal distribution, the chi-square distribution with 3 degrees of freedom. (We will use this distribution again in Chapter 12.)

(a)   Simulate 100 observations from the chi-square distribution and construct displays as follows:

```
RANDOM 100 C10;
   CHISQUARE 3.
DOTPLOT C10
HISTOGRAM C10
```

(b)   How does the shape of this chi-square distribution compare to the normal curve?

(c)   Again use the chi-square distribution with 3 degrees of freedom, but now simulate 100 samples, each of size 5, into C1–C5. Then use RMEAN to put the means of these 100 samples into C6. Now answer the questions in parts (c)–(e) of Exercise 8-3.

8.2   Confidence Interval for μ When σ Is Known

Suppose we want to estimate the mean, μ, of a population. We might take a sample, calculate the sample mean, $\bar{x}$, and use this as an estimate of μ. Exhibit 8.2 shows how close the population mean and sample mean are likely to be. A confidence interval uses this idea in a more formal way.

A 95% confidence interval is an interval, calculated from the sample, that is very likely to cover the unknown mean, μ. To be more precise, if we were to use the formula for a 95% confidence interval on many, many sets of data, then 95% of our intervals would cover the unknown mean we are trying to estimate and 5% would not cover it. Unfortunately, in practice we can never know which intervals are the successful ones and which are the failures. The value 95% is called the confidence level of the interval. The confidence levels most commonly used in practice are 95%, 99%, and 90%.

Exhibit 8.3   Confidence Interval When σ Is Known

```
MTB > set c1
DATA>    4.9   4.7   5.1   5.4   4.7   5.2   4.8   5.1
DATA> end
MTB > zinterval 0.2 c1

THE ASSUMED SIGMA = 0.200

              N       MEAN     STDEV   SE MEAN    95.0 PERCENT C.I.
c1            8      4.9875   0.2532   0.0707   (  4.8489,   5.1261)
```

First, suppose we know the standard deviation of the population. In practice this is rare, but making this assumption will help us understand the concept of a confidence interval. We will then look at the more realistic case, in which the population standard deviation is unknown.

As an example, suppose we want to find a 95% confidence interval for μ based on a sample of eight observations and suppose we know that $\sigma = .2$. Exhibit 8.3 uses Minitab's ZINTERVAL command to calculate the interval.

The 95% confidence interval goes from 4.849 to 5.126. This set of data happens to be one we generated ourselves using the procedures in Chapter 7, so we know $\mu = 5$. Since 5 is inside our confidence interval, we have a successful interval. We know that in the long run, 95% of our intervals will be successful and 5% will be failures. Here we got one of the successful ones.

The value 0.0707, labeled SE MEAN, is the standard error of the sample mean. This is just another name for the standard deviation of $\bar{x}$. It is calculated as $\sigma_{\bar{x}} = \sigma/\sqrt{n}$. In the preceding section, we saw that $\bar{x}$ and μ rarely differed by more than $2\sigma_{\bar{x}}$. This is the essence of the 95% confidence interval. If we go up and down from $\bar{x}$ by $2\sigma_{\bar{x}}$, we get

$$\bar{x} - 2\sigma_{\bar{x}} = 4.987 - 2(.071) = 4.845$$
$$\bar{x} + 2\sigma_{\bar{x}} = 4.987 + 2(.071) = 5.129$$

The interval printed by ZINTERVAL differs slightly from this. It uses 1.96 instead of 2 as a multiplier for $\sigma_{\bar{x}}$. The 1.96 is the proper value for a 95% confidence interval because 95% of the area under the standard normal curve is located between -1.96 and $+1.96$.

ZINTERVAL [K% confidence] sigma = **K**, for data in **C...C**

For each column, ZINTERVAL calculates and prints a confidence interval for the population mean, μ, using the formula

$$\bar{x} - z\,(\sigma/\sqrt{n}\,) \qquad \text{to} \qquad \bar{x} + z\,(\sigma/\sqrt{n}\,)$$

Here σ is the known value of the population standard deviation, $\bar{x}$ is the mean of the sample, n is the number of observations in the sample, and z is the value from the standard normal distribution corresponding to K% confidence.

If the confidence level is not specified, 95% is used.

Stat > Basic Stat > 1-Sample Z, choose option Confidence interval

---

Exercises

8-5   The output in Exhibit 8.3 gives a 95% confidence interval for μ. Use this output to calculate, by hand, a 90% confidence interval.

8-6   Suppose $n = 9$ men are selected at random from a large population. Assume the heights of the men in this population are normal, with $\mu = 69$ inches and $\sigma = 3$ inches. Simulate the results of this selection 20 times and in each case find a 90% confidence interval for μ. The following commands may be used:

```
RANDOM 9 C1-C20;
  NORMAL 69 3.
ZINTERVAL .90 3 C1-C20
```

(a)   How many of the intervals contain μ?

(b)   Would you expect all 20 of the intervals to contain μ? Explain.

(c)   Do all the intervals have the same width? Why or why not?

(d)   Suppose you took 95% intervals instead of 90%. Would they be narrower or wider?

(e)   How many of your intervals contain the value 72? The value 70? The value 69?

(f)   Suppose you took samples of size $n = 100$ instead of $n = 9$. Would you expect more or fewer intervals to cover 72? 70? 69? What about the width of the intervals for $n = 100$? Would they be narrower or wider than for $n = 9$?

(g) Suppose you calculated 90% confidence intervals for 20 sets of real data. About how many of these intervals would you expect to contain μ? Could you tell which intervals were successful and which were not? Why or why not?

8-7   In this exercise, we will simulate random samples into rows rather than columns. Using rows will make it easier to plot the confidence intervals for the samples, even though we will not be able to use the ZINTERVAL command because it expects data in columns.

(a) The following program first simulates observations into each of C1–C5. Each row contains one sample of five observations from a normal distribution with $\mu = 30$ and $\sigma = 2$. Next, a 90% confidence interval is calculated for each row. The lower endpoint of the interval is in C21 and the upper endpoint is in C22. The program then plots the 25 intervals. Run this program:

```
RANDOM 50 C1-C5;
  NORMAL 30 2.
RMEAN C1-C5 C10
LET K1 = 1.645 * 2/SQRT(5)
LET C21 = C10 - K1
LET C22 = C10 + K1
SET C23
  1:50
END
MPLOT C21 C23   C22 C23
```

(b) Examine the plot. Are the intervals all of the same width? How many intervals missed the value of $\mu = 30$? With 90% confidence intervals, about how many of the intervals would you expect to miss?

8-8   In a z interval, you must assume you know σ. This exercise takes a brief look at what happens if you specify an incorrect value.

(a) Repeat Exercise 8-7, above, but use 1 for σ instead of 2 when you calculate the endpoints of the interval. (Use $\sigma = 2$, however, in the NORMAL subcommand to RANDOM.) Did you still have about 90% of your intervals covering μ?

(b) Repeat part (a), but now use 4 for σ. Now what percent of your intervals cover μ?

(c) Would you say that knowing σ precisely is important or unimportant to the proper use of a z confidence interval?

8-9   The theory used in developing z confidence intervals assumes that your data come from a normal distribution. In this exercise, we will look at what happens when that assumption is violated—that is, when the data come from some other distribution.

(a)   Simulate 50 random observations from a uniform distribution on 0 to 1 into C1–C5. This distribution has a mean of .5 and a standard deviation of $1/\sqrt{12} = .2887$.

(b)   Compute and plot the 90% confidence intervals as in Exercise 8-7.

(c)   What percent of your intervals were successful? The uniform distribution is a fairly extreme departure from normality. What can you conclude about the sensitivity of z intervals to nonnormality? That is, does the 90% hold up fairly well or not?

8-10   *Approximate binomial confidence intervals.* The normal approximation to the binomial distribution (see Section 6.3) can be used as the basis for computing approximate confidence intervals. For binomial data, the observed proportion of successes p provides an estimate of p. The standard deviation of p is $\sqrt{p(1-p)/n}$, which can be estimated by $\sqrt{\hat{p}(1-\hat{p})/n}$. Thus the normal approximation indicates that we can compute a 90% confidence interval for p with the formula $\hat{p} \pm 1.645 \sqrt{\hat{p}(1-\hat{p})/n}$.

(a)   The following commands simulate 50 binomial observations with $n = 20$ and $p = .3$, compute p, put the upper and lower confidence limits into C3 and C4, and then plot these intervals. Run this program:

```
RANDOM 50 C1;
  BINOMIAL 20 .3.
LET C2 = C1/20
LET C10 = 1.645*SQRT(C2*(1-C2)/20)
LET C3 = C2 - C10
LET C4 = C2 + C10
SET C5
  1:50
END
MPLOT C3 C5   C4 C5
```

(b)   Examine the plot. Do all the intervals have the same width? Explain how many intervals were successful in catching the correct value of p.

Exhibit 8.4   Confidence Interval for OTIS Scores

```
MTB > set c1
DATA>    79    82   123   106   125    98    95   129    90   111
DATA>    99   116   106   107   100   124    98   124    84    91
DATA>   118   102    95    90    86   104   111   105   110    80
DATA>    78   120   110   107   125   117   126    98   111   110
DATA>   120   114   117   105    97    86   111    93   115   102
DATA>   111    82   117
DATA> end
MTB > tint .90 c1

             N      MEAN    STDEV   SE MEAN    90.0 PERCENT C.I.
c1          53    104.91    13.88      1.91   ( 101.71,   108.10)
```

8.3   Confidence Interval for μ When σ Is Not Known

Usually we do not know the standard deviation of the population and must estimate it from the data on hand. Then we can use the Student's t confidence interval procedure and the TINTERVAL command.

The Cartoon data set contains OTIS scores (similar to IQ scores) for a sample of 53 professional hospital workers. Suppose we want to find a 90% confidence interval for the mean OTIS score of all such people. Exhibit 8.4 shows the results.[1] Based on this output we might say, "We estimate the mean to be about 104.91, and we are 90% confident that it is somewhere between 101.71 and 108.10."

Here, SE MEAN is the estimated standard error (or standard deviation) of $\bar{x}$. It is calculated by using $s/\sqrt{n}$, where s is the sample standard deviation.

Exercises

8-11   (a)   Get a histogram for the SAT verbal scores in sample A of the Grades data (described in Appendix A). Calculate $\bar{x}$ and get a 95% confidence interval for μ = mean score. How does the confidence interval compare to the histogram?

[1]In the exhibit we used SET to enter the data. You could also use RETRIEVE to enter the entire CHOLEST data set, then use the COPY command, described in Section 1.4, to get the subset of professionals (the rows in which ED = 1).

TINTERVAL [K% confidence] for data in C...C

For each column, TINTERVAL calculates and prints a t confidence interval for the population mean, using the formula

$$\bar{x} - t\,(s/\sqrt{n}) \quad \text{to} \quad \bar{x} + t\,(s/\sqrt{n})$$

Here $\bar{x}$ is the sample mean, s is the sample standard deviation, n is the sample size, and t is the value from the t distribution, using $(n-1)$ degrees of freedom and K percent confidence.

If the confidence level is not specified, 95% is used.

Stat > Basic Stat > 1-Sample T, choose option Confidence interval

---

(b)   Also get a 90% and then a 99% confidence interval for mean SAT verbal scores. How do they compare with the 95% confidence interval in part (a)? Do they have the same centers? Do they have the same widths?

8-12   Repeat the simulation of Exercise 8-6, but now assume σ is unknown and use the TINTERVAL command. Get a total of twenty 90% intervals.

(a)   How many of the twenty intervals contain μ?

(b)   Would you expect all of the intervals to contain μ? Explain.

(c)   Do all of the intervals have the same width?

(d)   Compare the t intervals of this exercise to the z intervals of Exercise 8-6. On the average, which kind of interval seems to be wider?

(e)   Suppose you calculated 95% t intervals instead of 90%. Would they be narrower or wider?

(f)   How many of your intervals contain 72? 70? 69?

(g)   Suppose you took samples of size $n = 100$ instead of $n = 9$. Would you expect more or fewer intervals to cover 72? 70? 69? What about the size of the intervals for $n = 100$? Would they be longer or shorter than those for $n = 9$?

(h)   Suppose you calculated twenty 90% t intervals for real data. About how many would you expect to contain the true μ? Could you tell which?

8-13   Repeat Exercise 8-7, but use t confidence intervals rather than z confidence intervals. You will need to use RSTDEV to calculate the standard deviation for each row and use the appropriate value from a t table. The command INVCDF can be used to get the t table value.

8-14   The theory used in developing t confidence intervals assumes your data come from a normal distribution. In this exercise, we will look at what happens when that assumption is violated—that is, when the data come from some other distribution.

(a) Simulate 50 random observations from a uniform distribution on 0 to 1 into C1–C5.

(b) Compute and plot the 90% t confidence intervals as in Exercise 8-13. What is the appropriate value to use from the t table?

(c) What percent of your intervals were successful? The uniform distribution is a fairly extreme departure from normality. What can you conclude about the sensitivity of t intervals to nonnormality? That is, did the 90% hold up fairly well in this simulation?

8.4   Test of Hypothesis for μ When σ Is Known

Suppose a machine is set to roll aluminum into sheets that are 40.0 thousandths of an inch thick. To check that the machine stays in adjustment, the operator periodically takes five measurements of sheet thickness. On the latest occasion, the results were:

40.1   39.2   39.4   39.8   39.0

This sample has a mean of 39.5, which is below the desired mean of 40.0. Is this discrepancy just due to random fluctuation or does it indicate that the machine is not adjusted correctly? To answer this, we must have some idea of how much the thickness naturally varies from sheet to sheet. Suppose we know from past experience that the standard deviation of sheet thickness is $\sigma = .4$. Then we can use Minitab's ZTEST command to test the null hypothesis that the machine is properly adjusted, $H_0 : \mu = 40.0$, versus the alternative hypothesis that the machine is misadjusted, $H_1 : \mu \neq 40.0$. The results are shown in Exhibit 8.5.

The number −2.80, labeled Z in the output, is the quantity used to do the test. It is defined as

$$ z = \frac{\bar{x} - \mu}{\sigma/\sqrt{n}} = \frac{39.5 - 40}{.4\sqrt{5}} = -2.80 $$

If the null hypothesis is true—that is, if the population mean is 40—then z follows a standard normal distribution ($\mu = 0$ and $\sigma = 1$). Values of a standard normal smaller than −2 rarely occur. A value as small as −2.8 is very

Exhibit 8.5   Test of $\mu = 40$, with Known $\sigma = 0.4$

```
MTB > set c1
DATA>    40.1 39.2 39.4 39.8 39.0
DATA> end
MTB > ztest 40 .4 c1

TEST OF MU = 40.000 VS MU N.E. 40.000
The assumed sigma = 0.400

            N      MEAN    STDEV   SE MEAN        Z    P VALUE
c1          5     39.500   0.447     0.179    -2.80     0.0053
```

unusual. You could check this statement out by simulating data from a standard normal.

The p-value (also called significance level) says how unusual such a discrepancy is. Here the p-value = 0.0053, or 0.53%. This means that if μ = 40, then 0.53% of the time $\bar{x}$ will be this far or farther from μ. The rest of the time, $\bar{x}$ will be closer than this to μ. Therefore, $\bar{x}$ = 39.5 is quite unusual for a perfectly adjusted machine.

The value of the test statistic, z, and the p-value are related. Here we are using a two-sided test—that is, a test in which the alternative hypothesis is $H_1 : \mu \neq K$. In such cases the p-value equals twice the area beyond z, under a standard normal curve.

In most textbooks, statistical tests are done by specifying what is called an a level. Once we know the p-value, however, we can determine the results of a test for any value of a. If the p-value is less than a, we reject H_0; if it is greater than a, we do not reject H_0. For example, if we did a test here using a = .01, we would reject H_0, because .0053 is less than .01. If we used a = .001, we would not reject H_0.

Exercises

8-15   Imagine choosing n = 16 women at random from a large population and measuring their heights. Assume that the heights of the women in this population are normal, with μ = 64 inches and σ = 3 inches. Suppose you then test the null hypothesis H_0: μ = 64 versus the alternative that H_1:$\mu \neq$ 64,

---

ZTEST [mu = **K**] sigma = **K** for data in **C...C**

ALTERNATIVE = K

For each column, ZTEST tests the null hypothesis, $H_0 : \mu = K$, where μ is the hypothesized population mean. The default alternative hypothesis is $H_1 : \mu \neq K$. If μ is not specified on ZTEST, $H_0 : \mu = 0$ is tested. The command calculates the test statistic

$$z = \frac{\bar{x} - K}{\sigma/\sqrt{n}}$$

Here $\bar{x}$ is the sample mean, n is the sample size, σ is the known population standard deviation, and K is the hypothesized value of the population mean, μ.

The subcommand ALTERNATIVE allows you to do a one-sided test. Use $K = -1$ for $H_1 : \mu < K$ and $K = +1$ for $H_1 : \mu > K$.

Stat > Basic Stat > 1-Sample Z, choose option Test mean

---

using $a = .10$. Assume σ is known. Simulate the results of doing this test 20 times as follows:

```
RANDOM 16 C1-C20;
  NORMAL 64 3.
ZTEST 64 3 C1-C20
```

(a) In how many tests did you fail to reject H_0? That is, how many times did you make the "correct decision"?

(b) How many times did you make an "incorrect decision" (that is, reject H_0)? On the average, how many times out of 20 would you expect to make the wrong decision?

(c) What is the attained significance level (p-value) of each test? Are they all the same?

(d) Suppose you used $a = .05$ instead of $a = .10$. Does this change any of your decisions to reject or not? Should it in some cases?

8-16 As in Exercise 8-15, simulate choosing 16 women at random, measuring their heights, and testing H_0: $\mu = 64$ versus H_1: $\mu \neq 64$, but this time assume that the population really has a mean of $\mu = 63$, instead of 64. Thus, use the subcommand NORMAL with $\mu = 63$ and $\sigma = 3$ to simulate the samples. Use $a = .10$ and assume σ is known. Do this for a total of 20 tests.

(a)   In how many tests did you reject H_0? That is, how many times did you make the "correct decision"? How many times did you make an "incorrect decision"?

(b)   What we are investigating is called the power of a test—that is, how well the test procedure does in detecting that the null hypothesis is wrong, when indeed it is wrong. Repeat the above simulation, but now assume the true population mean is $\mu = 62$. (Continue to use $\sigma = 3$ and the same null hypothesis.) How often did you make the correct decision (reject) in these 20 tests? On the average, would you expect to make more "correct" decisions if the true mean were 62 or if the true mean were 63?

8-17   In this exercise, we will use simulation to study hypothesis tests. We will do our arithmetic with rows of data rather than with columns because it is easier this way. We will not be able to use the ZTEST command, because ZTEST expects the data in columns.

(a)   The following program first simulates 200 samples into C1–C5. Each row contains one sample of five observations from a normal distribution with $\mu = 30$ and $\sigma = 2$. Next, the z statistic for testing $H_0: \mu = 30$ versus $H_1: \mu \neq 30$ is calculated for each row. The program then makes a dotplot of these 200 z values. Run this program:

```
RANDOM 200 C1-C5;
  NORMAL 30 2.
RMEAN C1-C5 C10
LET C11 = (C10-30)/(2/SQRT(5))
DOTPLOT C11
```

(b)   Suppose we do the test using $a = .05$. Then we reject H_0 if the z value is less than -1.96 or greater than 1.96. For (approximately) what percent of your 200 samples do you reject H_0? (It should be roughly 5%.) This illustrates the fact that when you do hypothesis tests, you can make the wrong decision. In fact, if you do tests using $a = .05$ and the null hypothesis is true, you will make the wrong decision 5% of the time, on the average.

8.5 Test of Hypothesis for μ When σ Is Not Known

Suppose we once again test whether or not the aluminum sheet machine mentioned in the previous section is correctly adjusted; that is, whether μ = 40. However, this time let's not assume we know the value of σ. The test used is called the Student's t test and is performed by the TTEST command. Exhibit 8.6 shows the results. Again, the p-value is small. Therefore, we suspect that the rolling machine is not perfectly adjusted.

The number −2.50, labeled T in the output, is the quantity used to do the test. It is defined as

$$t = \frac{\bar{x} - \mu}{s/\sqrt{n}} = \frac{39.5 - 40}{.447/\sqrt{5}} = -2.50$$

If the null hypothesis is true—that is, if the population mean is 40— then t follows a t-distribution with $n - 1 = 4$ degrees of freedom. Values of this t-distribution smaller than −2 rarely occur. You can check this out by simulating data from a t-distribution with 4 degrees of freedom.

Practical Significance Versus Statistical Significance

It is rare that practical significance and statistical significance coincide precisely. Here our best estimate is that the process mean is about 39.5. TTEST showed that this result is statistically significant if $\alpha = .01$. This in itself gives us no clue whatsoever as to whether the process mean has shifted seriously off target. Is 39.5 far from 40.0 in a practical sense? It depends on the use of the rolled aluminum, the costs of production, and so on.

In some applications, tolerances are very tight. Even results as close as 39.9 may still be too far off to be acceptable. In other situations, it may not

Exhibit 8.6 Test of μ = 40 When σ Is Not Known

```
MTB > set c1
DATA>    40.1 39.2 39.4 39.8 39.0
DATA> end
MTB > ttest 40 c1

TEST OF MU = 40.000 VS MU N.E. 40.000

            N      MEAN     STDEV    SE MEAN         T    P VALUE
c1          5    39.500     0.447      0.200     -2.50      0.067
```

TTEST [mu = K] for data in C...C

ALTERNATIVE = K

For each column, TTEST tests the null hypothesis, $H_0 : \mu = K$, where μ is the hypothesized population mean. The default alternative hypothesis is $H_1 : \mu \neq K$. If μ is not specified on TTEST, $H_0 : \mu = 0$ is tested. The command calculates the Student's t test statistic

$$t = \frac{\bar{x} - K}{s/\sqrt{n}}$$

Here $\bar{x}$ is the sample mean, n is the sample size, s is the sample standard deviation, and K is the hypothesized value of the population mean, μ.

The subcommand ALTERNATIVE allows you to do a one-sided test. Use K = −1 for $H_1 : \mu < K$ and K = +1 for $H_1 : \mu > K$.

Stat > Basic Stat > 1-Sample t, choose option for Test mean

---

make much practical difference as long as the sheets are at least as thick as 39.3. In other words, a statistical test of significance tells us only whether or not the observed data are unusual under the hypothesized situation. It tells us nothing about what practical course of action we should take. This is the heart of the reason why we prefer confidence intervals over tests of significance. A 95% confidence interval for these data goes from 39.09 to 39.91. Thus, we are fairly confident that the true mean thickness is between 39.09 and 39.91. If all we really need is a mean thickness of 39.3 or more, then for practical purposes, the machine may be all right.

Exercise

8-18   Do a test to see whether there is evidence that the preprofessionals who participated in the Cartoon experiment (described in Appendix A) have OTIS scores that differ significantly from the national norm of 100. First, use COPY to select the data for the preprofessionals. Then get a histogram. What do you think? Now do the appropriate test, using $a = .05$. The following commands may be used:

```
COPY C1-C9  C1-C9;
  USE 'ED' = 0.
HISTOGRAM 'OTIS'
TTEST 100 'OTIS'
```

8.6 Departures from Assumptions

Many procedures in statistical inference, including those described in this chapter, are based on the assumption that the data are a random sample from a normal distribution. Here we take a quick look at how you can spot departures from these assumptions. We treat the most serious problem first—lack of randomness.

Nonrandomness

Most data are not a random sample from any population. In particular, observations taken close together are often more alike than those taken further apart. For example:

People from the same part of the country tend to be more alike than people from different regions.

Light bulbs manufactured on the same day tend to be more alike than those manufactured on different days.

Adjacent pieces from the same roll of tape tend to have more nearly the same degree of stickiness than do pieces of tape made on different days or by different machines.

The measurements made by one inspector are often more alike than measurements made by different inspectors.

Observations taken in close proximity, either in time or in space, are often correlated. Observations that are correlated do not form a simple random sample. One of the best ways to look for correlations is to make plots. For example, if measurements are made one after the other, plot them in that order. If they are made in bunches, plot them so the bunches are readily identifiable. If you see a clear pattern, you probably do not have a random sample.

Exhibit 8.7 gives six examples. In each case, the observations, y, are plotted versus "order." Here order could be time order, spatial order, or some other order that helps us see the nonrandom nature of the data. The first plot shows no evidence of nonrandomness, but all the other plots have very clear patterns.

Exhibit 8.7   Plots of Normal Data with Various Departures from Randomness

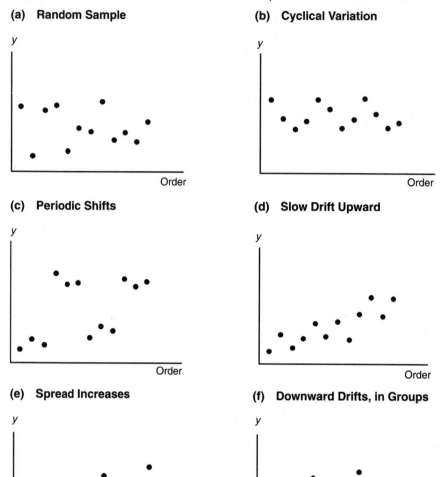

(a)   Random Sample

(b)   Cyclical Variation

(c)   Periodic Shifts

(d)   Slow Drift Upward

(e)   Spread Increases

(f)   Downward Drifts, in Groups

Nonnormality

The techniques in this chapter, and in many of the chapters that follow, require one more condition if they are to be exactly valid: The population from which our sample is drawn must have a normal distribution. This would seem to rule out most practical applications of these techniques, since no real population is ever exactly normal. But, in fact, it does not. Techniques based on the normal distribution give very good, though approximate, answers even if the production is nonnormal.

For example, suppose we construct a 95% t confidence interval for μ using 20 observations. If the population is nonnormal, then the true confidence will not be 95%. It might be 94% or 90% or 96%, depending on what the actual shape of the population is, but it is not likely to be very far from 95%. Similarly, suppose we do a t-test using these data and get a p-value of, say, .031. If the population is nonnormal, the correct p-value is not .031. It might be .038 or .047, or .026, but it will not be very far from .031.

We should mention that even though these procedures usually work well for nonnormal populations, in some cases the nonparametric methods in Chapter 13 are better (or more powerful, in statistical terminology). The confidence interval constructed by a nonparametric procedure may be shorter than the t confidence interval. A shorter interval gives us a more precise estimate of μ. A nonparametric test may be more likely to reject the null hypothesis when the null hypothesis is indeed false.

There are many methods that can help us decide whether a population is normal. Some of the sophisticated methods use formal tests. Here we will look at two simple graphical techniques.

Suppose we take a sample from the population and get a histogram. If the population is normal, the sample histogram should have approximately the shape of a normal curve. If the sample size is large, this approximation should be very good. Of course, if we have just a few observations, say 10 or 15, then it will be difficult to see a clear shape in the sample histogram; however, we still may be able to spot gross departures from normality.

Another plot, called a normal probability plot, is a useful supplement to histograms in checking for nonnormality. It plots the sample versus the values we would get, on the average, if the sample came from a normal population. This plot is approximately a straight line if the sample is from a normal population, but exhibits curvature if the population is not normal.

Minitab can make a normal probability plot. If C1 contains the data, use the two commands

```
NSCORES C1 C2
PLOT C1 C2
```

Exhibit 8.8   Normal Probability Plot of OTIS Scores from the Cartoon Data

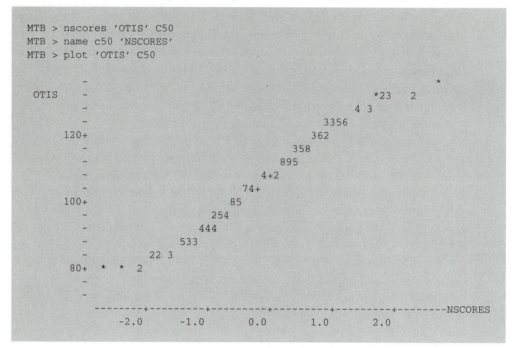

```
MTB > nscores 'OTIS' C50
MTB > name c50 'NSCORES'
MTB > plot 'OTIS' C50
```

Exhibit 8.8 gives a probability plot of the OTIS scores for all the participants in the Cartoon experiment. Again we are faced with the problem of
judging a picture. Does this plot look straight? There certainly is some curvature. But is it curved enough to doubt the normality of the OTIS scores?
Good judgment of graphs comes from experience. Some find it easier to
learn how to judge probability plots than histograms. Some of the exercises
at the end of this section show how you can use simulation to improve your
judgment of the nonnormality of histograms and normal probability plots.

Exercises

8-19   This exercise is designed to give you some idea of what a normal probability plot can look like. It will help you learn to decide whether a particular plot looks definitely nonnormal.

NSCORES of C put in **C**

Calculates the normal scores of a set of data. Here are the normal scores for a data set with six observations. The data set is in C1 and the normal scores are in C2.

C1	C2
1.1	−0.20
1.9	1.28
0.8	−1.28
1.3	0.64
1.2	0.20
0.9	−0.64

Loosely speaking, the normal scores for this data set can be defined as follows: The number −1.28 in C2 is the smallest value you would get, on the average, if you took samples of size $n = 6$ from a standard normal population ($\mu = 0$, $\sigma = 1$). This number is placed next to the smallest value in C1. The number −.64 in C2 is the second smallest value you would get, on the average, in samples of size 6 from a standard normal. It is placed next to the second smallest value in C1. This is continued for all six values. (More precisely, the ith smallest normal score is found by computing the inverse cdf of $(i − 3/8)/(n + 1/4)$.)

Release 8: **Calc > Functions and Statistics > Functions**, choose button for Normal scores
Windows: **Calc > Functions**, choose button for Normal scores

(a) Use the RANDOM command to simulate 20 observations from a normal distribution with mean 0 and standard deviation 1. Make a normal probability plot. Repeat this a total of five times.

(b) Repeat (a), but use $n = 50$ observations.

8-20 (a) Simulate 50 observations from a normal population with $\mu = 0$, $\sigma = 1$. Get a histogram and a probability plot.

(b) Repeat part (a) until you have five histograms and five probability plots. How do the histograms and plots compare?

8-21 Simulate five normal probability plots to compare with the OTIS scores. (Choose appropriate values for n, μ, and σ.) Compare these with the plot in

Exhibit 8.8. Is there any evidence that the OTIS scores may not be a random sample from a normal population?

8-22   Simulate 100 observations from a normal distribution with mean 0 and standard deviation 1 into C1. Then make a normal probability plot and comment on the appearance of the plot under each of the following circumstances:

(a)   Use the entire set of 100 simulated observations.

(b)   Use only the observations between −1 and 1, selected with the following instruction:

```
COPY C1 C2;
   USE C1 = -1:1.
```

(c)   Repeat (b), but select the observations between −2 and 2.

(d)   Append two "outliers" with the following commands.

```
COPY C1 C3
INSERT C3
   -4 4
END
```

(e)   Repeat (d), but put the outliers at −5 and 5. Then do (d) again, with the outliers at −6 and 6.

9

Comparing Two Means: Confidence Intervals and Tests

Many important questions involve comparisons between two sets of data.

Did the region with the high advertising budget have substantially greater sales than the region with the low budget?

How much stronger are the bonds made with the new glue than those made with the old glue?

Is there any evidence that women are paid less than men for the same jobs? If so, how much less?

How much difference is there between the survival rate for patients on the new drug and the survival rate on the standard treatment?

In this chapter we introduce some methods by which two sets of data can be compared. The remarks in Chapter 8 concerning random samples and normal populations apply here also. That is, although the methods we present are "exact" only if the populations have normal distributions, they still work quite well for most populations. However, the nonparametric methods of Chapter 13 are more powerful for some nonnormal populations.

We begin by discussing the important distinction between paired data and independent samples.

9.1   Paired and Independent Data

A Pennsylvania medical center collected some data on the blood cholesterol levels of heart-attack patients. A total of 28 heart-attack patients had
their cholesterol levels measured 2 days after their attacks, 4 days after,
and 14 days after. In addition, cholesterol levels were recorded for a control
group of 30 people who had not had heart attacks. The data set is given in
Table 9.1.

This data set contains both paired and independent variables. The columns headed "2 Days After" and "4 Days After" are paired because the
numbers in a given row are related—they are both for the same patient. For
example, the first patient had a "2 Days After" score of 270 and a "4 Days
After" score of 218. Because the observations are paired, we can determine
how much each patient's cholesterol changed during that period.

Now let's look at an example of independent or unpaired data. The
cholesterol levels in the last column of Table 9.1 are for a control group of
30 people. The people in this group were not linked in any known way to
any of the patients in the experimental group. Thus, we say that the "2 Days
After" data and the control group data are independent.

The distinction between paired and independent data is relatively clear
in this example, as it is in most data sets. However, it is an important concept you should keep in mind because it determines which statistical procedures are appropriate for a given set of data.

Exercises

9-1   For the cholesterol data in Table 9.1, determine which of the following
       variables are paired and which are not:
       (a)   The "2 Days After" and "14 Days After."
       (b)   The "4 Days After" and "Control Group."
       (c)   The "4 Days After" and "14 Days After."
       (d)   The "14 Days After" and "Control Group."

9-2   In each of the following problems, indicate whether the data are paired or
       independent samples. Justify your answer.
       (a)   A national fast-food chain used a special advertising campaign at half
             of its stores. It then compared sales at those stores with sales at its
             other stores.
       (b)   A total of 24 sets of mice were used in a study of a nutrition supple-
             ment. Each set consisted of two mice from the same litter. In each set,

Table 9.1   Blood Cholesterol After a Heart Attack

Experimental Group			Control Group
2 Days After	4 Days After	14 Days After	
270	218	156	196
236	234	*	232
210	214	242	200
142	116	*	242
280	200	*	206
272	276	256	178
160	146	142	184
220	182	216	198
226	238	248	160
242	288	*	182
186	190	168	182
266	236	236	198
206	244	*	182
318	258	200	238
294	240	264	198
282	294	*	188
234	220	264	166
224	200	*	204
276	220	188	182
282	186	182	178
360	352	294	212
310	202	214	164
280	218	*	230
278	248	198	186
288	278	*	162
288	248	256	182
244	270	280	218
236	242	204	170
			200
			176

one mouse was given the supplement and one was not. After three weeks, the weight gain of each mouse was recorded for analysis.

(c)   A tool manufacturer took a sample of 10 new wrenches. He treated 5 of them with a special chemical. The other 5 were left untreated. The manufacturer then compared the strength of the treated and the untreated tools.

(d)   A total of 50 people were used in a study to compare two treatments for the rash produced by poison ivy. A small patch of skin on the left arm of each person was exposed to poison ivy. Once a rash had developed, treatment A was used. The improvement was assessed and recorded. Two months later, the same 50 people were again exposed to poison ivy. This time, treatment B was used.

9-3   A number of data sets are described in Appendix A. Find three examples of paired data and three examples of unpaired data. (Do not use the examples that were used in this chapter.)

9-4   In the Cartoon data (described in Appendix A), which of the following comparisons involve paired data and which involve independent samples?

(a)   Difference between the colors and black and white, as measured by the immediate cartoon score, and by the immediate realistic score.

(b)   Difference between the OTIS scores in hospital A and hospital B.

(c)   Difference between the OTIS scores of preprofessionals and professionals in hospital C.

(d)   Difference between immediate cartoon and realistic scores for those participants who saw color slides.

(e)   Difference between immediate and delayed cartoon scores for those participants who saw color slides.

9.2 Difference Between Two Means: Paired Data

A shoe company wanted to compare two materials, A and B, for use on the soles of boys' shoes. We could design an experiment to compare the two materials in two ways. One way might be to recruit 10 boys (or more if our budget allowed) and give 5 of the boys shoes made with material A and give the other 5 boys shoes made with material B. Then after a suitable length of time, say three months, we could measure the wear on each boy's shoes. This would lead to independent samples. Now, you would expect a certain

Table 9.2 Wear for Boys' Shoes,
Using a Paired Design

Boy	Material A	Material B
1	13.2	14.0
2	8.2	8.8
3	10.9	11.2
4	14.3	14.2
5	10.7	11.8
6	6.6	6.4
7	9.5	9.8
8	10.8	11.3
9	8.8	9.3
10	13.3	13.6

variability among 10 boys—some boys wear out shoes much faster than others. A problem arises if this variability is large. It might completely hide an important difference between the two materials.

The other method, a paired design, attempts to remove some of this variability from the analysis so we can see more clearly any differences between the materials we are studying. Again, suppose we started with the same 10 boys, but this time had each boy test both materials. There are several ways we could do this. Each boy could wear material A for three months, then material B for three months. Or we could give each boy a special pair of shoes with the sole on one shoe made from material A and the other from material B. This latter procedure produced the data shown in Table 9.2.

First we entered the three variables into C1–C3 and named them 'Boy', 'Mat-A', and 'Mat-B'. Then we used MPLOT to display the data. Exhibit 9.1 shows the results. The plot shows that wear varied greatly from boy to boy. Boy 6 had the lowest wear—both shoes were well below 8. Boy 4 had the highest wear—both shoes were well above 14. For a given boy, however, both shoes had about the same amount of wear. Using statistical terminology, we would say, "The among-boy variability was high but the within-boy variability was low."

One important feature of the plot is the order of the letters: in 6 out of 10 cases, the As are clearly lower than the Bs, and in the remaining 4 cases the two materials are very close, within the resolution of the plot.

Exhibit 9.1 Plot to Compare Wear for Boys' Shoes

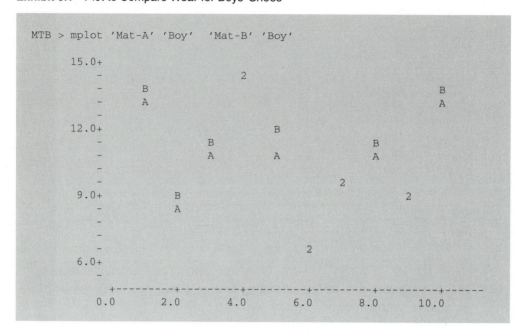

Confidence Intervals

In addition to looking at the original data, we can calculate a new quantity for each boy: the difference between the two materials. Both confidence intervals and tests for paired analyses use this difference.

Exhibit 9.2 calculates the difference $(B - A)$, then prints it and gets a confidence interval. We see that $(B - A)$ has two negative values, $-.2$ and $-.1$, and eight positive values. These two negative values are from the two boys who had more wear with material A. The positive values are from the boys who had more wear with B.

The mean of the differences is 0.410. That is, on the average, material B did 0.41 wear units worse than A, or equivalently A did 0.41 wear units better than B. To get some idea of the uncertainty in this estimate, we look at the confidence interval. This interval says we are 95% confident that the mean difference is between 0.133 and 0.687.

These numbers are a little hard to relate to. Suppose we express them as percentages. We use

```
LET K1 =   .41 / MEAN('Mat-A')
```

Exhibit 9.2   Confidence Interval for Paired Data

```
MTB > let c4 = c3-c2
MTB > name c4 'B-A'
MTB > print c1-c4

 ROW    Boy   Mat-A   Mat-B       B-A

   1     1    13.2    14.0    0.80000
   2     2     8.2     8.8    0.60000
   3     3    10.9    11.2    0.30000
   4     4    14.3    14.2   -0.10000
   5     5    10.7    11.8    1.10000
   6     6     6.6     6.4   -0.20000
   7     7     9.5     9.8    0.30000
   8     8    10.8    11.3    0.50000
   9     9     8.8     9.3    0.50000
  10    10    13.3    13.6    0.30000

MTB > tinterval 'B-A'

              N     MEAN    STDEV   SE MEAN    95.0 PERCENT C.I.
B-A          10    0.410    0.387    0.122    (   0.133,    0.687)
```

to calculate (average difference)/(average wear for B) = 0.04 = 4%. That is, on the average, material A resulted in 4% less wear than material B. What's the "best case"? According to the confidence interval, the most we could expect the difference to be is 0.687. We convert this to percentages with

```
LET K1 =   .687 / MEAN('Mat-A')
```

which gives 0.0646, about 6.5%. So at best, material A results in 6.5% less wear.

The paired-data analysis helped us see three things: There is a difference between the two materials. The difference is relatively consistent—in 8 of 10 cases, material A was better than material B. The difference is very small, no more than 6.5%.

Paired *t*-Test

We just saw that a confidence interval for paired data could be done by first computing the changes, then computing a *t* interval based on the changes. The *t*-test for paired data can be done in a similar manner. For example,

Exhibit 9.3   Paired *t*-Test on the Shoe Data

```
MTB > ttest 'B-A'

TEST OF MU = 0.000 VS MU N.E. 0.000

              N       MEAN     STDEV    SE MEAN        T     P VALUE
B-A          10      0.410     0.387      0.122     3.35     0.0086
```

Exhibit 9.3 shows how to test the null hypothesis that the average difference between the two materials for shoes is zero.

We see, as before, that the estimate of the mean change is 0.410. The
p-value for the test is 0.0086. With such a small *p*-value, we have strong
evidence that the mean change in the population is not zero. We would
reject our null hypothesis if we had used $a = 0.05$ or $a = 0.01$, or any value
of a down to 0.0086.

Exercises

9-5   Use the output in Exhibit 9.2 to calculate by hand a 90% confidence interval
for 'B–A'. Also calculate a 99% confidence interval by hand.

9-6   (a)   Use the Cholesterol data set stored in CHOLEST to estimate how
much change in cholesterol level there is between the second and
fourth day following a heart attack. Also do a plot like the one in
Exhibit 9.1.

(b)   Estimate the amount of change between the second and fourteenth
day following the attack. Notice that there are some missing observations in the data. Minitab commands omit missing cases from the
analysis. What effect might this have on your conclusions?

9-7   The 179 participants in the Cartoon experiment (described in Appendix A)
each saw cartoon and realistic slides.

(a)   Do an appropriate test to see whether there is any difference between
the two types of slides. Use the immediate scores.

(b)   Estimate the difference between the two types of slides with a 90%
confidence interval. Use HISTOGRAM to display the difference.

9.3   Difference Between Two Means: Independent Samples

The data in Table 9.3 are taken from a study on Parkinson's disease, a disease that, among other things, affects a person's ability to speak. Eight of the people in this study had received one of the most common operations to treat the disease. This operation seemed to improve the patients' condition overall, but how did it affect their ability to speak? Each patient was given several tests. The results of one of these tests are shown in Table 9.3. The higher the score, the more problems with speaking.

Exhibit 9.4 uses DOTPLOT to display these data. The SAME subcommand tells Minitab to use the same scale for the two samples so that we can compare them. The samples overlap quite a bit. At this point, it is not clear if the operation makes a difference.

Now suppose we want to do a formal test to see whether there is a significant difference and calculate a confidence for that difference. The command TWOSAMPLE, in Exhibit 9.5, does both of these things. Now we can see that the treatment is significant. The p-value for the test is 0.0073, and we are 95% confident that the difference is between 0.19 and 1.07.

Table 9.3   Speaking Ability for Patients with Parkinson's Disease

Patients Who Had Operation	Patients Who Did Not Have Operation
2.6	1.2
2.0	1.8
1.7	1.8
2.7	2.3
2.5	1.3
2.6	3.0
2.5	2.2
3.0	1.3
	1.5
	1.6
	1.3
	1.5
	2.7
	2.0

Exhibit 9.4   Dotplot for the Parkinson's Disease Data

```
MTB > name c1 = 'Op' c2 = 'No Op'
MTB > dotplot c1 c2;
SUBC>    same.
```

Exhibit 9.5   Two-Sample *t* Procedure Using Parkinson's Disease Data

```
MTB >  twosample c1 c2

TWOSAMPLE T FOR Op VS No Op
           N      MEAN      STDEV    SE MEAN
Op         8     2.450      0.411       0.15
No Op     14     1.821      0.556       0.15

95 PCT CI FOR MU Op - MU No Op: ( 0.19,   1.07)

TTEST MU Op = MU No Op (VS NE): T= 3.02   P=0.0073   DF=  18
```

Let's look for a moment at the practical significance of these results. The test says there is statistically significant evidence that the two populations differ. If we look at the sample means in Exhibit 9.5 and the dotplot in Exhibit 9.4, we see the direction of the difference. The patients who had the operation had more severe speech problems than those who did not have the operation. This, of course, does not prove that the operation causes difficulty in speaking. Perhaps the operation was performed only on patients who were already in poor condition.

The procedure used by TWOSAMPLE is slightly different from that described in many textbooks. Textbooks often use the pooled *t*, a procedure that assumes both populations have the same variance or, equivalently, the same standard deviation. The name *pooled* is used because the standard deviations from the two samples are pooled to get an estimate of the common standard deviation.

Suppose there are n_1 observations in the first sample, with mean $\bar{x}_1$ and standard deviation s_1. Suppose n_2, $\bar{x}_2$, and s_2 are the corresponding values for the second sample. Then the confidence interval goes from

$$(\bar{x}_1 - \bar{x}_2) - t\sqrt{\frac{s_1^2}{n_1} + \frac{s_2^2}{n_2}} \quad \text{to} \quad (\bar{x}_1 - \bar{x}_2) + t\sqrt{\frac{s_1^2}{n_1} + \frac{s_2^2}{n_2}}$$

where t is the value from a t-table corresponding to 95% confidence and degrees of freedom defined below.

The t-statistic used in the test is

$$t = \frac{(\bar{x}_1 - \bar{x}_2)}{\sqrt{\dfrac{s_1^2}{n_1} + \dfrac{s_2^2}{n_2}}}$$

The degrees of freedom is based on the following approximation:

$$\text{d.f.} = \frac{((s_1^2/n_1) + (s_2^2/n_2))^2}{\dfrac{(s_1^2/n_1)^2}{(n_1 - 1)} + \dfrac{(s_2^2/n_2)^2}{(n_2 - 1)}}$$

When we use the pooled method, the pooled estimate of the common variance is calculated by

$$s_p^2 = \frac{(n_1 - 1)s_1^2 + (n_2 - 1)s_2^2}{n_1 + n_2 - 2}$$

The pooled confidence interval goes from

$$(\bar{x}_1 - \bar{x}_2) - ts_p\sqrt{\frac{1}{n_1} + \frac{1}{n_2}} \quad \text{to} \quad (\bar{x}_1 - \bar{x}_2) + ts_p\sqrt{\frac{1}{n_1} + \frac{1}{n_2}}$$

where t is the value from a t-table corresponding to $n_1 + n_2 - 2$ degrees of freedom.

The pooled test statistic is

$$t = \frac{(\bar{x}_1 - \bar{x}_2)}{s_p\sqrt{\dfrac{1}{n_1} + \dfrac{1}{n_2}}}$$

If you use the pooled procedure when it is not appropriate—that is, when the standard deviations of the two populations are not equal—you could be seriously misled. For example, you might falsely claim to have evidence that the two populations differ when they really do not. Of course,

---

TWOSAMPLE [K% confidence] samples in **C C**

    POOLED
    ALTERNATIVE K

TWOSAMPLE gives the results of a *t*-test and a confidence interval to compare two independent samples. If the confidence level is not specified, 95% is used.

The default method estimates the standard deviation for each population separately. If you use the subcommand POOLED, then TWOSAMPLE assumes the two populations have the same variance (and thus the same standard deviation) and uses a pooled estimate of this common standard deviation.

The default hypothesis for the test is $H_0 : \mu_1 = \mu_2$ versus $H_1 : \mu_1 \neq \mu_2$. The subcommand ALTERNATIVE allows you to do a one-sided test. Use K = -1 for $H_1 : \mu_1 < \mu_2$ and K = $+1$ for $H_1 : \mu_1 > \mu_2$.

Stat > Basic Stat > 2-Sample t, choose option for Samples in different columns

---

you always have a chance of making such an error (called a Type I error) when you do a statistical test. In fact, this is exactly what a measures. When you do a test at $a = .05$, you are supposed to have a 5% chance of claiming there is a difference when in actuality there is none. But if you use the pooled procedure when the population standard deviations are not equal, your chances of a Type I error may be very different from 5%. How different depends on how unequal the standard deviations and the sample sizes are.

    TWOSAMPLE has a subcommand, POOLED, that allows you to do the pooled analysis. If you do not use POOLED when you safely could have (i.e., when the standard deviations of the two populations are equal), then, on the average, your analysis will be slightly conservative. That is, you will get a slightly larger confidence interval and you will be slightly less likely to reject a true null hypothesis. This conservatism essentially disappears with moderately large sample sizes (say, if both n_1 and n_2 are greater than 30). So, in most cases, and especially for large sample sizes, it's better not to use the POOLED subcommand. If the standard deviations are equal, you've lost little, and if they're unequal, you may have gained a lot.

Data in a Different Format

In TWOSAMPLE, the data for the two samples are in separate columns. There is another way to organize data, a way that is more commonly used in statistics software. Table 9.4 shows the Parkinson's disease data in both forms. The first form, which we'll call unstacked, has each group in a separate column. It is the form used by TWOSAMPLE. The second puts all the data for wear (this is often called the response variable) in one column, then uses a second column to say which group each observation belongs to.

Minitab has a second command, TWOT, to do a two-sample test and confidence interval. TWOT uses data in stacked form. The methods and output are exactly the same as for TWOSAMPLE; only the form of the data is different.

Table 9.4   The Parkinson's Disease Data Organized in Two Ways

Unstacked Data		Stacked Data	
Operation	*No Operation*	*Speaking Ability*	*Operation*
2.6	1.2	2.6	1
2.0	1.8	2.0	1
1.7	1.8	1.7	1
2.7	2.3	2.7	1
2.5	1.3	2.5	1
2.6	3.0	2.6	1
2.5	2.2	2.5	1
3.0	1.3	3.0	1
	1.5	1.2	2
	1.6	1.8	2
	1.3	1.8	2
	1.5	2.3	2
	2.7	1.3	2
	2.0	3.0	2
		2.2	2
		1.3	2
		1.5	2
		1.6	2
		1.3	2
		1.5	2
		2.7	2
		2.0	2

Exhibit 9.6 Pooled *t*-Test to Compare Male and Female Pulse Rates

```
MTB > boxplot 'PULSE1';
SUBC>    by 'SEX'.

SEX

                                    -------------
1              *        ----------I    +   I-------------- * *
                                    -----------

                                    -------------------
2                             ---------I        +        I---------------
                                    -------------------
                 ----+----------+----------+----------+----------+----------+--
PULSE1
                  50         60         70         80         90        100


MTB > TWOT 'PULSE1' 'SEX';
SUBC>    pooled.

TWOSAMPLE T FOR PULSE1
SEX    N       MEAN       STDEV    SE MEAN
1     57      70.42        9.95        1.3
2     35      76.9        11.6         2.0

95 PCT CI FOR MU 1 - MU 2: ( -11.0,   -1.9)

TTEST MU 1 = MU 2 (VS NE): T= -2.82   P=0.0058   DF=  90
```

In Appendix A, most data sets that contain two independent samples have the data in stacked form. Let's look at the PULSE experiment. Suppose we are interested in knowing how much difference, on the average, there is between the pulse rates of males and females. We will use PULSE1 because these pulse rates were taken before anyone exercised.

The assumption of equal standard deviations seems plausible in this case, so we will use POOLED to do the pooled *t* procedure. In Exhibit 9.6, BOXPLOT, with the subcommand BY, was used to compare these two groups visually; then TWOT did the test and confidence interval. The subcommand BY told Minitab to create a separate boxplot for each value in SEX and to put these on the same scale so we could compare the groups.

From the display, the women seem to have a higher pulse rate, but there is an overlap. The test, however, has a *p*-value of 0.0058. This tells us there

---

TWOT [**K**% confidence] data in **C**, groups in **C**

> **POOLED**
> **ALTERNATIVE K**

TWOT gives the results of a *t*-test and a confidence interval to compare two independent samples. The observations from the two samples are in the first column. The second column specifies which sample each observation belongs to. If the confidence level is not specified, 95% is used.

The output, the method, and the two subcommands are the same as for TWOSAMPLE. Only the form of the input data is different.

Stat > Basic Stat > 2-Sample t, choose option for Samples in one column

---

is statistically significant evidence that the pulse rates of the males and females are different even if we use an a as small as 0.01. We also note that the observed standard deviations of the males and females are reasonably close: 9.95 and 11.6. Thus the data do not seem to contradict our assumption of equal standard deviations.

Exercises

9-8   The following questions refer to the output in Exhibit 9.5.
  (a)   Use this output to calculate, by hand, a 95% confidence interval for the mean speaking ability for people who did not have the operation. Use Minitab to verify your answer.
  (b)   Using the formulas in the box for TWOSAMPLE, find a 90% confidence interval for the mean difference between people who had the operation and those who did not. *Note:* Round the degrees of freedom to the nearest value that is in your *t* table.

9-9   Repeat the analysis of the Parkinson's disease data using pooled procedures. How do the results change?

9-10 Physicists are constantly trying to obtain more accurate values of the fundamental physical constants, such as the mean distance from the earth to the sun and the force exerted on us by gravity. These are very difficult measurements, and require the utmost ingenuity and care. Here are some results obtained in a Canadian experiment to measure the force of gravity. The first group of 32 measurements was made in August, and the second group in December of the following year. (*Note:* The force of gravity, measured in centimeters per second squared, can be obtained from these numbers by first dividing each number by 1000, then adding 980.61; thus, the first measurement converts to 980.615.)

Measurements Made in August 1958

5	20	20	25	25	30	35
15	20	20	25	25	30	35
15	20	25	25	30	30	
15	20	25	25	30	30	
20	20	25	25	30	30	

Measurements Made in December 1959

20	30	35	40	40	45	55
25	30	35	40	40	45	60
25	30	35	40	45	50	
30	30	35	40	45	50	
30	35	40	40	45	50	

(a)   Use the data from August to estimate the force of gravity. Also find a 95% confidence interval.

(b)   Repeat (a) for the December measurements.

(c)   Now compare the measurements made in August to those made in December. Do a dotplot and an appropriate test using $a = .05$. Is there a statistically significant difference between the two groups? What is the practical significance of this result?

(d)   In between these two sets of measurements, it was necessary to change a few key components of the apparatus. Might this have made a difference in the measurements? If so, can you draw any guidelines for sound scientific experimentation? What if the experimenters had done all their measurements at one time? Might they have misled themselves about the accuracy of their results? What if still other parts of the apparatus were changed? Might the measurements change even more?

9-11   A small study was done to compare how well students with different majors do in an introductory statistics course. Seven majors were tested: biology, psychology, sociology, business, education, meteorology, and economics. At the end of the course, the students were given a special test to measure their understanding of basic statistics. Then a series of t-tests were performed to compare every pair of majors. Thus, biology and psychology majors were compared, biology and sociology majors, psychology and sociology majors, and so on, for a total of 21 t-tests. Simulate this study assuming that all majors do about the same. Assume there are 20 students in each major and that scores on the test have a normal distribution with $\mu = 12$ and $\sigma = 2$. Use

```
RANDOM 20 C1-C7;
  NORMAL 12 2.
```

This puts a sample for each of the seven majors into a separate column.

(a) What is the null hypothesis?

(b) What are the 21 pairs of majors for the 21 t-tests?

(c) Do the 21 t-tests. Unfortunately, you will need to type TWOSAMPLE 21 times.

(d) In how many of the tests did you reject the null hypothesis at $\alpha = 10$?

(e) Because this study was simulated, the true situation is known—there are no differences. However, you probably did find a significant difference in at least one pair of majors. This illustrates the "hazards" of doing a lot of comparisons without making proper adjustments in the procedure. Try to think of some other situations where you might do a lot of statistical tests. For example, suppose a pharmaceutical firm had 16 possible new drugs that they wanted to try out in the hope that at least one was better than the present best competing brand. What are some of the consequences of doing a lot of statistical tests?

9-12   A seventh-grade student (named Kristy) was given an experiment to run. She was to complete the same basic task a number of times. Half the time she was to do the task in a quiet room, and half the time she was to do it while chewing gum, with the radio and TV turned up loud.

	Time to Complete Task (Seconds)									
Quiet (Trials 1–10)	60	50	40	21	28	17	19	14	12	12
Noisy (Trials 11–20)	36	30	35	21	20	15	18	9	10	10

(a) Use DOTPLOT and DESCRIBE to compare the two sets of times. Discuss the results. Is one environment appreciably better than the other?

(b) Do an appropriate test.

(c) The 20 trials were done in the order indicated: first trial 1, then trial 2, and so on. Use TSPLOT to display the data. What do you see?

(d) Discuss the planning of the experiment and how it could have been improved. Would you change your conclusions for part (a) after seeing the plot in (b)? Can you offer any advice to future analysts who might be comparing data like these?

9-13 Comparisons between two groups can be made for several variables in the Pulse data (in Appendix A).

(a) Make stem-and-leaf displays of the first pulse rates, using the subcommand BY to separate the smokers from the nonsmokers.

(b) Repeat part (a), using boxplots.

(c) Does there seem to be any overall difference between the smokers and nonsmokers? Explain.

(d) Do an appropriate test.

10

Analysis of Variance

10.1 Analysis of Variance with One Factor

One-way analysis of variance is used to compare data from several populations. We will start with an example.

The flammability of children's sleepwear has received a lot of attention over the years. There are standards to ensure that manufacturers don't sell children's pajamas that burn easily. But a problem always arises in cases like this. How do you test the flammability of a particular garment? The shape of the garment might make a difference. How tightly it fits might also be important. Of course, the flammability of clothing can't be tested on children. Perhaps a metal mannequin could be used. But how should the material be lighted? Different people putting a match to identical cloth will probably get different answers. The list of difficulties is endless, but the problem is real and important.

One procedure that has been developed to test flammability is called the Vertical Semirestrained Test. The test has many details, but basically it involves holding a flame under a standard-size piece of cloth that is loosely held in a metal frame. The dryness of the fabric, its temperature, the height of the flame, how long the flame is held under the fabric, and so on are all carefully controlled. After the flame is removed and the fabric stops burning, the length of the charred portion of the fabric is measured and recorded.

Once we have a proposed way to test flammability, one important question is "Will the laboratories of the different garment manufacturers all be able to get about the same results if they apply the same test to the same

Table 10.1   Data from an Interlaboratory Study of Fabric Flammability

	Laboratory				
	1	*2*	*3*	*4*	*5*
	2.9	2.7	3.3	3.3	4.1
	3.1	3.4	3.3	3.2	4.1
	3.1	3.6	3.5	3.4	3.7
	3.7	3.2	3.5	2.7	4.2
Length of	3.1	4.0	2.8	2.7	3.1
Charred	4.2	4.1	2.8	3.3	3.5
Portion of	3.7	3.8	3.2	2.9	2.8
Fabric	3.9	3.8	2.8	3.2	3.5
	3.1	4.3	3.8	2.9	3.7
	3.0	3.4	3.5	2.6	3.5
	2.9	3.3	3.8	2.8	3.9
Sample Means	3.34	3.30	3.30	3.00	3.65

fabric?" A study was conducted to answer this question. A small part of the data is shown in Table 10.1.

   We want to look at displays of these data and use one-way analysis of variance to study the differences among laboratories. Our first task, however, is to enter the data in a format that is appropriate for displays and for doing analysis of variance. The following commands will do that:

```
NAME C1 'Charred'  C2 'Lab'
SET C1
   2.9   3.1   3.1   3.7   3.1   4.2   3.7   3.9   3.1   3.0   2.9
   2.7   3.4   3.6   3.2   4.0   4.1   3.8   3.8   4.3   3.4   3.3
   3.3   3.3   3.5   3.5   2.8   2.8   3.2   2.8   3.8   3.5   3.8
   3.3   3.2   3.4   2.7   2.7   3.3   2.9   3.2   2.9   2.6   2.8
   4.1   4.1   3.7   4.2   3.1   3.5   2.8   3.5   3.7   3.5   3.9
END
SET C2
   (1:5)11
END
```

   The first SET command puts all 55 measurements into one column, C1; first the eleven measurements for lab 1, then the eleven measurements for lab 2, then the eleven measurements for lab 3, and so on. Next we create a column, C2, for the labs. The first eleven measurements came from lab 1, so we put eleven 1s into C2. The next eleven measurements came from lab 2, so we put eleven 2s into C2, and so on. Notice that instead of typing eleven 1s, then eleven 2s, . . . , we used the patterned data feature of SET.

Exhibit 10.1   Dotplots of the Flammability Data

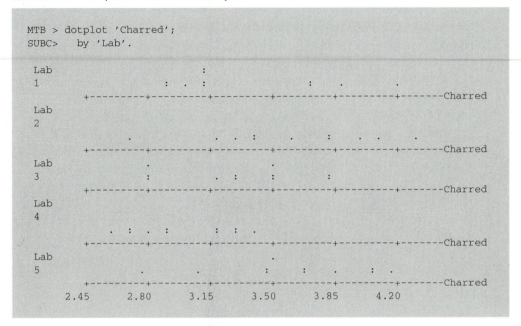

The abbreviation 1:5 is used to represent the list of integers 1, 2, 3, 4, 5. We then enclose this abbreviation in parentheses and put an 11 after the close parenthesis (with no blanks in between). The 11 says to repeat each number in the list eleven times. The patterned-data feature of SET, explained in Chapter 1, is especially useful for inputting analysis of variance data.

There are a number of ways we could display the data: histograms, dotplots, boxplots, and scatterplots are a few. In Exhibit 10.1, we used dotplots. The BY subcommand says to make a separate dotplot for each laboratory.

This display tells us a lot by itself. Even within a given laboratory, the char-length varies from specimen to specimen. All of the specimens are supposedly identical because they were cut from the same bolt of cloth, but there were probably still some differences between them. In addition, the test conditions surely differed slightly from specimen to specimen—the tautness of the material, the exact length of time the flame was held under the specimen, and so on. We must expect some variation in results even when all measurements are made in the same laboratory, under apparently identical conditions. The variation within a laboratory can be thought of as random error. There are several other names for this variation: within-group variation, unexplained variation, residual variation, or variation due to error.

There is a second source of variation in the data: variation due to differences among labs. This variation is called among-group variation, variation due to the factor, or variation explained by the factor. Here the factor under study is laboratory. (In statistics, the word *treatment* is sometimes used instead of *factor*.) If we look at the dotplots again, we see some variation among the five labs. Lab 4 got mostly low values, whereas labs 2 and 5 got mostly high values. Instead of using dotplots, we could compare the sample means for the five labs. These are given in Table 10.1, above. The five sample means do vary from lab to lab, but again, some variation is to be expected. The question is, "Do the five sample means differ any more than we would expect just from random variation?"

Put another way, "Is the variation we see among the groups significantly greater than the variation we would expect to see given the amount of variation within groups?" Analysis of variance is a statistical procedure that gives an answer to this question.

The One-Way Analysis of Variance Procedure

In one-way analysis of variance, we want to compare the means of several populations. We assume that we have a random sample from each population, that each population has a normal distribution, and that all of the populations have the same variance, σ^2. In practice, the normality assumption is not too important, the equal-variances assumption is not important (provided the number of observations in each group is about the same), but the assumption of a random sample is very important. If we do not have a random sample from the population, or something that is very close to it, our conclusion can be far from the truth.

The main question is, "Do all of the populations have the same mean?" Suppose that a is the number of populations we have, and that μ_1 is the mean of the first population, μ_2 is the mean of the second population, μ_3 is the mean of the third, and so on. Then the null hypothesis of no differences is

$$H_0 : \mu_1 = \mu_2 = \mu_3 = \ldots = \mu_a$$

To test this null hypothesis, we can use the Minitab command ONE-WAY. Exhibit 10.2 shows the results of analyzing the data in Table 10.1. The first part of the output is called an analysis of variance table. In it, the total sum of squares is broken down into two sources—the variation due to

Exhibit 10.2 Output from ONEWAY Using the Flammability Data

```
MTB > oneway 'Charred' 'Lab'

ANALYSIS OF VARIANCE ON Charred
SOURCE      DF        SS        MS        F         p
Lab          4     2.987     0.747     4.53    0.003
ERROR       50     8.233     0.165
TOTAL       54    11.219
                                      INDIVIDUAL 95% CI'S FOR MEAN
                                      BASED ON POOLED STDEV
LEVEL       N      MEAN      STDEV   --+---------+---------+---------+-----
   1       11    3.3364     0.4523              (------*------)
   2       11    3.6000     0.4604                   (------*------)
   3       11    3.3000     0.3715            (------*------)
   4       11    3.0000     0.2864   (------*------)
   5       11    3.6455     0.4321                      (------*------)
                                    --+---------+---------+---------+-----
POOLED STDEV =    0.4058            2.80      3.15      3.50      3.85
```

the factor (here it is differences among the five labs) and the variation due to random error (here it is variation within labs). Thus, (SS TOTAL) = (SS Lab) + (SS ERROR). In this example, $11.219 = 2.987 + 8.233$. Each sum of squares has a certain number of degrees of freedom associated with it. These are used to do tests. The degrees of freedom also add up: (DF TOTAL) = (DF Lab) + (DF ERROR).

Formulas for the Three Sums of Squares. In the following table, x_{ij} is the jth observation in the sample from population i, x_i, and n_i are the sample mean and sample size for sample i, a is the number of populations, n is the total number of observations, and $\bar{x}$ is the mean of all n observations.

Source	DF	SS
Factor	$a - 1$	$\Sigma_i n_i (\bar{x}_i - \bar{x})^2$
Error	$n - a$	$\Sigma_i \Sigma_j (x_{ij} - \bar{x}_i)^2$
Total	$n - 1$	$\Sigma_i \Sigma_j (x_{ij} - \bar{x})^2$

Exhibit 10.2 gives mean squares (abbreviated MS). Each mean square is just the corresponding sum of squares divided by its degrees of freedom. The next column gives the quotient: F = (MS FACTOR)/(MS ERROR). This F-ratio is a useful test statistic. It is very large when MS FACTOR is

---

ONEWAY on data in **C**, levels in **C**

Performs a one-way analysis of variance. The first column contains the data; the second says what level (or treatment) each observation belongs to. The numbers used for levels must be integers.

ONEWAY has subcommands to do multiple comparison procedures. These are described in Chapter 16.

Stat > ANOVA > Oneway

---

much larger than MS ERROR—that is, when the variation among the labs is much greater than the variation due to random error. In such cases, we would reject the null hypothesis that the labs all have the same average char-length. Is this F-ratio large? That's what the p-value tells us, as it did in the t-tests discussed in Chapters 8 and 9. Here the p-value is 0.003. Thus, for $a = 0.05$ or even 0.01, we would reject the null hypothesis and conclude we have evidence that there are some differences among the five labs.

The next table in Exhibit 10.2 summarizes the results separately for each lab. The sample size, sample mean, sample standard deviation, and a 95% confidence interval are given for each lab. Each confidence interval is calculated by the formula

$$\bar{x}_i - ts_p / \sqrt{n_i} \quad \text{to} \quad \bar{x}_i + ts_p / \sqrt{n_i}$$

Here $\bar{x}_1$ and n_i are the sample mean and sample size for level i, s_p = POOLED STDEV = $\sqrt{\text{MS ERROR}}$ is the pooled estimate of the common standard deviation, σ, and t is the value from a t-table corresponding to 95% confidence and the degrees of freedom associated with MS ERROR. These intervals give us some idea of how the population means differ.

Exercises

10-1   Use DOTPLOT and ONEWAY to see whether the OTIS scores differed by a statistically significant amount for the three education groups—prepro-fessional, professional, student—in the Cartoon experiment (described in Appendix A).

10-2   Plywood is made by cutting a thin continuous layer of wood off a log as it is spun around. Several of these thin layers are then glued together to make plywood sheets.

Thirty logs were used in a study. A chuck was inserted into each end of a log. The log then was turned and a sharp blade was used to cut off a thin layer of wood. Three temperature levels were used. Ten logs were tested at each temperature. The torque that could be applied to the log before the chuck spun out was measured. Use DOTPLOT and ONEWAY to see how the temperature level affects the amount of torque that can be applied.

Temperature	Torque									
60°F	17.5	16.5	17.0	17.0	15.0	18.0	15.0	22.0	17.5	17.5
120°F	15.5	17.0	18.5	15.5	17.0	18.5	16.0	14.0	17.5	17.5
150°F	16.5	17.5	15.5	16.5	16.0	13.5	16.0	13.5	16.0	16.5

10-3 In a simple pendulum experiment, a weight (bob) is suspended at the end of a length of string. The top of the string is supported by some sort of stable frame. Under ideal conditions, the time (T) required for a single cycle of the pendulum is related to the length (L) of the string by the equation

$$T = \frac{2\pi}{g}\sqrt{L}$$

where, as usual, $\pi = 3.1415 \ldots$ and g is the pull (acceleration) of gravity. If the time required for a cycle and the length of the pendulum μ are measured, this equation can be solved to give an estimate of g. In theory, neither the length of the pendulum nor the type of bob should have any effect on the estimate of g. In practice, however, things that are not supposed to make any difference often do. So it is often a good policy, when doing an experiment, to vary, in a carefully balanced manner, things that are not supposed to matter. Therefore, the experiment was run using four different lengths of string and two types of bobs.

The following estimates of g, in centimeters per second squared, were obtained by college professors during a short course on the use of statistics in physics and chemistry courses.

	Length of Pendulum (cm)			
	60	70	80	90
Heavy Bob	924	994	970	1000
	973	969	975	1017
	955	968	970	1055
Light Bob	966	973	985	960
	949	997	999	1041
	955	988	994	962

(a)   Use DOTPLOT to display the data and do a one-way analysis of variance to see whether length affects the estimate of g. Use only the data
obtained with the heavy bob.

(b)   Repeat part (a), using only the data obtained with the light bob.

10-4   In the Restaurant survey (described in Appendix A), the various restaurants
were classified by the variable TYPEFOOD as fast food, supper club, and
other. Use appropriate displays and ONEWAY to determine the following:

(a)   Do these three groups spend approximately the same percent of their
sales on advertising? (Use the variable ADS.)

(b)   Do they spend the same percent on wages? (Use the variable
WAGES.)

(c)   Do they spend the same percent on the cost of goods? (Use the variable COSTGOOD.)

10-5   Very small amounts of manganese are important to a good diet. Unfortunately, measurement of these small amounts is quite difficult. To help researchers evaluate their ability to measure such small amounts, the
National Institute of Standards and Technology (NIST; formerly the National Bureau of Standards) sells samples of cow liver together with an
accurate chemical analysis of the amount of manganese in each sample.
The data given here are from one part of the evaluation. Eleven pieces were
taken from one cow's liver. The experimenters wanted to know whether the
amount of manganese varied from piece to piece. Of course, even if the
pieces were exactly the same, there would still be some differences in the
recorded amounts just because of errors in making the measurements.
Therefore, the question posed was "Do the 11 recorded amounts vary more
than you would expect from measurement error alone?" To get some idea
of how large measurement error was, the experimenters measured each
piece twice. The amount of manganese (in parts per million) is given in the
table below. Use ONEWAY to see whether there is a statistically significant
difference among the 11 pieces.

					Piece					
1	*2*	*3*	*4*	*5*	*6*	*7*	*8*	*9*	*10*	*11*
10.02	10.41	10.25	9.41	9.73	10.07	10.09	9.85	10.02	9.92	9.7
10.03	9.79	9.80	10.17	10.75	9.76	9.38	9.99	9.51	10.01	10.0

10-6   Steel makers find that small differences in the amount of oxygen in steel
make an important difference in the quality of the steel they produce. Since
small amounts of oxygen are difficult to measure accurately, NIST agreed
to make very careful measurements on some homogeneous material, then

to sell the samples to steel manufacturers that could use them to check whether their own instruments were making accurate measurements.

The researchers at NIST took a long "homogeneous" steel rod and cut it into 4-inch pieces. They randomly selected 20 of these pieces and labeled them 1 through 20. Then they made two very careful measurements of the amount of oxygen in each piece. Measurements were made over a five-day period in January. On each day, the measurements were made in the order shown below. The data are given below and stored in the file STEEL.

Bear in mind that finding and measuring quantities as small as 5 parts of oxygen in 1 million parts of steel is very difficult and is subject to a wide variety of unsuspected sources of error.

Day	Piece Number	Amount of Oxygen (parts per million)	Day	Piece Number	Amount of Oxygen (parts per million)
1	17	5.6	4	13	5.4
1	11	5.9	4	18	6.1
1	10	6.8	4	2	5.7
1	15	7.5	4	10	6.1
1	5	4.7	4	4	5.7
1	6	4.0	4	20	4.6
1	19	4.4	4	7	5.7
2	4	6.6	4	6	4.4
2	7	4.9	4	12	4.1
2	1	5.5	5	16	6.3
2	13	4.9	5	15	7.5
2	3	6.3	5	8	6.1
2	18	4.2	5	14	6.4
2	9	3.3	5	3	5.1
2	14	4.8	5	20	5.7
3	16	6.1	5	5	3.8
3	12	5.3	5	1	3.8
3	8	5.2			
3	2	4.3			
3	11	4.0			
3	19	6.2			
3	17	5.1			
3	9	3.3			

(a) Is there any evidence that the 20 pieces are not homogeneous? Use ONEWAY.

(b) Is there any evidence of day-to-day variation?

(c) There is a major unsuspected source of error in these measurements. Can you find it?

10.2 Analysis of Variance with Two Factors

Table 10.2 presents data from an experiment to study the driving abilities of two types of drivers, inexperienced and experienced. This is our first factor. Twelve drivers of each type took part in the study. Three road types were used: first class, second class, and dirt. Road type is our second factor. Four of the inexperienced drivers were assigned at random to each road type, and four of the experienced drivers were assigned at random to each road type. This gives a total of $2 \times 3 = 6$ treatment combinations or cells in our design, with four observations per cell. Each subject drove a one-mile section of road. The number of steering corrections the driver had to make was recorded. This is the dependent, or response, variable.

Again, our first task is to enter the data in a form that is appropriate for displays and two-way analysis of variance. The following commands do that:

```
NAME C1='corrects' c2='expernc'  c3='road'
SET 'corrects'
   4  18   8  10    23  15  21  13    16  27  23  14
   6   4  13   7     2   6   8  12    20  15   8  17
END
SET 'expernc'
  (0:1)12
END
SET 'road'
  (1:3)4      (1:3)4
END
```

Table 10.2 Driving Performance

	Road Type		
	First Class	*Second Class*	*Dirt*
	4	23	16
Inexperienced	18	15	27
Driver	8	21	23
	10	13	14
	6	2	20
Experienced	4	6	15
Driver	13	8	8
	7	12	17

Exhibit 10.3   Using TABLE to Display the Driving Performance Data

```
MTB > table 'expernc' 'road';
SUBC>    data 'corrects'.

 ROWS: expernc     COLUMNS: road

             1        2        3

 0      4.000   23.000   16.000
       18.000   15.000   27.000
        8.000   21.000   23.000
       10.000   13.000   14.000

 1      6.000    2.000   20.000
        4.000    6.000   15.000
       13.000    8.000    8.000
        7.000   12.000   17.000

 CELL CONTENTS --
         corrects:DATA
```

The first SET puts the response variable into C1. A second column, C2, is used for driving experience; 0 = inexperienced and 1 = experienced. We use the patterned data feature of SET to enter twelve 0s followed by twelve 1s. A third column, C3, is used for road condition; 1 = first class, 2 = second class, and 3 = dirt. We use SET to enter four 1s, four 2s, and four 3s for the inexperienced drivers, then another four 1s, four 2s, and four 3s for the experienced drivers.

Exhibit 10.3 uses the TABLE command to display the data in a form similar to that of Table 10.2. This makes it easy to check for typing errors. Exhibit 10.4 uses TABLE to calculate means. This table gives us a good idea of how the drivers did on each road type. On the average, inexperienced drivers made 16.000 steering corrections, whereas experienced drivers made only 9.833. Further, if we look at the individual cell means, we notice that the inexperienced drivers did worse than the experienced drivers on all three road types.

Suppose we now look at road type. On the average, the number of steering corrections goes up as the road quality goes down, from 8.750 to 12.500 to 17.500. The pattern does not quite hold up if we look at the individual cells. The experienced drivers did slightly better on second-class roads than on first-class roads. However, the decrease is small and could be due merely to random variation.

Exhibit 10.4   Using TABLE to Calculate Means

```
MTB > table 'expernc' 'road';
SUBC>    means 'corrects'.

 ROWS: expernc      COLUMNS: road

              1         2         3       ALL

   0    10.000    18.000    20.000    16.000
   1     7.500     7.000    15.000     9.833
 ALL     8.750    12.500    17.500    12.917

 CELL CONTENTS --
          corrects:MEAN
```

Additive Models

Now, just as we did in one-way analysis of variance, we can express the total variation in the data as the sum of the variation from several sources:

$$\text{(Total variation in data)} = \text{(variation due to driving experience)}$$
$$+ \text{(variation due to road type)}$$
$$+ \text{(variation due to random error)}$$

If the variation due to driving experience is much greater than the variation due to random error, we will have statistically significant evidence of a difference between the two types of drivers. Similarly, if the variation due to road type is much greater than the variation due to random error, we will have statistically significant evidence of a difference among the three road types. We will give formulas for these when we discuss the model with interaction in the next section.

The ANOVA output in Exhibit 10.5 gives the breakdown of the total variation. Thus (SS Total) = (SS expernc) + (SS road) + (SS Error). As in ONEWAY, each sum of squares has a certain number of degrees of freedom associated with it and these also add up. Each mean square is the corresponding sum of squares divided by its degrees of freedom. The value of F = (MS for the factor)/(MS Error). And the p-value gives the statistical significance of the corresponding F-statistic. Both p-values are small. Thus there is evidence to say that both factors, driving experience and road type, have an effect on steering corrections.

Exhibit 10.5   Output from ANOVA for the Driving Performance Data

```
MTB > anova 'correct' = 'expernc'  'road'

Factor       Type Levels Values
expernc      fixed    2    1    2
road         fixed    3    1    2    3


Analysis of Variance for corrects

Source       DF         SS        MS       F      P
expernc       1     228.17    228.17    8.56  0.008
road          2     308.33    154.17    5.78  0.010
Error        20     533.33     26.67
Total        23    1069.83
```

The first table in Exhibit 10.5 lists the factors, what type each is, the number of levels each has, and the actual values of the levels. There are two types of factors, fixed and random, but in this chapter we will use only fixed factors. Random factors are discussed in Chapter 16.

Look at the form of the ANOVA command in Exhibit 10.5. The response is listed first, then an equal sign, then the two factors. ANOVA has some special syntax rules that are different from those for other Minitab commands: First, you cannot use any extra text, except after a # (see p. 9). Second, the equal sign is required. Third, you do not have to put quotes around variable names.

Models with Interactions

There is another source of variations that we have not fully discussed: interaction between the two factors. The driving performance data set does not have a significant interaction, and we will have you verify that in an exercise. Now we will look at another experiment.

Table 10.3 presents data from an experiment designed to study the effects of two factors on the quality of pancakes. The two factors were the amount of whey and whether or not a supplement was used. There are four levels of whey (0%, 10%, 20%, and 30%) and two levels of supplement (used and not used), giving a total of $4 \times 2 = 8$ treatment combinations or cells. Three pancakes were baked using each treatment combination. Each pancake was then rated by an expert; the three ratings were averaged to give

Table 10.3 Quality of Pancakes

	Amount of Whey			
	0%	10%	20%	30%
No Supplement	4.4	4.6	4.5	4.6
	4.5	4.5	4.8	4.7
	4.3	4.8	4.8	5.1
Supplement	3.3	3.8	5.0	5.4
	3.2	3.7	5.3	5.6
	3.1	3.6	4.8	5.3

one overall quality rating. The higher the quality rating, the better the pancake. This was done three times for each treatment combination, giving a total of $3 \times 8 = 24$ overall quality ratings.

First we entered the data. Quality ratings were all put in one column. A second column was used for supplement (0 = no, 1 = yes), and a third column for the amount of whey (0, 10, 20, 30). Exhibit 10.6 uses TABLE to calculate cell means. First, let's look at the average effect of each factor. These are often called the main effects.

Pancake quality increases, on the average, as the percentage of whey increases. The six pancakes with 0% whey have a mean quality of 3.8; the six with 10% whey have a mean quality of about 4.2; for 20% the mean is 4.9; and for 30% it is 5.1. The average effect of using the supplement is not as great as the effect of whey: the mean quality of the 12 pancakes with no supplement is just a little higher than the mean quality for the 12 with the supplement.

We also can look at the individual cells in Exhibit 10.6. We used these cell means to draw a plot by hand, shown in Exhibit 10.7. Now the effect of the supplement is not clear: Supplement increases quality when whey is 20% or 30%, but seems to lower quality when whey is 0% or 10%. Thus, whether or not the supplement increases quality *depends* on the amount of whey. When the effect of one factor depends on the level of another, we say the two factors interact. Here, supplement and whey appear to interact. If the two factors did not interact, the two lines in the plot of Exhibit 10.7 would be roughly parallel.

Exhibit 10.8 shows the output from ANOVA. This is the same command we used for the driving performance data, but now we have added an extra term on the command line. This is the term for the interaction between supplement and whey. Notice the form: the name of the first factor, then an asterisk, then the name of the second factor. We did not use quotes around

Exhibit 10.6   TABLE Output for Pancake Data

```
MTB > table 'Supplmnt' 'Whey';
SUBC>    mean 'Quality'.

 ROWS: Supplmnt       COLUMNS: Whey

              0         10        20        30       ALL

   0    4.4000    4.6333    4.7000    4.8000    4.6333
   1    3.2000    3.7000    5.0333    5.4333    4.3417
 ALL    3.8000    4.1667    4.8667    5.1167    4.4875

   CELL CONTENTS --
          Quality:MEAN
```

Exhibit 10.7   Hand-Drawn Plot of Cell Means for Pancake Data

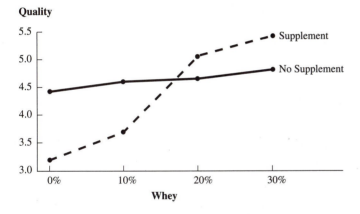

Exhibit 10.8   Output from ANOVA for Pancake Data

```
MTB > anova Quality = Supplmnt  Whey  Supplmnt*Whey

Factor       Type Levels Values
Supplmnt     fixed      2      0    1
Whey         fixed      4      0    10   20    30

Analysis of Variance for Quality

Source             DF         SS        MS       F        P
Supplmnt            1     0.5104    0.5104   17.01    0.001
Whey                3     6.6912    2.2304   74.35    0.000
Supplmnt*Whey       3     3.7246    1.2415   41.38    0.000
Error              16     0.4800    0.0300
Total              23    11.4063
```

the variable names, since they are optional with the ANOVA command. You can, as always, use the column numbers instead of the names. Thus we could use

```
ANOVA C1 = C2 C3 C2*C3.
```

The total variation is now partitioned into four sources:

(Total variation) = (variation due to rows)
+ (variation due to column)
+ (variation due to interaction)
+ (variation due to error)

Notice that the sums of squares (and the degrees of freedom) in Exhibit 10.8 also add up as in this formula.

Now let's look at the tests. We always test for interaction first. If interaction is present, then we must be careful how we interpret the other two tests. The p-value for interaction is 0.000, so there is strong evidence that the effect of the supplement depends on the amount of whey used. Or, put another way, the effect of whey depends on whether or not the supplement is used.

Let's look at the two other tests. These are often called tests for the average or main effects of the two factors; here they are supplement and whey. The p-value for supplement is very small, 0.001, so there is significant evidence of an overall effect of supplement. The p-value for whey is also very small, so there is significant difference due to whey.

Now let's look at Exhibit 10.6 again. Notice that the overall mean for no supplement is 4.6333, which is larger than the overall mean of 4.3417 for supplement. This is consistent with our test: supplement has an effect. But the test does not tell the whole story. In fact, it tends to distort the real story. If we look at the individual cells, we see that sometimes supplement is better than no supplement and sometimes it is worse. On the average, no supplement is better than supplement. In general, whenever there is a significant interaction, you should look at the individual cells to see what's really going on. Otherwise, the tests on the main effects may be very misleading.

Formulas for the Sums of Squares. We will use the following notation: a is the number of levels of the first factor (that is, the number of rows in the table of data); b is the number of levels of the second factor (that is, the number of columns in the table of data); n is the number of observations in each cell; x_{ijk} is the kth observation in cell (i, j) (i.e., in the cell in row i and column j); $\bar{x}_{ij.}$ is the mean of the n observations in cell (i, j); $\bar{x}_{i..}$ is the mean

---

ANOVA C = termlist

Does a one-way and a two-way analysis of variance. Designs with two factors must be balanced; that is, each cell must contain the same number of observations. The level numbers must be integers. The equal sign is required. Put the response before the equal sign; the factors and interaction term (if you give one), after the equal sign. No extra text may be used on the ANOVA command except after a #.

Examples:

```
ANOVA  C1 = C2          # one-way analysis of variance
ANOVA  C1 = C2 C3       # two-way additive model
ANOVA  C1 = C2 C3 C2*C3 # two-way model with interaction
```

ANOVA can also analyze more complicated designs. See Chapter 16 for a discussion.

Stat > ANOVA > Balanced Anova

---

of the bn observations in row i; $\bar{x}_{.j}$ is the mean of the an observations in column j; and x is the mean of all abn observations in the table.

Source	DF	SS
Rows	$a - 1$	$bn\Sigma_i(\bar{x}_{i..} - \bar{x})^2$
Columns	$b - 1$	$an\Sigma_j(\bar{x}_{.j.} - \bar{x})^2$
Interaction	$(a-1)(b-1)$	$n\Sigma_i \Sigma_j (\bar{x}_{ij.} - \bar{x}_{i..} - \bar{x}_{.j.} + \bar{x})^2$
Error	$ab(n-1)$	$\Sigma_i \Sigma_j \Sigma_k (x_{ijk} - \bar{x}_{ij.})^2$
Total	$abn - 1$	$\Sigma_i \Sigma_j \Sigma_k (x_{ijk} - \bar{x})^2$

The additive model we fit to the driving performance uses the same formulas for the rows, columns, and total that are used in this table. Since SS always add up, the SS for the error term in the additive model is (SS for Interaction) + (SS for Error) in this table. The same is true for the DF. In general, if you omit any term from a model in analysis of variance, the SS and DF for that term are added into the error term.

ANOVA Command. The ANOVA command can also do one-way analysis of variance, just as the command ONEWAY can. ONEWAY has some special features, which is why we explained it in Section 10.1. ANOVA can also analyze up to nine factors and has many special features of its own. We will explain some of these in Chapter 16.

Exercises

10-7   Use ANOVA to see whether there is a significant interaction in the Driving Performance data in Table 10.2.

10-8   An experiment was done to see whether the damage to corn, wheat, dried milk, and other stored crops by a beetle, *Trogoderma glabrum*, could be reduced. The idea was to lure male adult beetles to inoculation sites where they would pick up disease spores, which they would then carry back to the remaining population for infection. The experiment consisted of two factors, "sex attractant or none" and "disease pellet present or not." The dependent variable was the fraction of larvae failing to reach adulthood.

   Analyze the data presented in the table below. Use TABLE to get cell means; plot these means by hand. Use ANOVA to do an analysis of variance. Interpret your results.

	Disease	
Attractant	*Yes*	*No*
Yes	.979	.222
	.743	.189
	.775	.143
	.885	.262
No	.312	.239
	.293	.253
	.188	.159
	.388	.241

10-9   The pendulum data of Exercise 10-3 involve two factors, length and bob. Do a two-way analysis of variance on the data. Use TABLE to get cell means. Plot these by hand.
   (a)   Does length seem to affect the estimate of g? Does the type of bob used affect the estimate? Is there a significant interaction between length and bob?
   (b)   Suppose you wanted to estimate the acceleration of gravity using a pendulum. What do the results in part (a) tell you about designing such an experiment?

10-10 A substantial percentage of the potatoes raised in this country never have a chance to reach the table. Instead, they fall victim to potato rot while being stored for later use. To find out what could be done to reduce this loss, an experiment was carried out at the University of Wisconsin. Potatoes were injected with bacteria known to cause rot, and were then stored under a variety of conditions. After five days, the diameter of the rotted portion of each potato was measured.

Three factors were varied in this experiment: (1) the amount of bacteria injected into the potato (1 = low amount, 2 = medium amount, 3 = high amount); (2) the temperature during storage (10°C, 16°C); (3) the amount of oxygen during storage (2%, 6%, 10%). Data are given below and are stored in the file POTATO.

Bacteria	Temperature	Oxygen	Rot	Bacteria	Temperature	Oxygen	Rot
1	1	1	7	2	2	1	17
1	1	1	7	2	2	1	18
1	1	1	9	2	2	1	8
1	1	2	0	2	2	2	3
1	1	2	0	2	2	2	23
1	1	2	0	2	2	2	7
1	1	3	9	2	2	3	15
1	1	3	0	2	2	3	14
1	1	3	0	2	2	3	17
1	2	1	10	3	1	1	13
1	2	1	6	3	1	1	11
1	2	1	10	3	1	1	3
1	2	2	4	3	1	2	10
1	2	2	10	3	1	2	4
1	2	2	5	3	1	2	7
1	2	3	8	3	1	3	15
1	2	3	0	3	1	3	2
1	2	3	10	3	1	3	7
2	1	1	2	3	2	1	26
2	1	1	4	3	2	1	19
2	1	1	9	3	2	1	24
2	1	2	4	3	2	2	15
2	1	2	5	3	2	2	22
2	1	2	10	3	2	2	18
2	1	3	4	3	2	3	20
2	1	3	5	3	2	3	24
2	1	3	0	3	2	3	8

These data probably should be analyzed using a three-way analysis of variance, but you can discover most of what is going on by using several two-way analyses and some plots and tables.

(a)   Do a two-way analysis of variance on all the data, using bacteria and temperature as the two factors (ignore oxygen for the moment). Use TABLE to get cell means. Plot these means by hand.

(b)   Next, do a two-way analysis of variance on all the data, using bacteria and oxygen as the two factors (ignore temperature in this analysis). Use TABLE to get cell means. Plot these means by hand.

(c)   How do each of the three factors influence potato rot? Do any of the factors appear to interact?

(d)   The factors that can be controlled, to some extent, by the food supplier are temperature and oxygen. Analyze these two factors at each of the three levels of bacteria. Do temperature and oxygen seem to have the same effect at each level?

(e)   What recommendations would you make to someone who was storing potatoes?

10.3 Randomized Block Designs

In a randomized block design (abbreviated RBD) there is one factor we want to study, just as in a one-way design. But now we try to reduce some of the variability in the data by grouping the material, people, locations, time periods, or whatever into relatively homogeneous blocks. The treatments we wish to compare are then assigned at random within each block, with each treatment appearing exactly once in each block.

The data in Table 10.4 give the elasticity of billiard balls made under three different conditions. The experiment was carried out as follows: Ten batches of melted plastic were prepared. (These are the 10 blocks of material.) Each batch was divided into three equal portions. One portion was chosen at random and set aside as a control. A second portion was chosen

Table 10.4   Elasticity of Billiard Balls

Batch	Control	Additive A	Additive B
1	51.000	75.000	39.000
2	45.000	89.000	43.000
3	49.000	73.000	51.000
4	66.000	84.000	34.000
5	53.000	66.000	54.000
6	41.000	85.000	43.000
7	58.000	73.000	42.000
8	56.000	71.000	41.000
9	60.000	78.000	37.000
10	63.000	65.000	44.000

Exhibit 10.9   Using TABLE to Calculate Means for the Billiard Ball Data

```
MTB > name c1='Elastic' c2='Batch'  c3='Additive'
MTB > set 'Elastic'
DATA>   51 45 49 66 53 41 58 56 60 63
DATA>   75 89 73 84 66 85 73 71 78 65
DATA>   39 43 51 34 54 43 42 41 37 44
DATA> end
MTB > set 'Batch'
DATA>   3(1:10)
DATA> end
MTB > set 'Additive'
DATA>   (0:2)10
DATA> end
MTB > table 'Batch' 'Additive';
MTB >   means 'Elastic'.

 ROWS: Batch     COLUMNS: Additive

              0         1         2        ALL

    1     51.000    75.000    39.000    55.000
    2     45.000    89.000    43.000    59.000
    3     49.000    73.000    51.000    57.667
    4     66.000    84.000    34.000    61.333
    5     53.000    66.000    54.000    57.667
    6     41.000    85.000    43.000    56.333
    7     58.000    73.000    42.000    57.667
    8     56.000    71.000    41.000    56.000
    9     60.000    78.000    37.000    58.333
   10     63.000    65.000    44.000    57.333
  ALL     54.200    75.900    42.800    57.633

 CELL CONTENTS --
         Elastic:MEAN
```

at random from the remaining two, and was mixed with additive *A*. The third portion was mixed with additive *B*. In this way, the experimenters hoped to balance out any variations in the plastic from batch to batch. Elasticity was measured on a scale from 0 to 100, with the higher numbers representing greater elasticity. (Higher elasticity is considered more desirable.)

Exhibit 10.9 enters the data. The first SET puts elasticity into C1. A second column, C2, was used for batch number. We used the patterned data feature of SET to enter the numbers 1 through 10 three times. A third col-

Exhibit 10.10   Analysis of Variance for Billiard Ball Data

```
MTB > anova 'Elastic' = 'Batch' 'Additive'

Factor      Type Levels Values
Batch       fixed     10    1    2    3    4    5    6    7    8    9   10
Additive    fixed      3    1    2    3


Analysis of Variance for Elastic

Source      DF         SS         MS        F        P
Batch        9      82.30       9.14     0.12    0.999
Additive     2    5654.87    2827.43    36.62    0.000
Error       18    1389.80      77.21
Total       29    7126.97
```

umn, C3, was used for additive; 0 = no additive, 1 = additive A, and 2 = additive B. Then we produced a table of means. Since there is just one number in each cell, the cell means are the data. The column means gives us the average for each treatment. Additive A seems to have improved elasticity over the control, but additive B seems to have reduced it.

Now, just as we did in one-way analysis of variance, we can express the total variation in the data as the sum of the variation from several sources:

$$(\text{Total variation in data}) = (\text{variation due to batch})$$
$$+ (\text{variation due to additive})$$
$$+ (\text{variation due to random error})$$

If the variation due to additive is much greater than the variation due to random error, we will have statistically significant evidence of a difference among the three levels of additive.

Minitab's ANOVA command in Exhibit 10.10 gives the breakdown of the total variation. Thus (SS Total) = (SS batch) + (SS additive) + (SS Error). As in ONEWAY, each sum of squares has a certain number of degrees of freedom associated with it, and these also add up. Each mean square is the corresponding sum of squares divided by its degrees of freedom. The value of F = (MS for the factor)/(MS Error). And the value of p gives the statistical significance of the corresponding F-statistic. The p-value for additive is small, 0.000. Thus, there is evidence that additive matters.

ANOVA C = C C

ANOVA analyzes a randomized block design (as well as many other models). The level numbers must be integers from -10000 to $+10000$. The equal sign is required. Put the response before the equal sign; the block and treatment factors, after the equal sign. No extra text may be used on the ANOVA command except after a #. .

Stat > ANOVA > Balanced Anova

The *p*-value for batch is not small. In general, we really don't care whether the blocking variable is significant. We do blocking to reduce variation so that we are more likely to detect differences in treatment. The fact that batch is not significant tells us we probably could have run this experiment as a completely randomized one-way analysis and still discovered that additive is statistically significant.

Formulas for the Sums of Squares. In the table below, a is the number of treatments, b is the number of blocks, x_{ij} is the observation in block i given treatment j, $\bar{x}_{i.}$ is the mean of all r observations in block i, $\bar{x}_{.j}$ is the mean of all b observations given treatment j, and $\bar{x}$ is the mean of all ab observations.

Source	DF	SS
Blocks	$b-1$	$a\Sigma_i\,(\bar{x}_{i.}-\bar{x})^2$
Treatments	$a-1$	$b\Sigma_j\,(\bar{x}_{.j}-\bar{x})^2$
Error	$(b-1)(a-1)$	$\Sigma_i\,\Sigma_j\,(x_{ij}-\bar{x}_{i.}-\bar{x}_{.j}+\bar{x})^2$
Total	$ba-1$	$\Sigma_i\,\Sigma_j\,(x_{ij}-\bar{x})^2$

Exercises

10-11　An experiment was performed at the University of Wisconsin to compare the yield of six varieties of alfalfa. Four fields were used. Each field was divided into six plots, one plot for each variety. The four fields can be considered to be four blocks. They are analogous to the 10 batches in the billiard ball experiment. The total yield from each combination was recorded. The data are stored in the file ALFALFA.

Use appropriate displays and tests to analyze the data.

Alfalfa Variety	Field			
	1	*2*	*3*	*4*
Atlantic	3.22	3.31	3.26	3.25
Buffalo	3.04	2.99	3.27	3.20
Culver	3.06	3.17	2.93	3.09
Lohontar	2.64	2.75	2.59	2.62
Narragansett	3.19	3.40	3.11	3.23
Rambler	2.49	2.37	2.38	2.37

10-12   An experiment was done to compare different methods of freezing meat loaf. Meat loaf was to be baked, then frozen for a time, and finally compared by expert tasters. Eight loaves could be baked in the oven at one time. However, some parts of an oven are usually hotter or differ in some other important way from other parts. If this is the case, loaves baked in one part of the oven might taste better than those baked in another part, and differences in freezing methods might be masked. Therefore, a preliminary test was conducted to see whether there were any noticeable differences among the eight oven positions used in the study.

The data shown below came from this preliminary experiment. One batch of eight loaves was mixed; one loaf was assigned at random to each oven position. The loaves then were baked and analyzed. A second batch of eight loaves was mixed, assigned at random to the eight oven positions, baked, and analyzed. A third batch was tested in the same manner. Each batch is a block, and there are eight treatments (oven positions) within each block. Each loaf was analyzed by measuring the percentage of drip loss (that is, the amount of liquid that dripped out of the meat loaf during cooking divided by the original weight of the loaf).

Analyze the data, using appropriate tests and displays. Make sure you take into account all the information you are given about this experiment.

Oven Position	Batch		
	1	*2*	*3*
1	7.3300	8.1100	8.0600
2	3.2200	3.7200	4.2800
3	3.2800	5.1100	4.5600
4	6.4400	5.7800	8.6100
5	3.8300	6.5000	7.7200
6	3.2800	5.1100	5.5600
7	5.0600	5.1100	7.8300
8	4.4400	4.2800	6.3300

Oven Position on the Shelf

5	6	7	8
4	3	2	1

Front of Shelf

Note: Thermometers to measure temperature were inserted into loaves 1, 4, 5, 7, and 8.

10.4 Residuals and Fitted Values

Analysis of variance can be viewed in terms of fitting a model to the data. This leads to a fitted value and a residual for each observation. A fitted value is our best estimate of the underlying population mean corresponding to that observation. The residual is the amount by which an observation differs from its fitted value. (The concepts of fitted values and residuals are discussed in more depth in Chapter 11, on regression models.)

One-Way Designs

In one-way analysis of variance, the fitted values are just the group or cell means. A residual is then the difference between an observed value and the corresponding group mean. Consider, for example, the Fabric data. The observations, fitted values, and residuals for the first two laboratories are as follows:

Laboratory 1				*Laboratory 2*		
Obs	*Fit*	*Residual*		*Obs*	*Fit*	*Residual*
2.9	3.34	−0.44		2.7	3.60	−0.90
3.1	3.34	−0.24		3.4	3.60	−0.20
3.1	3.34	−0.24		3.6	3.60	0.00
3.7	3.34	0.36		3.2	3.60	−0.40
3.1	3.34	−0.24		4.0	3.60	0.40
4.2	3.34	0.86		4.1	3.60	0.50
3.7	3.34	0.36		3.8	3.60	0.20
3.9	3.34	0.56		3.8	3.60	0.20
3.1	3.34	−0.24		4.3	3.60	0.70
3.0	3.34	−0.34		3.4	3.60	−0.20
2.9	3.34	−0.44		3.3	3.60	−0.30

It is good practice to make plots of the residuals to check for unequal variances, nonnormality, dependence on other variables, and so on. We often make the following plots:

Plot of the residuals versus the fitted values.
Histogram of residuals.
Plot of the residuals versus the order in which the data were collected.
Plot of the residuals versus other variables (when data on other variables are available).

Two-Way Designs with an Interaction Term

In two-way analysis of variance with an interaction term, the fitted values are the cell means. A residual is then the difference between the observed value and the corresponding cell mean. We will use the Pancake data in Table 10.3 as an example. Exhibit 10.6 gave the cell means. The observations, fitted values, and residuals from the cells in which whey = 0% and 10% are as follows:

	0% Whey			10% Whey		
	Obs	*Fit*	*Residual*	*Obs*	*Fit*	*Residual*
No Supplement	4.4	4.40	0.00	4.6	4.63	−0.03
	4.5	4.40	0.10	4.5	4.63	−0.13
	4.3	4.40	−0.10	4.8	4.63	0.17
Supplement	3.3	3.20	0.10	3.8	3.70	0.10
	3.2	3.20	0.00	3.7	3.70	0.00
	3.1	3.20	−0.10	3.6	3.70	−0.10

Additive Models and Randomized Block Designs

A two-factor design and a randomized block design are conducted differently, but when you analyze the data in Minitab, you use the ANOVA command for both. Also, if you fit an additive model to the two-factor data, the fitted values and residuals are calculated in exactly the same way.

$$\text{Fitted value for cell } (i, j) = (\text{mean of data in row } i)$$
$$+ (\text{mean of data in column } j)$$
$$- (\text{mean of all data})$$

ANOVA   C = termlist

FITS   **C**
RESIDUALS C

Does a one-way and a two-way analysis of variance with or without interaction, and a randomized block design. You can store the fitted values by
using the subcommand FITS, and the residuals by using the subcommand
RESIDUALS.

Stat > ANOVA > Balanced Anova

Consider the Driving Performance data in Table 10.2. Means are given
in Exhibit 10.4. The fitted value for the first cell, inexperienced drivers on
first-class roads, is $16.000 + 8.750 - 12.917 = 11.833$. The fitted value for
the next cell, inexperienced drivers on second-class roads, is $16.000 +
12.500 - 12.917 = 15.583$.

	First-Class Roads			*Second-Class Roads*		
	Obs	*Fits*	*Residuals*	*Obs*	*Fits*	*Residuals*
	4	11.833	−7.833	23	15.583	7.417
Inexperienced	18	11.833	6.166	15	15.583	−0.583
Drivers	8	11.833	−3.833	21	15.583	5.417
	10	11.833	−1.833	13	15.583	−2.583

11

Correlation and Regression

Some of the most interesting problems in statistics occur when we try to find a model for the relationship among several variables. The data in Table 11.1 illustrate a common situation. Two tests were given to 31 individuals. We often want to answer questions such as the following: What is the correlation between the scores on the two tests? If you know someone's score on the first test, does that help you at all in predicting that person's score on the second test? What is a good prediction of the second score for a person who scored 70 on the first test? In this chapter, we will see how questions like these can be answered.

11.1 Correlation

There are several ways to measure the association between two variables. The most common measure is the Pearson product moment correlation coefficient, or just the *correlation* for short. This is usually designated by the letter r.

Table 11.1 presents some data and Exhibit 11.1 shows a plot and the output from the command CORRELATION. In this example, Minitab printed the result $r = 0.703$.

The correlation coefficient is always between -1 and $+1$. We will denote the two variables by x and y. Here x would be the first score and y the second. The correlation coefficient is positive if y tends to increase as x increases—that is, if a plot of y versus x slopes upward. Conversely, the correlation is negative if y tends to decrease as x increases—that is, if a plot

Table 11.1   Two Scores for 31 People

Subject Number	First Test Score	Second Test Score
1	50	69
2	66	85
3	73	88
4	84	70
5	57	84
6	83	78
7	76	90
8	95	97
9	73	79
10	78	95
11	48	67
12	53	60
13	54	79
14	79	79
15	76	88
16	90	98
17	60	56
18	89	87
19	83	91
20	81	86
21	57	69
22	71	75
23	86	98
24	82	70
25	95	91
26	42	48
27	75	52
28	54	44
29	54	51
30	65	73
31	61	52

Exhibit 11.1   Plot and Correlation of Data in Table 11.1

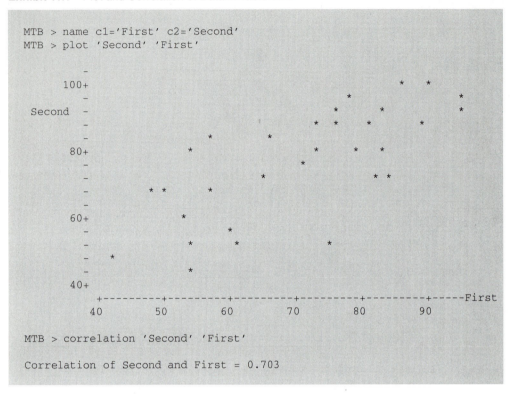

```
MTB > name c1='First' c2='Second'
MTB > plot 'Second' 'First'
```

```
MTB > correlation 'Second' 'First'

Correlation of Second and First = 0.703
```

of y versus x slopes downward. If the points fall exactly on a straight line, then $r = +1$ if the points slope upward and -1 if the trend is downward. The closer the points are to forming a straight line, the closer r is to $+1$ or -1. The closer r is to $+1$ or -1, the easier it is to predict y from x.

If there is almost no association between x and y, then r will be near 0. The converse, however, is not true. There are cases in which r is near 0 but there is still a clear association between x and y. We will see some examples in the exercises. The correlation coefficient measures one type of association—how closely the points fall on a straight line. Fortunately, this is the most common type of association. But a correlation coefficient can be quite misleading at times, and plots may give you a better view of what is really going on.

CORRELATION coefficients for **C...C**

Computes the correlation between all pairs of columns listed in the commands. Minitab calculates the usual Pearson product moment correlation coefficient, as given by

$$r = \frac{\Sigma(x - \bar{x})(y - \bar{y})}{\sqrt{\Sigma(x - \bar{x})^2 \, \Sigma(y - \bar{y})^2}}$$

For example,

```
CORRELATION C1 C2 C3
```

calculates three correlations: the correlation between C1 and C2, that between C1 and C3, and that between C2 and C3.

Stat > Basic Stat > Correlation

Exercises

11-1 Listed below are four variables:

X1	Y1	X2	Y2
1	4	1	1
2	5	2	4
3	6	3	7
4	7	4	7
5	8	5	4
6	2	6	1

(a) Calculate the correlation coefficient between X1 and Y1. Do you think there is any association between these two variables? Plot Y1 versus X1. Now what do you think?

(b) Repeat part (a) for the variables X2 and Y2.

11-2 There are three variables in the Trees data (described in Appendix A): diameter, height, and volume. Plot volume versus height, volume versus diameter, and height versus diameter. Find the correlation associated with each of the three plots. Which relationship seems strongest? Which variable would probably be the best predictor of volume? Is there any evidence to make you doubt the reasonableness of the *linear* (straight-line) correlation coefficients you computed?

11-3   Listed below are the price and number of pages for each of 15 books that
were reviewed in the February 1982 issue of the journal *Technometrics*.
Make a plot of price versus book length. Compute the correlation between
book length and price.

Pages	Price ($)
302	30
425	24
526	35
532	42
145	25
556	27
426	64
359	59
465	55
246	25
143	15
557	29
372	30
320	25
178	26

11-4   In this exercise, we first simulate data under various conditions. We then
plot the data, guess the correlations, and check our guesses.

(a)   Use RANDOM 50 C1 C2 to simulate two independent random normal samples into columns C1 and C2. Make a plot of C1 versus C2
and guess the correlation coefficient. Then compute the correlation
coefficient and check your guess. Repeat this process four more
times.

(b)   A trick allows us to simulate correlated samples. Repeat (a), but this
time use the commands below. Plot C1 versus C3, guess the correlation, and check it. Repeat four more times.

```
RANDOM 50 C1 C2
LET C3 = C1 + C2
```

(c)   Repeat (b), but this time use

```
RANDOM 50 C1 C2
LET C3 = C2 - C1
```

(d)   Repeat (b), but use

```
RANDOM 50 C1 C2
LET C3 = 5*C1 + C2
```

(e)   Repeat (b), but use

```
RANDOM 50 C1 C2
LET C3 = C2 - 5*C1
```

11-5 (a)   Exhibit 4.1 is a plot of gas consumption per hour versus the outside
            temperature. Guess the correlation of these data. Compute the corre-
            lation coefficient and check your guess.
      (b)   Repeat (a) for the plot of cartoon scores versus OTIS scores in Ex-
            hibit 4.2.

11-6 Use SET to put the integers 1 to 50 into C1. Now compute

```
C2 = C1*C1
C3 = SQRT(C1)
C4 = 3*C1
C5 = (C1-25.5)**2
```

In each case, there is an exact relationship with C1. Plot C2 versus C1, C3
versus C1, and so on. Look at the relationships. Now compute the correlations among C1–C5. Explain why the correlation coefficients are not all 1.

11.2 Simple Regression: Fitting a Straight Line

Correlation tells us how much association there is between two variables;
regression goes farther. It gives us an equation that uses one variable to help
explain the variation in another variable. In this section we will show how
to use the REGRESS command to fit straight lines.

Look at the data in Table 11.1 again. Suppose we wanted an equation
that relates the second test scores to those from the first. From Exhibit 11.1
we see that if a person scored 70 on the first test, we might expect a score
of about 75 on the second test. If another person scored 90 on the first test,
we might expect about a 90 on the second test. If still another scored 55 on
the first test, we might expect a second score of about 65.

In fact, it looks as if we might be able to draw a straight line on the plot
and use it to predict scores on the second test. Suppose we decide to do that.
How would we draw the line? Any straight line we draw will miss some of
the points. Exhibit 11.2 gives another copy of this plot, with several lines
drawn. Which one looks best?

A method known as least squares is often used to decide which line to
choose. Suppose we look at the deviation between what a given line
suggests and what actually happened for each person. If the line suggests a second score of 79 and the actual score was 88, then the deviation

Exhibit 11.2   Several Possible Lines for Predicting Second Scores from First Scores

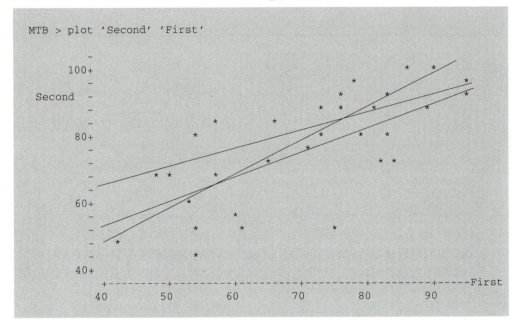

```
MTB > plot 'Second' 'First'
```

is 88 − 79 = 9. Suppose we compute all the deviations for a line, square them, and add them up.

Intuitively, a line with large deviations is not as good as a line with smaller deviations. The least-squares criterion says we should use the line that gives us the smallest sum of squared deviations. The surprising thing is that it is relatively easy to find the equation for that line.

If we are trying to use x to determine y, any straight line can be written in the form $y = a + bx$, where a and b are two numbers that tell us which line we are using. For example, $y = 3 + 2x$ and $y = -5 + 0.8x$ specify two different straight lines. For the least-squares line, a and b can be found with the formulas

$$b = \frac{\Sigma(x - \bar{x})(y - \bar{y})}{\Sigma(x - \bar{x})^2} \quad \text{and} \quad a = \bar{y} - b\bar{x}$$

Minitab will give values for a and b, and many other useful quantities, when the REGRESS command is used.

Suppose we want to use REGRESS to relate the second test scores to the first scores. All we need to do, once the data have been put in the worksheet, is type

```
REGRESS 'SECOND' 1 'FIRST'
```

In the REGRESS command, you must type the number of predictors before you type the predictor variables. Here there is just one predictor, the score on the first test, so we type the number 1.

Exhibit 11.3 shows the full output that may be obtained from RE-GRESS. Ordinarily an abbreviated version is given; we use the full output here to facilitate our explanation. To get the full output, type the command BRIEF 3 (p. 310). Then all REGRESS commands that follow will give the full output.

Look at some of the simpler parts of this output. The regression equation is Second = 22.5 + 0.755 First, or $y = 22.5 + 0.755x$. This equation corresponds to one of the lines in Exhibit 11.2. (Can you determine which one?) Since this is the least-squares line, it is impossible to find any straight line that gives a smaller sum of squared deviations than this one. The sum of squared deviations for this line is 3844. It is printed in the Analysis of Variance table in Exhibit 11.3 as SS Error. SS is an abbreviation for *sum of squares* and Error is another name for *deviation*.

The next block of output again gives the values of $a = 22.47$ and $b = 0.7546$, with some additional information. Next comes the Analysis of Variance table, which gives, as we will see later, a breakdown of the variation in the data.

The last block of output lists each x value and y value, then gives the fitted y value from the equation. For example, the first person scored 50 on the first test and 69 on the second test. The fitted score was 60.20, which was computed by calculating $22.47 + 0.7546(50)$. The corresponding deviation or residual is thus $y - (\text{fitted } y) = 69 - 60.2$, which gives 8.8.

R-Squared (Coefficient of Determination)

The output also gives R-sq = 49.4%. Whenever a straight line is fit to a set of data, R^2 is just the square of the ordinary correlation coefficient we discussed in Section 11.1. There we found $r = .703$, which, when squared, gives .494, or 49.4%, the same value given here in the REGRESS output.

R^2 also has two other, more general, interpretations. It is the square of the correlation between the observed y values and the fitted y values. It also is the fraction of the variation in y that is explained by the fitted equation. Look at the Analysis of Variance table in Exhibit 11.3. The total sum of

Exhibit 11.3   REGRESS Output for Predicting the Second Score from the First Score

```
MTB > brief 3
MTB > regress 'Second' 1 'First'

The regression equation is
Second = 22.5 + 0.755 First

Predictor        Coef        Stdev      t-ratio          p
Constant        22.47        10.22         2.20       0.036
First          0.7546       0.1417         5.32       0.000

s = 11.51        R-sq = 49.4%      R-sq(adj) = 47.7%

Analysis of Variance

SOURCE          DF           SS          MS          F          p
Regression       1        3757.4      3757.4      28.35      0.000
Error           29        3844.0       132.6
Total           30        7601.4

Obs.    First      Second      Fit   Stdev.Fit    Residual    St.Resid
  1     50.0       69.00      60.20       3.58        8.80        0.80
  2     66.0       85.00      72.27       2.17       12.73        1.13
  3     73.0       88.00      77.55       2.09       10.45        0.92
  4     84.0       70.00      85.85       2.80      -15.85       -1.42
  5     57.0       84.00      65.48       2.83       18.52        1.66
  6     83.0       78.00      85.10       2.71       -7.10       -0.63
  7     76.0       90.00      79.81       2.20       10.19        0.90
  8     95.0       97.00      94.15       4.02        2.85        0.26
  9     73.0       79.00      77.55       2.09        1.45        0.13
 10     78.0       95.00      81.32       2.32       13.68        1.21
 11     48.0       67.00      58.69       3.82        8.31        0.77
 12     53.0       60.00      62.46       3.24       -2.46       -0.22
 13     54.0       79.00      63.21       3.14       15.79        1.43
 14     79.0       79.00      82.08       2.38       -3.08       -0.27
 15     76.0       88.00      79.81       2.20        8.19        0.72
 16     90.0       98.00      90.38       3.44        7.62        0.69
 17     60.0       56.00      67.74       2.56      -11.74       -1.05
 18     89.0       87.00      89.62       3.32       -2.62       -0.24
 19     83.0       91.00      85.10       2.71        5.90        0.53
 20     81.0       86.00      83.59       2.54        2.41        0.21
 21     57.0       69.00      65.48       2.83        3.52        0.32
 22     71.0       75.00      76.04       2.07       -1.04       -0.09
 23     86.0       98.00      87.36       3.00       10.64        0.96
 24     82.0       70.00      84.34       2.62      -14.34       -1.28
 25     95.0       91.00      94.15       4.02       -3.15       -0.29
 26     42.0       48.00      54.16       4.56       -6.16       -0.58
 27     75.0       52.00      79.06       2.16      -27.06       -2.39R
 28     54.0       44.00      63.21       3.14      -19.21       -1.73
 29     54.0       51.00      63.21       3.14      -12.21       -1.10
 30     65.0       73.00      71.51       2.22        1.49        0.13
 31     61.0       52.00      68.50       2.48      -16.50       -1.47

R denotes an obs. with a large st. resid.
```

squared deviations, written Total SS, is a measure of the variation of y about its mean. Here it is 7601. The Regression SS is the amount of this variation that is explained by the regression line. Here it is 3757. The fraction of variation explained is 3757/7601, which is again .494. It is sometimes more convenient to convert this to a percentage and to say the regression equation explains 49.4% of the variation in y.

REGRESS C on 1 predictor in C

Computes the regression equation. Output includes the equation and other information. More details about REGRESS are given later in this chapter.

Stat > Regression > Regression

Exercises

11-7   Refer to the output from REGRESS in Exhibit 11.3.
  (a)   Based on the fitted equation, if a person scored 54 on the first test, what score would you predict for the second test?
  (b)   How many individuals got 54 on the first test? What did each of them get on the second test? Find the residual (deviation) for each.
  (c)   If a person got 100 on the first test, what score would you predict on the second test? (Always check your answers for reasonableness.)

11-8   The following are the mean Scholastic Aptitude Test scores (SAT scores) for the years 1967–1981.

Year	Verbal	Math		Year	Verbal	Math
1967	466	492		1975	434	472
1968	466	492		1976	431	472
1969	463	493		1977	429	470
1970	460	488		1978	429	468
1971	455	488		1979	427	467
1972	453	484		1980	424	466
1973	445	481		1981	424	466
1974	444	480				

(a)   Plot the verbal SAT scores versus year. Repeat for math scores.

(b)   Fit a regression line using year to predict the verbal SAT scores.

(c)   Fit a regression line using year to predict the math SAT scores.

(d)   If you have a copy of the plot on paper, draw in the two regression lines (by hand).

(e)   Do both verbal and math scores seem to be changing at about the same rate (same number of points per year)?

(f)   What is the predicted average SAT score for math in 1970? In 1985? In 1990? In 2000? Which of these seem to make sense? Repeat for verbal scores.

(g)   Here are the actual data for some of the years since 1981. What do you think of your predictions? It is generally considered risky to forecast much beyond the data you have. Was that a reasonable concern here?

Year	Verbal	Math
1984	426	471
1985	431	475
1987	430	476
1988	428	476
1989	427	476
1990	424	476
1991	422	474
1992	423	476

11-9   Given below are the winning times in seconds for the men's 1500-meter run in the Olympics from 1900 to 1992. (*Note:* No Olympics were held in 1916, 1940, and 1944 because of wars; in 1980 the United States and some other countries did not participate.)

Year	Time	Year	Time	Year	Time
1900	246.0	1932	231.2	1968	214.9
1904	245.4	1936	227.8	1972	216.3
1908	243.4	1948	229.8	1976	219.2
1912	236.8	1952	225.2	1980	218.4
1920	241.8	1956	221.2	1984	212.5
1924	233.6	1960	215.6	1988	216.0
1928	233.2	1964	218.1	1992	220.1

(a) Plot winning time versus year. Does winning time seem to be changing over the years? Is it changing according to a straight line?

(b) Fit a regression line for predicting winning time from year. If you have a plot of the data on paper, draw the regression line on the plot.

(c) On the average, how much has the winning time decreased in the four years between Olympics? Use the fitted equation.

(d) Are there any appreciable departures from the straight line that you might have anticipated from outside information? Explain.

11-10 The following data were collected to study the relationship between the temperature of a battery and its output voltage. The eight measurements were taken in the order shown.

	Reading							
	1	2	3	4	5	6	7	8
Temperature (Celsius)	10.0	10.0	23.1	23.1	34.0	34.0	45.6	45.6
Voltage	4290	4270	4470	4485	4723	4731	4920	4935

(a) Plot the data and use REGRESS to fit a straight line to estimate voltage as a function of temperature. Does the regression line seem to fit the data well?

(b) Do you spot any weakness in the order in which the readings were taken? Can you think of some better orders in which to make the eight readings if you had to do this experiment again? Discuss.

11-11 (a) How well can you predict the volume of a tree from its diameter? Use the Trees data (described in Appendix A) to find an equation for black cherry trees in Pennsylvania.

(b) Plot the data (volume versus diameter) and, if the plot is on paper, draw your regression line on it. How well does the line seem to fit the data? Do you see any problems?

(c) Repeat (a) and (b), but use height as the predictor of volume.

(d) Which is the better predictor, height or diameter? Which equation would be easier to use in the woods to predict the volume of a given tree? Why?

11.3 Making Inferences from Straight-Line Fits

Sometimes we will be content to use REGRESS just to fit an equation to the set of data we happen to have. On other occasions we may want to generalize from the data in hand to some larger population. Sometimes our data will be a random sample from some population. In such cases, we often will be more interested in the characteristics of the population than in the characteristics of the sample. At other times, our data will have resulted from a carefully designed and executed experiment. Then we will usually be more interested in the underlying relationship between the variables than in the values we happened to get in this particular experiment.

In other cases, our data will be produced more by happenstance. For example, we may have good estimates of the automobile fatality rate and of the average highway speeds of vehicles in each of the 50 states. Sometimes we may want to pretend, for a moment, that such data are a random sample from some population. We will know that in reality there are only 50 states in all, but pretending our 50 are a random sample allows us to use the procedures of statistical inference for general guidance. Then we might seek to answer questions such as "Is the relationship we observe between fatalities and speed one that could reasonably have occurred due to chance alone?"

Conditions for Inference

There are several conditions that must be met, at least approximately, before we can make reasonable statistical inferences:

1. In the underlying population, the relationship between x and y should be a straight line. Suppose that for each value of x we find the mean of all the corresponding y values. Then these means must, at least approximately, fall on a straight line. We will denote this straight line by $A + Bx$, where A and B are some fixed values. For example, suppose we could calculate the mean of all the ys in the population corresponding to $x = 1$, the mean of all the ys corresponding to $x = 2$, the mean of all the ys corresponding to $x = 3$, and so on. Then all these means should fall on a straight line. We will call this line the *population regression line*. The values a and b, discussed in Section 11.2, are our sample estimates of the population values A and B.

2. For each x, the amount of variation in the population of ys should be approximately the same. This variance is usually called the variance of y about the regression line, and is denoted by σ^2. Correspondingly, σ is called the standard deviation of y about the regression line.

3. For each value of x, the distribution of ys in the population should be approximately normal.

4. The y values that are actually obtained should be approximately independent. In other words, the amount by which a particular y value

differs from its mean should not be related to the amount by which any other y value differs from its mean.

In Sections 11.6 and 11.7, we will discuss some procedures for checking whether these conditions hold. Condition 4 is the most important and the most difficult to check. The best advice when doing surveys or experiments is to try to do them properly in the first place. For example, try to make sure that no observations are unduly related to one another. Whenever practical, use randomization to determine the order in which measurements are made. Try to avoid biases or systematic errors.

Interpreting the Output

The following all require that the conditions for inference be at least approximately satisfied:

Estimate of σ. We begin by looking at s in Exhibit 11.3. This often is called the standard deviation of y about the regression line, or the standard error of estimate. This quantity gives us an estimate of σ. It is computed by using the formula

$$s = \sqrt{\frac{\Sigma(y - \text{fitted } y)^2}{n - 2}}$$

The value s can be thought of as a measure of how much the observed y values differ from the corresponding average y values as given by the least-squares line. It has $(n - 2)$ degrees of freedom and is used in all the formulas for standard deviations. All t-tests and confidence intervals will be based on this s, and thus all will have $(n - 2)$ degrees of freedom. In our example, the number of degrees of freedom is $(31 - 2) = 29$.

Standard Deviation of Coefficients. Under the conditions for inference, the estimated coefficients a and b each have an approximately normal distribution. The estimated standard deviations of these coefficients are given in the column headed "Stdev" in Exhibit 11.3. The estimated standard deviation of a is 10.22, and the estimated standard deviation of b is 0.1417.

Confidence Intervals. The general formula for a t confidence interval is

(quantity) $\pm$ (value from t-table) $\times$ (estimated stdev of quantity)

The t-value corresponding to 29 degrees of freedom and 95% confidence is 2.045. Thus, a 95% confidence interval for A (the population value of a) is

$$a \pm t(\text{stdev of } a)$$

or

$$22.47 \pm 2.045(10.22)$$

This gives the interval from 1.57 to 43.37. A 95% confidence interval for B (the population value of b) is similarly given by

$$.7546 \pm 2.045(.1417)$$

This gives the interval from .46 to 1.04.

Tests of Significance. We often want to know if there is any statistically significant evidence of an association between x and y. Thus, a hypothesis we frequently want to test is that B is 0. We use the general formula

$$t = \frac{b - (\text{hypothesized value})}{(\text{estimated stdev of } b)}$$

Here we find that

$$t = \frac{7546 - 0}{.1417} = 5.33$$

This is given in the column headed "t-ratio." With 29 degrees of freedom, this value of t is highly significant, giving us evidence that B is probably not 0. This in turn implies that the score on the first test is at least slightly useful as a predictor of the score on the second test. (*Note:* This t-test is equivalent to testing whether the population correlation coefficient is 0.)

A similar t-test of the null hypothesis that $A = 0$ gives

$$t = \frac{22.47 - 0}{10.22} = 2.20$$

This is also statistically significant, although just barely. Thus, from a statistical standpoint, both a and b have been shown to be useful in the equation.

Standard Deviation of a Fitted y Value. The estimated standard deviations of the fitted y values are given in the column headed "Stdev.Fit" in Exhibit 11.3. These can be used to get confidence intervals for the population mean of all y values corresponding to a given value of x. For example, the first line shows that the fitted value for $x = 50$ is 60.20. The 95% confidence interval for the mean of all ys corresponding to $x = 50$ is given by

$$(\text{fit}) \pm t(\text{stdev of fit})$$

or
$$60.2 \pm 2.045(3.58)$$

This gives the interval 52.9 to 67.6. Thus, we can be 95% confident that for all persons who scored 50 on the first test, the mean score on the second test is between 52.9 and 67.6.

This tells us how well persons with a first test score of 50 will do on the average. But how about one particular individual who scores 50 on the first test?

Prediction Interval for a Single y Value. Since individuals are never as predictable as averages, we must expect more uncertainty in this prediction. The following calculations give us an interval that we can be 95% confident will contain the second test score of an *individual* who gets a 50 on the first test:

$$60.2 \pm 2.045\sqrt{(3.58)^2 + (11.51)^2}$$

This gives the interval 35.5 to 84.9. The 3.58 is, again, the estimated standard deviation of the fitted value, and $s = 11.51$ is the estimated standard deviation of an individual. Note that the interval for an individual is, as expected, larger than the interval for the average.

Most texts give the following, slightly different-looking, formula for a prediction interval corresponding to a given x, which we denote by x_0.

$$\text{(fitted } y) \pm ts \sqrt{\frac{1}{n} + \frac{(x_0 - \bar{x})^2}{\Sigma(x - \bar{x})^2} + 1}$$

We can rewrite this formula as

$$\text{(fitted } y) \pm t \sqrt{\text{(estimated stdev of fitted } y)^2 + s^2}$$

This is precisely the formula we used in the calculations above.

Predictions for New Values of x. The procedure we just used for $x = 50$ will work for any value of x that was in our set of data, such as $x = 66$, $x = 73$, or $x = 84$. But how about a prediction interval for a value of x that was not in our original set of data? For example, how could we find a prediction interval for $x = 68$? To get the predicted second score is not too difficult. All we have to do is substitute 68 into the regression equation. This gives

$$y = 22.47 + (0.7546)(68) = 73.78$$

REGRESS C on 1 predictor C

PREDICT for **E**

The PREDICT subcommand computes estimates for any values of *x*. It prints out a table of fitted *y*, the standard deviation of fitted *y*, a 95% confidence interval, and a 95% prediction interval.

E may be a constant, such as 68 or K3, or it may be a column containing a list of *x* values. Up to 10 PREDICT subcommands may be used with one REGRESS command.

Stat > Regression > Regression, then use Options button

The easiest way to get the estimated standard deviation of this predicted *y* value is to use the PREDICT subcommand. To use PREDICT for *x* = 68, use

```
REGRESS 'SECOND'  1 'FIRST';
  PREDICT 68.
```

Minitab will then print the following extra lines:

```
 Fit   Stdev.Fit        95% C.I.          95% P.I.
73.78        2.10  ( 69.49,   78.08) ( 49.84,   97.92)
```

Exercises

11-12 Refer to the output in Exhibit 11.3.
  (a) Calculate a 90% confidence interval for B, the slope of the underlying regression line.
  (b) Find a 90% confidence interval for A, the intercept of the underlying regression line.
  (c) Find a 90% confidence interval for the average of the second test scores of all persons in the population who scored 90 on the first test.

11-13 Refer to the output in Exhibit 11.3. Suppose a person scored an 86 on the first test. What score should be expected on the second? Find an interval that you are 95% confident will cover the second score for that individual.

11-14 Refer to the output in Exhibit 11.3. Test the null hypothesis, H_0: $A = .5$ versus the alternative hypothesis, H_1: $A \neq .5$.

11-15 Refer to the test data in Table 11.1.

(a)   Find a 90% confidence interval for the mean second test scores for all persons in the population who obtained a score of 77 on the first test. You will need to use the PREDICT subcommand, because 77 is not in the data set.

(b)   Find a 90% prediction interval for the second test score for an individual who achieved a score of 77 on the first test.

11-16 Refer to Exercise 11-9 for the 1500-meter race. Give 95% confidence intervals for the two regression coefficients.

11-17 (a)   We can simulate data from a regression model as follows. To choose a model, we need to specify three things: A, B, and σ. Suppose we use $A = 3$, $B = 5$, and $\sigma = .5$. Next, we must specify values for x. Suppose we take two observations at each integer from 1 to 10. First use SET to put the 20 values of x into a column. Then use LET to calculate $A + Bx$. Now simulate 20 observations from a normal distribution with $\mu = 0$ and $\sigma = .5$. Add these to $A + Bx$ to get the observed ys. Get a plot and do a regression for these simulated data. Record a, b, s, and R^2.

(b)   Repeat part (a), using $\sigma = 2.00$. Compare the results with those of part (a).

(c)   Repeat part (a), using $\sigma = 10.0$. Compare the results with those of parts (a) and (b).

11-18 Maple trees have winged fruit, called samara, which come spinning to the ground in the fall. A forest scientist was interested in the relationship between the velocity with which the samara fell and their "disk loading." The disk loading is a function of the size and weight of the fruit and is closely related to the aerodynamics of helicopters. Tests were run on samara from three trees. The results are given below and are stored in the worksheet MAPLE.

Tree 1		Tree 2		Tree 3	
Loading	Velocity	Loading	Velocity	Loading	Velocity
.239	1.34	.238	1.20	.192	0.91
.208	1.06	.206	1.06	.200	1.13
.223	1.14	.172	0.88	.175	1.00
.224	1.13	.235	1.24	.187	0.98
.246	1.35	.247	1.37	.181	0.96
.213	1.23	.239	1.37	.195	0.88
.198	1.23	.233	1.43	.155	0.81
.219	1.15	.234	1.32	.179	0.91
.241	1.25	.189	0.99	.184	1.00
.210	1.24	.192	1.00	.177	0.87
.224	1.34	.209	1.12	.177	1.02
.269	1.35			.186	0.94

(a)   Does velocity seem to be a straight line function of loading? Examine separately for each tree.

(b)   A scientist hypothesizes that the straight lines will go through the origin (the point $x = 0$, $y = 0$). Do they seem to do this, at least approximately?

(c)   Test $H_0: A = 0$ and compare the result to your answer in (b).

(d)   Is there any difference in the relationship between velocity and loading for the three different trees?

11.4 Multiple Regression

So far we have described how one variable can be used to help explain the variation in another—for example, how the score on one test relates to that on another test. But what if you have two or three, or even more variables that could help with the explanation? One technique you can use, and the one we will describe in this section, is multiple regression. We begin with an example.

Many universities use multiple regression to estimate how well the various applicants would do if they were admitted. An equation used at one major university was

(Freshman GPA) = .61813(HS GPA) + .00137(SAT verbal)
                    + .00063(SAT math) − .19787

This equation shows that the estimated grade point average (GPA) at the end of the freshman year is equal to .61813 times the high school grade point average, plus .00137 times the Scholastic Aptitude Test verbal score, plus .00063 times the Scholastic Aptitude Test mathematics score, minus .19787. This equation was an important criterion in deciding whom to admit to that university. The variables HS GPA, SAT verbal, and SAT math are often called *predictor variables*. Here they are used to predict Freshman GPA.

This equation was obtained by using multiple regression on the records of previous students. The university had the freshman-year GPAs for some past students, as well as the high school GPA and the two SAT scores for each. They asked, "Which equation best explains freshman GPA from these other variables?" The procedure they used is very similar to the one we used in Sections 11.2 and 11.3 for straight-line equations.

The Grades example (in Appendix A) gives some data from another university. Suppose we wanted to develop a similar equation for freshman GPAs at that university, using just the SAT verbal and SAT math scores. We used Minitab to find an equation based on a sample of 100 freshmen (sample *A* from Grades). The results are shown in Exhibit 11.4. Notice that you must type the number of predictor variables in the REGRESS command. Exhibit 11.4 gives the default output from REGRESS: Only those observations which are "unusual" because of their x values (marked by X on the output) or because of their residuals (marked by R on the output) are printed. If you want all 100 observations printed, type BRIEF 3 before you type REGRESS.

Interpreting the Output

The equation for Freshman GPA is

$$(\text{fitted GPA}) = .471 + (.00356)(\text{VERBAL}) + (.000158)(\text{MATH})$$

We can use this equation to estimate how well a student will do who scored 500 on both SATs. We compute

$$(\text{fitted GPA}) = .471 + (.00356)(500) + (.000158)(500)$$
$$= .471 + 1.78 + .08$$
$$= 2.33$$

Thus we forecast that the student's GPA will be 2.33.

For a student who scored 500 on the verbal test and 800 on the math test, we would predict

$$(\text{Estimated GPA}) = .471 + (.00356)(500) + (.000158)(800)$$
$$= 2.38$$

Exhibit 11.4   Multiple Regression Output for Predicting Freshman GPA from SAT Scores

```
MTB > regress 'gpa' 2 'verbal' 'math'

The regression equation is
GPA = 0.471 + 0.00356 VERBAL +0.000158 MATH

Predictor         Coef        Stdev      t-ratio          p
Constant        0.4706       0.5433         0.87      0.388
VERBAL       0.0035628    0.0007350         4.85      0.000
MATH         0.0001576    0.0008514         0.19      0.854

s = 0.5018      R-sq = 23.5%      R-sq(adj) = 22.0%

Analysis of Variance

SOURCE          DF           SS          MS          F          p
Regression       2        7.5137      3.7568      14.92      0.000
Error           97       24.4227      0.2518
Total           99       31.9364

SOURCE          DF       SEQ SS
VERBAL           1       7.5051
MATH             1       0.0086

Unusual Observations
Obs.    VERBAL         GPA       Fit   Stdev.Fit    Residual     St.Resid
  2        454      2.3000    2.1624      0.1544      0.1376        0.29 X
 40        490      1.2000    2.3269      0.1149     -1.1269       -2.31R
 54        592      2.4000    2.6493      0.1863     -0.2493       -0.54 X
 89        361      2.4000    1.8517      0.1682      0.5483        1.16 X

R denotes an obs. with a large st. resid.
X denotes an obs. whose X value gives it large influence.
```

Thus a student with scores of 500 and 500 and another student with scores of 500 and 800 have estimated GPAs that differ by only .05. This seems to indicate that SAT math scores are not very useful estimates of GPA at that university.

How good is this estimation equation as a whole? Put another way, how much might the GPAs for individual students vary from what we predict? One way to answer this question is to look at the value of R^2. It is 23.5%, which means that our equation explains only 23.5% of the variation in GPAs. The remaining 76.5% of the variation in freshman GPAs is left unexplained.

From these results, it appears that SAT scores have only limited value in forecasting who will succeed in college. We can give several possible reasons for this. First, we do not have results for a random sample of all students. All we have is students who applied to, were admitted to, and attended that university. Second, perhaps those students with low SAT scores were advised to take easier courses and thus received higher grades than they ordinarily would have, whereas those with higher SAT scores were encouraged to take more difficult courses. Third, it is possible that tests such as the Scholastic Aptitude Test simply do not do a very good job of measuring whatever it takes to get good grades in college.

Notation and Assumptions in Multiple Regression

Suppose there are two explanatory variables—call them x_1 and x_2. We assume the following:

1. The underlying population regression line is approximately $y = B_0 + B_1x_1 + B_2x_2$.
2. For all values of x_1 and x_2, the ys have approximately the same variance, σ^2.
3. For each x_1 and x_2, the ys have approximately normal distributions.
4. The ys are approximately independent.
5. We use b_0, b_1, b_2, and s for the estimated values of B_0, B_1, B_2, and σ, respectively.

Confidence Intervals and Tests

All the confidence intervals we calculated for straight lines in Section 11.3 can be calculated here in essentially the same way. For example, in Exhibit 11.4, the estimated standard deviation of b_2 is .0008514. Since there are 97 degrees of freedom, a 95% confidence interval for B_2 is .000158 ± (1.99)(.0008514), which gives the interval −.00154 to .00185.

We can also do t-tests on the coefficients just as we did in simple regression. For example, to test H_0: $B_2 = .005$, we form the t-ratio $t = (b_2 - .005)/(\text{estimated stdev of } b_2)$. This ratio has a t-distribution with 97 degrees of freedom.

The t-ratio for the hypothesis that $B_2 = 0$ is given on the output, and is just .19. Thus, b_2 was not statistically different from zero. This test indicates that SAT math scores were not statistically significant in explaining freshman GPAs in this sample of students.

Suppose we want a 95% confidence interval for the population mean value of y corresponding to an SAT verbal score of 454 and an SAT math score of 471. This pair of scores happens to be in the data we used in the regression; it is the second observation and it is shown in Exhibit 11.4. The fitted GPA is 2.1624, and the Stdev.Fit is .1544. This means that $2.1624 \pm (1.99)(.1544)$, or 1.86 to 2.47, gives a 95% confidence interval for the mean GPA of all students in the population who had an SAT verbal score of 454 and an SAT math score of 471.

For a new *individual* with these same test scores, the prediction interval can be computed, as on page 287, to be

$$2.1624 \pm (1.99)\sqrt{(.5018)^2 + (.1544)^2}$$

This gives the interval 1.66 to 2.67, emphasizing again our lack of ability to make precise forecasts of freshman GPAs.

Suppose we want to compute an estimate for scores that did not happen to be in the original set of data, say 600 and 750. We then can calculate

$$
\begin{aligned}
(\text{Pred GPA}) &= .471 + (.00356)(600) + (.000158)(750) \\
&= 2.73
\end{aligned}
$$

To obtain a confidence interval or prediction interval for values that are not in the original data set, we use the PREDICT subcommand; for example,

---

REGRESS C on K predictors C...C

PREDICT for E...E

Calculates and prints a multiple regression equation.

The PREDICT subcommand tells Minitab to compute predicted values, standard deviations, 95% confidence intervals, and 95% prediction intervals. The arguments on PREDICT can be columns or constants. Columns must all be of the same length. There should be as many arguments on the PREDICT subcommand as there are predictors on the REGRESS command. You may use up to ten PREDICT subcommands in one REGRESS command. Here is an example:

```
REGRESS C8 on 4 C1-C4;
  PREDICT 16, 20, 4, 30.
  PREDICT C11-C14;
  PREDICT C11 20 C13 30.
```

Stat > Regression > Regression, use the Options button to specify predicted values

```
REGRESS 'GPA' 2 'VERBAL' 'MATH';
  PREDICT 600 750.
```

This gives the following output:

```
   Fit   Stdev.Fit          95% C.I.            95% P.I.
 2.7265     0.0954    ( 2.5371,  2.9159)  ( 1.7125,  3.7405)
```

We see that a student with an SAT verbal score of 600 and an SAT math score of 750 has an expected freshman GPA of 2.73, as we calculated. In addition, we are 95% confident that the average GPA of all such freshmen at this university is between 2.5371 and 2.9159. The 95% prediction interval for an individual freshman is 1.7125 to 3.7405.

Exercises

11-19 Refer to the output in Exhibit 11.4.
(a) Get a 95% confidence interval for B_1, the coefficient of SAT verbal score.
(b) Is B_1 significantly different from zero (use $a = .05$)? What does this say about the relationship between SAT verbal score and GPA?
(c) Is your answer to (b) consistent with the low value of R^2? Explain.

11-20 (a) Use the data in sample B from the Grades data (described in Appendix A) to develop another equation for predicting GPA from SAT scores.
(b) How does this equation compare to the equation in Exhibit 11.4? Compare the estimates of B_0, B_1, B_2, σ, and R^2 for the two equations.

11-21 In Exercise 11-11, we fitted an equation for estimating volume of a black cherry tree from its diameter. Suppose we use height as a second explanatory variable.
(a) Find an equation for estimating volume from diameter and height. How much extra help does height seem to give you when you are predicting volume?
(b) Use REGRESS and PREDICT to calculate a 95% confidence interval for estimating the average volume of trees with diameter = 11 and height = 70.

11-22 In this exercise, we will fit a model to predict systolic blood pressure using the Peru data set. Read the description of this data set in Appendix A.
(a) Regress systolic blood pressure on years since migration. What is the relationship?
(b) Add in a second predictor, weight. How does this model compare to the one in part (a)?

(c) What do the results in parts (a) and (b) tell you about fitting models to data?

(d) Try adding a third predictor to the model in part (b). Can you find one that improves the model?

11.5 Fitting Polynomials

So far, we have talked about fitting straight lines (Sections 11.2 and 11.3) and about fitting equations with several variables (Section 11.4). Now we will show how to fit curved data such as those shown in Exhibit 11.5. In this section, we will describe how to use polynomial models. In Section 11.6 we will show how transformations give another, often preferable, method of analysis.

Polynomials are equations that involve powers of the x variable. The statistical model for a second-degree polynomial, or quadratic equation, is

$$y = B_0 + B_1x + B_2x^2$$

An Example

The data in Exhibit 11.5 were gathered in conjunction with an environmental impact study to find the relationship between stream depth and flow rate. Flow rate is the total amount of water that flows past a given point in a fixed amount of time. Data were collected on seven different streams. The data in Exhibit 11.5 are all from the same site on one stream. We are interested in estimating the flow rate from the depth of the stream.

The plot shows a gap in the data. There are no observations of stream depth between .5 and .7. The relationship between flow and depth would be clearer if there were no gap. But these are all the data we have, so we will do our best with the information on hand. The overall pattern in the plot is certainly not a straight line. There is an upward curve. Let's try a second- degree polynomial.

To fit a second-degree polynomial, all we need to do is compute x^2 and put it in another column. Then we simply regress y on the two variables x and x^2, just as we did in multiple regression. Exhibit 11.6 shows the appropriate commands.

The form of the output is essentially the same as for a straight line. The regression equation is

$$(\text{Flow}) = 1.68 - 10.9(\text{Depth}) + 23.5(\text{Depth})^2$$

The standard deviation about the regression line is .2794.

Exhibit 11.5 Stream Data and Plot

```
MTB > set c1
DATA>    0.34  0.29  0.28  0.42  0.29  0.41  0.76  0.73  0.46  0.40
DATA> end
MTB > set c2
DATA>    0.636 0.319 0.734 1.327 0.487 0.924 7.350 5.890 1.979 1.124
DATA> end
MTB > name c1='Depth'  c2='Flow'
MTB > plot 'Flow' 'Depth'


       -
  7.5+                                                              *
       -
  Flow -
       -
       -                                                        *
  5.0+
       -
       -
       -
       -
  2.5+
       -                                        *
       -                                 *
       -                            **
       -        *2        *
  0.0+
       ------+---------+---------+---------+---------+---------+Depth
           0.30      0.40      0.50      0.60      0.70      0.80
```

We can do a *t*-test of the null hypothesis that, in the population, the coefficient of the (Depth)^2 term is zero. The data give $t = (23.535 - 0)/4.274 = 5.51$. This is statistically significant, giving us evidence that the population coefficient for (Depth)^2 is not zero.

One might ask, "Just what is the population in this case?" To answer this question, we need to imagine the amount of water in this stream fluctuating up and down over a long period of time, but without changing the basic flow pattern of the stream. Then all the measurements of flow rate and depth that might be made over this long period of time are our population. We assume (but don't really believe) that our set of measurements is a random sample from this population. All we can really hope is that our

Exhibit 11.6   Fitting a Quadratic Polynomial to the Stream Data

```
MTB > brief 3
MTB > let c3 = 'Depth'**2
MTB > name c3='Depth-Sq'
MTB > regress 'Flow' 2 'Depth' 'Depth-Sq'

The regression equation is
Flow = 1.68 - 10.9 Depth + 23.5 Depth-Sq

Predictor         Coef        Stdev      t-ratio            p
Constant         1.683        1.059         1.59        0.156
Depth          -10.861        4.517        -2.40        0.047
Depth-Sq        23.535        4.274         5.51        0.000

s = 0.2794       R-sq = 99.0%       R-sq(adj) = 98.7%

Analysis of Variance

SOURCE           DF           SS           MS           F          p
Regression        2       54.105       27.053      346.50      0.000
Error             7        0.547        0.078
Total             9       54.652

SOURCE           DF       SEQ SS
Depth             1       51.739
Depth-Sq          1        2.367

Obs.     Depth        Flow          Fit    Stdev.Fit     Residual     St.Resid
  1      0.340      0.6360       0.7107       0.1029      -0.0747        -0.29
  2      0.290      0.3190       0.5123       0.1477      -0.1933        -0.82
  3      0.280      0.7340       0.4868       0.1634       0.2472         1.09
  4      0.420      1.3270       1.2727       0.1344       0.0543         0.22
  5      0.290      0.4870       0.5123       0.1477      -0.0253        -0.11
  6      0.410      0.9240       1.1860       0.1281      -0.2620        -1.05
  7      0.760      7.3500       7.0223       0.2138       0.3277         1.82
  8      0.730      5.8900       6.2961       0.1830      -0.4061        -1.92
  9      0.460      1.9790       1.6667       0.1575       0.3123         1.35
 10      0.400      1.1240       1.1040       0.1218       0.0200         0.08
```

measurements act just about the same as a random sample. If all our measurements were made in the spring, or early in the morning, or just after a storm, or within the same week, we could be fairly confident they would not act like a random sample. It is hoped that the experimenter took such factors into consideration when the data were collected. If not, our inferences may be seriously incorrect.

Exercises

11-23 In Exercise 11-11, we fitted the equation (volume) $= B_0 + B_1$ (diameter) to the Trees data. In Exercise 11-21, we added height as a second explanatory variable and fitted the equation (volume) $= B_0 + B_1$ (diameter) $+ B_2$ (height). In this case, use just diameter to predict volume, but now fit the quadratic equation (volume) $= B_0 + B_1$ (diameter) $+ B_2$ (diameter)^2.

(a) How well does this quadratic equation fit? Compare its fit to that of the straight line we fitted in Exercise 11-11.

(b) Does a quadratic equation seem a reasonable choice from a theoretical standpoint? Consider the geometry of diameter versus volume for a tree.

(c) Compare the quadratic equation in (a) to the equation in Exercise 11-21, which was based on height and diameter. How well does each fit?

(d) It is considerably more difficult to measure the height of a tree than its diameter. Based on this information, what equation would you most likely use in practice to estimate the volume of a tree?

11.6 Interpreting Residuals in Simple and Polynomial Regression

Whenever we fit a model to a set of data, we always should plot the data and the residuals. Recall that a residual is the difference between the observation and the fitted value determined by the regression equation. Thus residuals tell us how the model missed in fitting the data. For example, in Exhibit 11.6, for the first observation, the stream depth is 0.340, the observed flow rate is 0.6360, and the estimated flow rate is 0.7107. Thus the first residual is $(0.6360 - 0.7107) = -0.0747$, and the second residual is -0.1933. These, and all the other residuals, are listed in the column headed "Residual."

Exhibits 11.7 and 11.8 show what happens when we fit an inadequate model to the stream-flow data. The data in Exhibit 11.7 are curved, but we tried to represent them with a straight line anyway. Here it is easy to see that the equation is not a good fit. For low x values, the observed y values are above the straight line; for middle x values, the observed y values are below the straight line; and for the largest x value, the observed y is above the line. This means the residuals will be positive for low values of x, negative for middle values of x, and positive again for the high value of x. This pattern shows up nicely in the plot of residuals versus x shown in Exhibit 11.8.

Exhibit 11.7   Stream Data with a Straight-Line Regression Equation

```
MTB > name c20 'Resids'
MTB > Regress 'Flow' 1 'Depth';
SUBC>    residuals 'Resids'.

The regression equation is
Flow = -3.98 + 13.8 Depth

Predictor           Coef        Stdev     t-ratio           p
Constant         -3.9821       0.5430       -7.33       0.000
Depth             13.834        1.161       11.92       0.000

s = 0.6035       R-sq = 94.7%      R-sq(adj) = 94.0%

Analysis of Variance

SOURCE          DF           SS          MS          F          p
Regression       1       51.739      51.739     142.07      0.000
Error            8        2.913       0.364
Total            9       54.652

Obs.     Depth        Flow        Fit   Stdev.Fit    Residual    St.Resid
  1      0.340       0.636      0.721       0.222      -0.085       -0.15
  2      0.290       0.319      0.030       0.257       0.289        0.53
  3      0.280       0.734     -0.109       0.265       0.843        1.55
  4      0.420       1.327      1.828       0.192      -0.501       -0.88
  5      0.290       0.487      0.030       0.257       0.457        0.84
  6      0.410       0.924      1.690       0.194      -0.766       -1.34
  7      0.760       7.350      6.531       0.420       0.819        1.89
  8      0.730       5.890      6.116       0.389      -0.226       -0.49
  9      0.460       1.979      2.381       0.193      -0.402       -0.70
 10      0.400       1.124      1.551       0.196      -0.427       -0.75
```

To make it easy to do plots such as these, REGRESS has a subcommand that stores the residuals. We used this subcommand in Exhibit 11.7. Patterns such as the one in Exhibit 11.7 sometimes show up more clearly in plots of the residuals than in plots of the original data. The plots of the data and residuals help us see when the model we have fit to the data is seriously wrong. A strong pattern in the residual plot indicates that we probably have a poor model.

What happens when we fit a second-degree polynomial to the stream data? Exhibit 11.9 shows a residual plot for such a fit. Is there any pattern here? For example, do low x values tend to have low residuals? How about the residuals for middle and high values? Are there any patterns that would indicate that our model is not adequate? In this plot, we do not see any. It appears that the model we fit was all right, at least as far as this plot is concerned.

Exhibit 11.8   Plot of Residuals from Straight-Line Fit in Exhibit 11.7

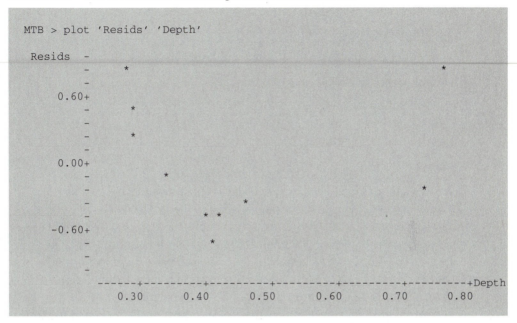

Exhibit 11.9   Residuals from Second-Degree Polynomial Fit to Stream Data

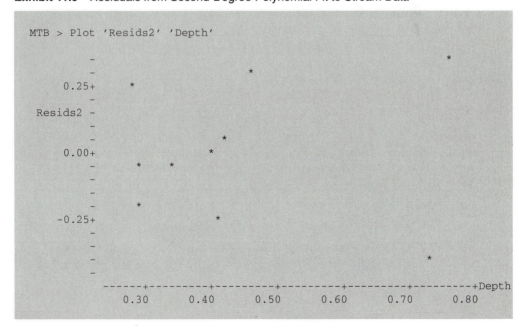

REGRESS C on K in C...C

RESIDUALS into **C**

The residuals are stored in the indicated column.

Stat > Regression > Regression, check the box for Residuals

Summary

If we fit a straight line to data when the basic relationship between x and y is a curve, a plot of the residuals versus x will be curved. The plot of y versus x also will be curved, but the curvature may not be as apparent. We might then fit a second-degree polynomial. If the residual plot after this fit is no longer curved (and no other problems are indicated), we may have a good fit. If the residual plot still shows curvature, we could try fitting a third-degree, or even a higher-degree, polynomial. As a general rule, however, it is better to use a transformation, as discussed in Section 11.7, than to use polynomials of degree higher than two.

Residual plots also help us spot outliers—observations that are far from the majority of the data or far from the fitted equation. Outliers should always be checked for possible errors or special causes. In some cases, they should be temporarily set aside to see whether they have any effect on the practical interpretation of the results of the analysis.

Exercises

11-24   Stream flow data from another site, site 3, are given below.

Flow Rate	.820	.500	.433	.215	.120	.172	.106	.094	.129	.240
Stream Depth	.96	.92	.90	.85	.84	.84	.82	.80	.83	.86

(a) Plot flow rate versus stream depth.
(b) Find a 95% confidence interval for the slope, B.
(c) Find a 95% confidence interval for the intercept, A.

11-25 (a) Plot residuals versus stream depth for the straight line fit in Exercise 11-24. Does the plot indicate any lack of fit, or any other problem? Explain.

(b)   Fit a quadratic model (second-order polynomial) to the data from site 3. Does it fit any better? Explain.

11-26 Stream flow data from another site, site 4, are given below.

Flow Rate	.352	.320	.219	.179	.160	.113	.043	.095	.278
Stream Depth	.71	.72	.64	.64	.67	.61	.56	.73	.72

(a)   Plot flow rate versus stream depth.
(b)   Fit a straight line to explain flow rate based on stream depth. Where does this line fall on the plot?
(c)   Plot the residuals against stream depth. Is there any indication that another model should be used? Is anything else indicated?
(d)   Fit a quadratic model. Does it fit any better? Explain.

11-27 Compare the analyses for sites 3 and 4 on the following points:
(a)   How well you can predict flow rate from depth.
(b)   Whether or not a quadratic model fits better than a straight line.
(c)   Whether similar relationships between flow rate and depth seem to hold for both sites.

11-28 A simple pendulum experiment from physics consists of releasing a pendulum of a given length (L), allowing it to swing back and forth for 50 cycles, and recording the time it takes to swing through these 50 cycles. Data from five trials of this experiment are given below.

	Trial				
	1	2	3	4	5
Length	175.2	151.5	126.4	101.7	77.0
Time for 50 Cycles	132.5	123.4	112.8	101.2	88.2

(a)   Let T be the average time per cycle. Compute T for each trial by dividing the time by 50. Plot T versus L and fit a straight line to explain T based on L. How well can you estimate time per cycle from pendulum length using a straight line?
(b)   Plot the residuals versus L. Are there any indications that a higher-degree polynomial should be used? Fit a better model if one seems needed.

11-29 In Exercise 11-17 we simulated data from a regression equation. Repeat those simulations, and each time plot the residuals versus x. This should give you some idea of how a residual plot looks when a correct model is fitted.

11-30 A statistician named Frank Anscombe constructed the data listed below to make an important point. The following steps should dramatically illustrate his point. (*Note:* y_1, y_2, and y_3 all use the same x values.) The data are shown below and are saved in the worksheet FA.

x_1, x_2, x_3	y_1	y_2	y_3	x_4	y_4
10	8.04	9.14	7.46	8	6.58
8	6.95	8.14	6.77	8	5.76
13	7.58	8.74	12.74	8	7.71
9	8.81	8.77	7.11	8	8.84
11	8.33	9.26	7.81	8	8.47
14	9.96	8.10	8.84	8	7.04
6	7.24	6.13	6.08	8	5.25
4	4.26	3.10	5.39	19	12.50
12	10.84	9.13	8.15	8	5.56
7	4.82	7.26	6.42	8	7.91
5	5.68	4.74	5.73	8	6.89

(a) Use REGRESS to fit a straight line to each pair of variables, y_i and x_i. Compare the regression output from the different data sets. Do they have anything in common? Based on the regression output, would you tend to think the data sets were pretty much alike?

(b) For each data set, make a plot of y versus x. Now think back to the regression output in part (a). Can you guess what important point Anscombe was trying to make?

11.7 Using Transformations

Rather than fit a polynomial to curved data, it is often preferable to transform them to see if a simpler model can be found. The use of transformations is new in some areas of research, but engineers and physical scientists have long used transformations to simplify relationships. In fact, they have a saying: "Anything is a straight line after using a logarithmic transformation." Of course, this statement is an exaggeration, but it does indicate the usefulness of transformations.

Suppose we have two variables, say x and y, and we want a simple way to describe the relationship between them. We start by plotting y versus x. If the plot is more or less a straight line, we have our simple description. But suppose the plot is a curve. Three curves frequently encountered in data analysis are shown in Exhibit 11.10. Plots (a) and (b) both have an upward trend—that is, as x increases, y also increases. But there is a difference: plot

(a) curves down, whereas (b) curves up. Plot (c) has a downward trend—as x increases, y decreases—and it curves up. Our objective is to transform x or y or both in order to get a straight line.

Three popular transformations are square root, log, and negative reciprocal. These can be applied to x or y or both. The arrows in Exhibit 11.10 indicate which ones to try. In panel (a), the arrow points down on the x axis. This tells us that if we pull the upper end of the curve in the direction of the arrow, we will tend to straighten the curve. Put another way, we need to compress the high values of x to straighten the curve.

We often start with $\sqrt{x}$. If this is not strong enough, we try $\log(x)$, and then $-1/x$, until we find a transformation that works. In panel (b), we need to compress the high values of y. So we start with $\sqrt{y}$, and then, if necessary, try $\log(y)$, and then $-1/y$. In panel (c), there are downward arrows on both axes. This means we should try transforming either x or y or both. For example, we might try y versus $\sqrt{x}$, or $\log(y)$ versus $\sqrt{x}$. Our goal in all cases is to get a straight line. Of course, this doesn't always happen. But it does work surprisingly often.

We can use the stream data from Exhibit 11.5 as an example. Plotting Flow versus Depth did not give us a straight line. The curvature in the plot of Flow versus Depth is similar to that in panel (b) of Exhibit 11.10. To achieve a straight line, we need to pull in the upper end of the Flow scale. The indicated transforms are thus $\sqrt{\text{Flow}}$, $\log(\text{Flow})$, and $-1/\text{Flow}$. These are plotted versus Depth in Exhibit 11.11. Also shown is $\log(\text{Flow})$ versus $\log(\text{Depth})$.

Several of these plots seem to be reasonably close to a straight line. Let's try $\sqrt{\text{Flow}}$ versus Depth. If we fit a straight line to this plot, the resulting equation would be

Exhibit 11.10   Three Common Types of Curves Relating Two Variables

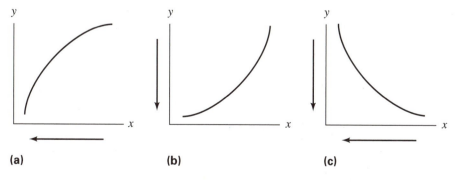

(a)   (b)   (c)

$$\sqrt{\text{Flow}} = a + b(\text{Depth})$$

To fit this model, we could use

```
NAME C11 = 'SQRTFLOW'
LET 'SqrtFlow' = SQRT('FLOW')
REGRESS 'SQRTFLOW' 1 'DEPTH'
```

Minitab tells us that the best equation is

$$\text{SqrtFlow} = -.558 + (4.16)(\text{Depth})$$

If Depth were .7, we would estimate SqrtFlow as

$$\text{SqrtFlow} = -.558 + 4.16(.7) = 2.35$$

Then to estimate Flow, at Depth = .7, we calculate

$$\text{Flow} = (2.35)^2 = 5.52$$

Another plot that looks fairly good is log(Flow) versus log(Depth). Fitting a straight line to this plot gives the equation

$$\log(\text{Flow}) = a + b \log(\text{Depth})$$

Either of these equations might do well enough for estimation, particularly within the range of depths for which we have data. In addition, they might give us some idea of a good theoretical model for the relationship between stream flow rate and depth. Using a polynomial might be equally good for estimation over the range of the data, but it probably would not work very well outside the range of the data, nor would it be likely to give us much theoretical insight.

Effect of Transformations on Assumptions

The conditions necessary for inference in regression have been listed in Section 11.3. Whenever we use a transformation, there will be some effect on the validity of the conditions. If we have an exactly straight line, and take the logarithm of x or y, we will not have an exactly straight line anymore. If y has exactly the same variance for all values of x, then $\log(y), \sqrt{y}$, and $-1/y$ will not. If y is exactly normally distributed, then $\log(y)$, $\sqrt{y}$, and $-1/y$ will not be.

On the other hand, if some of these conditions were not met before we transformed the data, they may be afterward. Sometimes a transformation will help with one of these conditions, but will cause problems with an-

Exhibit 11.11   Transformations of Stream Data

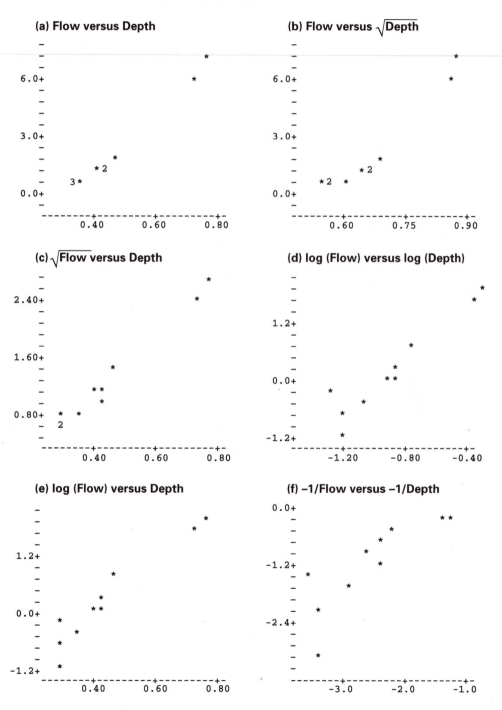

other. Surprisingly often, though, a transformation that helps with one condition also helps with others.

To summarize, our advice is to use transformations as you would use any other statistical technique—not blindly, but with a healthy skepticism that says, "I know none of these assumptions are met exactly, and I know that some are more important than others. I will check them all, particularly the most important ones, and always use the results of my analysis as guidance, not as gospel."

Exercises

11-31 In Exercise 11-28, you fitted a line to some pendulum data. A residual plot indicates a curve in the data—a curve that may not have been apparent in a data plot. You then probably fitted a quadratic. This gives a residual plot with no apparent pattern (of course, with just five observations, it's difficult to do a very precise analysis). It is known from physics that the correct relationship is $T = (2\pi/\sqrt{q})\sqrt{L}$, where $q = 981$ cm/sec^2 (at the latitude where the experiment was done). Thus if we fit a straight line of T versus $\sqrt{L}$, we should find $A = 0$ and $B = 2\pi/\sqrt{981} = .201$.

(a) Fit the model $T = A + B\sqrt{L}$. How well does it fit compared to a quadratic model? If you didn't know any physics, could you decide between these two models based on these data?

(b) Do your estimates of A and B seem close to the theoretical values?

(c) Now test $H_0: A = 0$ and $H_0: B = .201$ (use $\alpha = .05$). Are the results of these tests consistent with your answers in (b)? Explain.

11-32 An experiment was run in which a tumor was induced in a laboratory animal. The size of the tumor was recorded as it grew.

Number of Days After Induction	Size of Tumor (cc)
14	1.25
16	1.90
19	4.75
21	5.45
23	7.53
26	14.50
28	16.70
30	21.00
33	27.10
35	30.30
37	40.50
41	51.40

Investigate the relationship between time and tumor size. Is the relationship linear? Can it be "linearized" by an appropriate transformation?

11-33 In Exercise 11-22, we studied the Peru data (described in Appendix A) with the aim of predicting systolic blood pressure from other variables in the data set. We will continue that study here.

(a) One model in Exercise 11-22 used WEIGHT and YEARS to predict systolic blood pressure. Fit that model if you have not already done so.

(b) The researchers in this study created a new variable, fraction of life since migration, using FRACTION = YEARS/AGE. They reasoned that younger people adapt to new surroundings more quickly than older people. Therefore, FRACTION might be a better measure of how long a person lived in the new environment than YEARS. Combining variables in this way is a type of transformation. Use this new predictor along with WEIGHT to predict systolic blood pressure. How does this model compare to the one in part (a)?

(c) Use the three predictors YEARS, FRACTION, and WEIGHT to predict systolic blood pressure. How does this model compare to the ones in parts (a) and (b)?

(d) Can you find another predictor to add to the model in part (c) that improves things? Look at just the original predictors (there are seven remaining) in the data set.

(e) There is (at least) one major lesson in the analyses you have done using these data. What is it?

11-34 When a magnifying glass is used to view an object, there is a simple rela-
       tionship between the apparent size of the object as seen through the lens
       (image size) and the distance of the object from the lens (object distance).
       Here are some data that were obtained in an experiment.

Object Distance (Centimeters)	12	13	14	15	16	17	18	19	20	21	22
Image Size (Centimeters)		12.0	9.4	7.2	6.2	5.2	4.5	4.0	3.6	3.2	3.0	2.7

   (a) Plot image size versus object distance. How does image size vary
        with object distance?

   (b) Suppose you were to place an object at a distance of 12.5 centimeters
        from the lens. What would you estimate the image size to be? Sup-
        pose you were to place it at 23 centimeters. Approximately what im-
        age size would you expect?

   (c) Now let's investigate the relationship between these two variables. It
        is obviously some sort of curve, but what curve? Compare the plot in
        part (a) with the sketches in Exhibit 3.9. Which does it match?

   (d) See if you can find a way to transform one of the variables so that the
        resulting plot is approximately a straight line.

11.8 Some Additional Features of REGRESS

BRIEF output at level K

BRIEF controls the amount of output from all REGRESS commands that
follow. K is an integer from 1 to 3; the larger the integer, the more output.

K = 1 The regression equation, a table of coefficients, s, R^2 and R^2-adjusted,
       and the first part of the analysis of variance table (regression, error,
       and total sums of squares) are printed.

K = 2 In addition, the second part of the analysis of variance table (pro-
       vided there are two or more predictors) and the "unusual" observa-
       tions in the table of data, fits, stdev fit, and so on are printed. An
       observation is marked as unusual if its predictors are unusual (i.e.,
       far away from most other predictors) or if its standardized residual
       is large (specifically, more than 2). In both cases, you should check
       the observation to make sure it is correct. This is the usual output.

K = 3 In addition, the full table of data, fits, stdev. fit, and so on are printed.

Releases 8 and 9 Windows: There is no menu entry for BRIEF. You must type it.

REGRESS C K C...C [store standardized residuals in **C** [fits in **C**]]

PREDICT E...E
RESIDUALS into **C**
COEFFICIENTS into **C**
NOCONSTANT

REGRESS has many options and subcommands. In this chapter, we described a few of the elementary ones. See Chapter 15 to learn about the others.

You can store the standardized residuals and fits by listing two storage columns on the REGRESS line. Standardized residuals also are printed in the output. They are calculated by (residual)/(standard deviation of this residual). The standard deviation is different for each residual and is given by the formula

$$\text{(standard deviation of residual)} = \sqrt{\text{MSE} - (\text{standard deviation of fit})^2}$$

PREDICT was described in Section 11.4, and RESIDUALS in Section 11.6.

COEFFICIENTS stores the coefficients $b_0, b_1, \ldots, b_k$ down the column.

NOCONSTANT fits a model without a constant term—that is, the model

$$y = b_1 x_1 + b_2 x_2 + \ldots + b_k x_k$$

There is also a main command NOCONSTANT. It acts like a switch. Once you use it, all REGRESS models from then on are fitted without a constant term. To turn it off, you use CONSTANT.

Releases 8 and 9 Windows: **Stat > Fit Intercept**, uncheck this to get NOCONSTANT.

11.9 Optional Material on High-Resolution Graphs

In this section, we will show you how to draw a fitted regression equation on a data plot. We will use the stream data of Exhibit 11.5 for the examples.

Releases 7 and 8

The two exhibits for this section were done on a Macintosh. The graphs are similar on other computers but may have slightly different scales.

Straight Lines. If the regression equation is a straight line, then we can use the fitted values to draw the line. We add two columns to the REGRESS command. The standardized residuals are stored in the first column and the fitted values in the second.

```
NAME C11='Fits1'
REGRESS 'Flow' 1 'Depth' C10 'Fits1'
```

Fits1 and Depth contain the *x*- and *y*-coordinates of the regression line. You can use GPLOT with the LINE subcommand to create the plot, which is shown in Exhibit 11.12.

```
GPLOT 'Flow' 'Depth';
  LINE 'Fits1' 'Depth'.
```

Another way to draw the regression line was shown in Section 4.5, where we added a line to the plot of the Furnace data.

Quadratic Equations. The method we use when the equation is a straight line will not work, in general, when the regression equation is a quadratic or any other curve. In these cases, we need to use a fine grid to get a smooth curve, and we need to have the points for the *x*-coordinates in order. Therefore, we do the calculations ourselves. Suppose we want to draw the quadratic equation, shown in Exhibit 11.6, that we fit to the stream data. The equation is

$$\text{Flow} = 1.68 - 10.9(\text{Depth}) + 23.5(\text{Depth-Sq})$$

First we use SET to put *x*-values into a column. Then we use LET to calculate the points for the quadratic, using the coefficients in the equation above. Finally, we use GPLOT to plot the data and LINE to draw the quadratic equation. The necessary commands are given below and the plot is shown in Exhibit 11.13.

```
NAME C10='x'  C11='quad'
SET 'x'
  0.28:0.76/.001
END
LET 'quad' = 1.68 - 10.9*'x' + 23.5*'x'**2
GPLOT 'Flow' 'Depth';
  LINE 'quad' 'x'.
```

Release 9 Windows and Commands

Straight Lines. If the regression equation is a straight line, we can use the fitted values to draw the regression line on the plot. First we store the fitted values using the subcommand FITS:

Exhibit 11.12   Stream Data with Regression Line

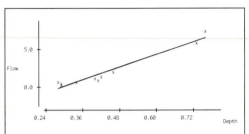

Exhibit 11.13   Stream Data with Quadratic Fit

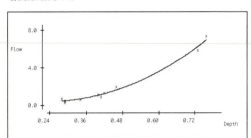

```
NAME C11='Fits1'
REGRESS 'Flow' 1 'Depth';
  FITS 'Fits1'.
```

Fits1 and Depth contain the *x*- and *y*-coordinates of the regression line. You can use PLOT with the Lines option to create the plot, which is shown in Exhibit 11.14. In Release 9 Windows, the Lines option is under Annotations. This option was discussed in Chapter 4 and its dialog box shown in Exhibit 4.17. The command to produce the plot is

```
PLOT 'Flow'*'Depth';
  LINE 'Depth' 'Fits1'.
```

Another way to draw the regression line was shown in Section 4.5, where we added a line to the plot of the Furnace data.

Quadratic Equations. In general, the method we use when the equation is a straight line will not work when the regression equation is a quadratic or any other curve. In these cases, we need to use a fine grid to get a smooth curve, and we need to have the points for the *x*-coordinates in order. Therefore, we use SET and LET to do the calculations ourselves. Suppose we want to draw the quadratic equation, shown in Exhibit 11.6, that we fit to the stream data. The equation is

$$\text{Flow} = 1.68 - 10.9(\text{Depth}) + 23.5(\text{Depth-Sq})$$

First we use SET to put *x*-values into a column. Then we use LET to calculate the points for the quadratic, using the coefficients in the equation above. Finally, we use PLOT to plot the data and LINE to draw the quadratic equation. Output is shown in Exhibit 11.15. Here are the commands to do the plot.

Exhibit 11.14   Stream Data with
Regression Line

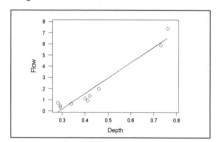

Exhibit 11.15   Stream Data with
Quadratic Fit

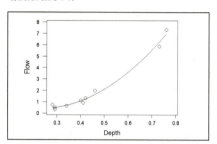

```
NAME C10 'x'  C11 'quad'
SET 'x'
  0.28:0.76/.001
END
LET 'quad' = 1.68 - 10.9*'x' + 23.5*'x'**2
PLOT Flow*Depth;
  LINE x quad.
```

12

Chi-Square Tests and Contingency Tables

Many times we count the number of occurrences of an event. For example, we count how many fatal accidents there are on a holiday weekend or how many people in different age groups support restrictions on international trade. We may then be interested in whether or not our counts are in agreement with some theory. In this chapter we show how Minitab may be used to test agreement in such cases.

12.1 Chi-Square Goodness-of-Fit Test

Several years ago, an article in the *Washington Post* described a high-school boy named Edward who made 17,950 coin flips and "got 464 more heads than tails and so discovered that the United States Mint produces tail-heavy pennies." Is this result statistically significant? The statistician W. J. Youden called Edward and asked how he had done this experiment. Edward explained that he had tossed five pennies at a time and his younger brother had recorded the results as Edward called them out. The results are in Table 12.1.

If the coins are symmetrical, the probability of a head and the probability of a tail should both be .5. Thus, if you toss five coins, the number of heads should follow a binomial distribution with $n = 5$ and $p = .5$. This is our null hypothesis. To test this hypothesis, we use a chi-square goodness-of-fit test. Minitab does not have a command to do this test directly,

Table 12.1   Results from Coin-Tossing Data

Number of Heads in Five Tosses	Number of Times Found
0	100
1	524
2	1080
3	1126
4	655
5	105
Total Tosses	3590

but we can use Minitab as a "calculator" and do it by hand. Exhibit 12.1 shows the calculations.

First we enter the data into C1 and C2, and assign names to all the columns we plan to use. There are six cases, or "cells"; 0 heads in 5 tosses, 1 head in 5 tosses, 2 heads in 5 tosses, and so on. The PDF command calculates the proportion of the time each case would occur if H_0 were true. Next, we multiply each proportion by the total number of observations (the number of times Edward tossed the five coins). This gives the number we would expect to get in each case. These are printed at the end of Exhibit 12.1.

The Observed and Expected numbers do not agree exactly. Even if our hypothesis were exactly true, we would not expect perfect agreement. Some amount of chance variation will always be present. But, on the other hand, if our hypothesis were true, we should not expect "too much" difference between the Observed and Expected counts. How much difference is "too much"? The chi-square test gives us a way of doing a formal test. The chi-square statistic is

$$\chi^2 = \Sigma \, [(\text{Observed} - \text{Expected})^2/\text{Expected}]$$

If the Observed and Expected numbers are very different, the value of χ^2 will be large. In Exhibit 12.1, we first calculated the values (Observed − Expected)^2/Expected. Then we added them up to get $\chi^2 = 21.568$. Now we need the p-value associated with this statistic. CDF gave the area up to 21.568 as .9994. We want the tail area, the area above this, which is $1 - .994 = .006$. This is very significant. Edward's pennies do not follow a binomial with $n = 5$ and $p = .5$.

Exhibit 12.1   Commands to Do the Calculations for a Chi-Square Test

```
MTB > read c1 c2
DATA>    0    100
DATA>    1    524
DATA>    2   1080
DATA>    3   1126
DATA>    4    655
DATA>    5    105
DATA> end
MTB > name c1 = 'x' c2 = 'Obs' c3 = 'Exp' c4 = 'Chisq' c10 = 'P(x)'
MTB > pdf 'x' 'P(x)';
SUBC>     binomial 5 0.5.
MTB > let 'Exp' = 3590*'P(x)'
MTB > let 'Chisq' = ('Obs' - 'Exp')**2/'Exp'
MTB > sum 'Chisq'
   SUM      =        21.568
MTB > cdf 21.568;
SUBC>    chisq 5.
   21.5680     0.9994

MTB > print c1-c4 c10

   ROW    x     Obs      Exp    Chisq      P(x)
     1    0     100   112.19    1.3240   0.03125
     2    1     524   560.94    2.4323   0.15625
     3    2    1080  1121.87    1.5630   0.31250
     4    3    1126  1121.87    0.0152   0.31250
     5    4     655   560.94   15.7732   0.15625
     6    5     105   112.19    0.4605   0.03125
```

Let's look just a bit further. We printed the individual contributions to the chi-square statistic in Exhibit 12.1. One number stands out, far out; the number 15.7732, corresponding to 4 heads. Look at the counts for this case. There are many more Observed counts than expected. There may be some mistake. Perhaps many of the tosses with 4 tails were recorded as 4 heads.

Exercises

12-1 The following table gives accidental deaths from falls, by month, for the year 1979.

Month	Number of Deaths from Falls
Jan	1150
Feb	1034
Mar	1080
Apr	1126
May	1142
Jun	1100
Jul	1112
Aug	1099
Sep	1114
Oct	1079
Nov	999
Dec	1181
Total	113,216

(a) Do accidental deaths due to falls seem to occur equally often in all 12 months? Do a chi-square goodness-of-fit test. What is an appropriate null hypothesis? Calculate the expected number of falls under this null hypothesis. Complete the test.

(b) Can you give some reasons for the result in part (a)? What patterns do you see in the data?

12-2 Suppose we count the number of days in a week on which at least .01 inch of precipitation falls. Can we model this by a binomial distribution? Below, we give some data collected in State College, Pennsylvania, during the years 1950–1969. All the observations are from the same month, February. This gives us (4 weeks in Feb.) × (20 years) = (80 weeks in all).

Number of Precipitation Days in a Week	0	1	2	3	4	5	6	7
Number of Weeks in Which This Occurred	3	12	17	25	14	5	4	0

(a) Use these data to estimate p = the probability of precipitation on a given day.

(b) Test to see whether a binomial distribution with $n = 7$ and p as estimated in part (a) fits the data. You can use the PDF command to help you get the expected counts for each group.

(c) Does your answer to part (b) agree with what you would expect of rainfall data? Explain.

12-3 The Missouri Department of Conservation wanted to learn about the migration habits of Canadian geese. They banded more than 18,000 geese from one flock and classified each goose into one of four groups according to its age and sex. Information on the bands asked all hunters who shot one of these geese to send the band to a central office and tell where they shot the goose. The number of geese that were banded in each of the four groups and the number of bands that were returned from geese that were shot outside the normal migration route for this flock are given below.

Group	Number of Geese Banded	Number of Bands Returned
Adult Male	4144	17
Adult Female	3597	21
Yearling Male	5034	38
Yearling Female	5531	36
Totals	18,306	112

(a)   Suppose the four groups were equally likely to stray from the normal migration routes. Calculate the number of bands you would expect to be returned in each of the four groups. Notice that you must take into account the fact that the number of geese banded in each group was not the same.

(b)   Are the four groups equally likely to stray? Do an appropriate test.

12.2 Contingency Tables

A researcher did a study to investigate the relationship between being an artist and believing in extrasensory perception (ESP). He asked 114 artists and 344 nonartists to classify themselves into one of three categories: (1) believe in ESP, (2) believe more-or-less, (3) do not believe. His results are given in a contingency table, Table 12.2.

One question that is frequently asked about such data is "Is there any association between being an artist and belief in ESP?" Another version of

Table 12.2 Contingency Table from a Study of ESP

	Believe in ESP	Believe More-or-Less	Do Not Believe	Total
Artists	67	41	6	114
Nonartists	129	183	32	344
Total	196	224	38	458

the same question is "Do artists and nonartists seem to have about the same degree of belief in ESP?" The corresponding null hypothesis can be phrased as:

H_0: The proportions in the three categories for belief in ESP are the same for artists as for nonartists

If we assume this null hypothesis is true, we can do the following calculations: Of the 458 people in the study, 196/458, or 42.79%, believe in ESP; 224/458, or 48.91%, believe more-or-less; and 38/458, or 8.30%, do not believe. There are 114 artists. Therefore, 42.79% of the 114 artists, or 48.79 people, should believe in ESP. Similarly, 48.91% of 114, or 55.76 artists, should believe more-or-less, and 8.30% of 114, or 9.46 artists, should not believe. The same proportions should hold for the nonartists. Thus 42.79% of 344 = 147.21 should believe, 48.91% of 344 = 168.26 should believe more-or-less, and 8.30% of 344 = 28.55 should not believe. These are the Expected counts if our null hypothesis is true.

Are the Observed and Expected counts close? To determine whether the differences are more than could reasonably be due to chance alone, we use a chi-square test. The test statistic is similar to the one we used in the preceding section:

$$\chi^2 = \Sigma[(\text{Observed} - \text{Expected})^2/\text{Expected}]$$

For the ESP data,

$$
\begin{aligned}
\chi^2 &= (67 - 48.79)^2/48.79 + (41 - 55.76)^2/55.76 + (6 - 9.46)^2/9.46 \\
&\quad + (129 - 147.21)^2/147.21 + (183 - 168.26)^2/168.26 \\
&\quad + (32 - 28.55)^2/28.55 \\
&= 15.94
\end{aligned}
$$

The number of degrees of freedom is computed as follows: Let r = the number of rows in the contingency table. Here there are two: artists and nonartists. Let c = the number of columns. There are three: believe, believe more-or-less, and do not believe. The number of degrees of freedom is $(r - 1) \times (c - 1) = (2 - 1) \times (3 - 1) = 2$.

Exhibit 12.2 shows how to do these calculations using Minitab's CHI-SQUARE command. Notice that we enter each row of the contingency table into a separate row of the worksheet. To finish the test, we compare the calculated chi-square value to a table of the chi-square distribution with 2 degrees of freedom, or we use Minitab's CDF command. For $a = .005$, CDF gives 7.88. Our observed value is much larger than this. Therefore,

Exhibit 12.2 Output from CHISQUARE for the ESP Data

```
MTB > read c1-c3
DATA>      67     41      6
DATA>     129    183     32
DATA> end
       2 rows read.
MTB > chisq c1-c3

Expected counts are printed below observed counts

             C1       C2       C3     Total
    1        67       41        6       114
          48.79    55.76     9.46

    2       129      183       32       344
         147.21   168.24    28.54

Total       196      224       38       458

ChiSq =   6.800 +   3.905 +   1.265 +
          2.254 +   1.294 +   0.419 = 15.936
df = 2
```

we have strong evidence that the distribution of ESP beliefs among artists
is different from that among nonartists.

Notice that the output also gives the cell-by-cell contributions to the
overall χ^2 value. In this case the largest contribution, 6.80, comes from the
"artists who believe in ESP" cell. The Observed count is 67, but the Expected count if the null hypothesis were true is only 48.8. Many more artists
than expected believe in ESP.

CHISQUARE analysis of the table in C...C

Does a chi-square test for association on the table of counts given in the
specified columns.

Stat > Tables > Chisquare Test

Exercises

12-4   Two researchers at Penn State studied the relationship between infant mortality and environmental conditions in Dauphin County, Pennsylvania. This
county has one large city, Harrisburg. The researchers recorded the season
in which each baby was born and whether or not it died before one year of
age. Data on infant deaths and births for Dauphin County, 1970, are given
below.

| | Season of Birth | | | |
	Jan Feb Mar	Apr May Jun	Jul Aug Sep	Oct Nov Dec
Died Before One Year	14	10	35	7
Lived One Year	848	877	958	990

(a)   Is an infant more likely to die if it is born in one season than in another? Which season has the highest risk? The lowest risk?

(b)   Newspapers reported severe air pollution, covering the entire East,
during the end of July. Air pollution, especially during the end of
pregnancy and in the first few days after birth, is suspected of increasing the risk of an infant's death. Is this theory consistent with the data
and the analysis in part (a)?

12-5   The data in Exercise 12-4 are for all of Dauphin County. However, environmental conditions as well as socioeconomic conditions are different for
the city of Harrisburg and for the surrounding countryside. Here we give
the data separately for these two regions. Do the analysis in Exercise 12-4
separately for each region. What are your conclusions now?

Data for Harrisburg

| | Season of Birth | | | |
	Jan Feb Mar	Apr May Jun	Jul Aug Sep	Oct Nov Dec
Died Before One Year	6	6	25	3
Lived One Year	306	334	347	369

Data for Dauphin County, Excluding Harrisburg

	Season of Birth			
	Jan Feb Mar	Apr May Jun	Jul Aug Sep	Oct Nov Dec
Died Before One Year	8	4	10	4
Lived One Year	542	543	611	621

12-6 A survey on car defects was done in Dane County, Wisconsin, based on a random sample of people who had just purchased used cars. Each owner was sent an invitation to bring the car in for a free safety inspection. Car owners who did not respond to the invitation were sent postcard reminders. Eventually about 56% of the owners had their cars inspected, giving a total of 8842 cars. Here we will look at four types of defects: brakes, suspension, tires, and lights. There is one contingency table for each type of defect. Each car was classified as to whether it was purchased from a car dealer or from a private owner.

Brakes	Defective	Not Defective
Dealer	931	2723
Private	1690	3498

Suspension	Defective	Not Defective
Dealer	1120	2534
Private	1171	3477

Tires	Defective	Not Defective
Dealer	1147	2507
Private	1765	3423

Lights	Defective	Not Defective
Dealer	602	3052
Private	1098	4090

(a) Is there a relationship between a car's having defective brakes and whether it was purchased from a dealer or a private owner? Do an appropriate test. What percent of dealer-sold cars have defective brakes? What percent of privately sold cars? (Calculate these percents by hand.) How do the percents compare?

(b) Repeat part (a) for defects in the suspension system.

(c) Repeat part (a) for defective tires.

(d) Repeat part (a) for defective lights.

(e) Compare the results from parts (a)–(d). Is there a common pattern?

(f) Try to think of some possible reasons for the results in parts (a)–(e).

12-7 Here we give more data from the Wisconsin car survey described in Exercise 12-6. The age of each car was also recorded. These ages were then grouped into three categories: cars under three years old, cars between three and six years old, and cars over six years old.

	Under Three Years	Three–Six Years	Over Six Years
Dealer	570	1792	1292
Private	513	1859	2816

Is there a relationship between the age of a car and whether it was purchased from a dealer or from a private owner? Do an appropriate test. Describe the relationship. Calculate (by hand) row percents to use in your description.

12-8 Mark Twain has been credited in numerous places with the authorship of 10 letters published in 1861 in the *New Orleans Daily Crescent*. The letters were signed "Quintus Curtius Snodgrass." Did Twain really write these letters? In a 1963 paper, Claude S. Brinegar used statistics to compare the Snodgrass letters to works known to have been written by Mark Twain. We present some of his very interesting analyses in this exercise. The data are stored in the worksheet TWAIN.

There are many statistical tests of authorship. The one Brinegar used is quite simple in concept—just compare the distributions of word lengths for the two authors. If these distributions are very different, we will have some evidence that the authors are probably different people. In using this type of test, we must assume that the distribution of word lengths is about the same in all works written by the same author. Parts (a), (b), and (c), below, attempt to provide some checks on this assumption.

The 10 Snodgrass letters were first divided into three groups. Then the number of two-letter words in each group was recorded, followed by the number of three-letter words in each group, the number of four-letter words, and so on. One-letter words were omitted. There are only two such words, "I" and "a," and the use of "I" tends to characterize content (work written in first person or not) more than an unconscious style. Data for Mark Twain were obtained from letters he wrote to friends at about the same time and from two works written at a later time. These will be used to see whether the word-length distribution remained about the same throughout Twain's writings.

(a) Compare the three groups of Snodgrass letters to see whether they are consistent in word-length distribution. First compare them graphically. To make the numbers comparable, change the frequencies into relative frequencies (proportions). Then plot each column versus word length. Put all three plots on the same axes by using MPLOT.

Do the distributions look similar? Next, do a chi-square test of the null hypothesis that all three sets of letters have the same word-length distribution. Is there any evidence that they differ?

(b)   Next, compare the three groups of Mark Twain letters. Do both a plot and a chi-square test, as in part (a). Is there any evidence that these three collections of writings differ in word-length distribution?

(c)   Now compare Mark Twain's writings over a large span of years. The samples from *Roughing It* and *Following the Equator* were taken for this purpose. First combine the three columns of Mark Twain letters into one column of "early works." Compare the "early," "middle," and "late" samples for Mark Twain. Do both a plot and a chi-square test, as in part (a). Is there any evidence that the word-length distribution changed over the years?

(d)   Finally, now that we've checked the authors for consistency, let's compare the Twain and Snodgrass works. As in part (c), combine the three columns of Twain's early works into one column. Also combine the three columns of Snodgrass letters into one column. Finally, compare Twain's letters with the Snodgrass letters. Do both a plot and a chi-square test. Do you think Mark Twain wrote the Snodgrass letters?

(e)   In what ways do the two authors differ, as far as word-length distribution is concerned? Examine both the plot and chi-square output from part (d) to find out. Does the chi-square output tell you anything the plot does not, or vice-versa?

Word Counts for Quintus Curtius Snodgrass Letters

Word Length	First Three Letters	Second Three Letters	Last Four Letters
2	997	831	857
3	1026	828	898
4	856	669	777
5	565	420	446
6	366	326	300
7	318	293	285
8	258	183	197
9	186	150	129
10	96	94	86
11	63	49	40
12	42	30	29
13 and over	25	25	11
Totals	4798	3998	4055

Word Counts for Known Mark Twain Writings

Word Length	Two Letters from 1858 and 1861	Four Letters from 1863	Letter from 1867	Sample from Roughing It, 1872	Sample from Following the Equator, 1897
2	349	1146	496	532	466
3	456	1394	673	741	653
4	374	1177	565	591	517
5	212	661	381	357	343
6	127	442	249	258	207
7	107	367	185	215	152
8	84	231	125	150	103
9	45	181	94	83	92
10	27	109	51	55	45
11	13	50	23	30	18
12	8	24	8	10	12
13 and over	9	12	8	9	9
Totals	1811	5794	2858	3031	2617

12.3 Making the Table and Computing Chi-Square

In the preceding section the contingency table was already prepared; all we had to do was compute the value of chi-square. In this section we show how Minitab can be used to make the table and then do the chi-square test all in one operation. The procedure is a simple extension of the one discussed in Chapter 5 for making tables.

Consider again the Restaurant data. In Section 5.1, we used the instruction

```
TABLE 'OWNER' BY 'SIZE'
```

to make a table that classified the restaurants by type of ownership and size. TABLE will also do a chi-square analysis if we add the CHISQUARE subcommand, as shown in Exhibit 12.3.

The 2 on the CHISQUARE subcommand says to print two values in each cell of the table. The first is the Observed count; the second is the Expected count—the count we would expect if the null hypothesis of no association were true. The calculated value of chi-square is 67.9. This is quite large for a table with $(3 - 1) \times (3 - 1) = 4$ degrees of freedom. For example, if we used $a = .005$, CDF gives 14.86. We therefore have strong evidence that the size of a restaurant is related to the type of ownership.

Exhibit 12.3   Chi-Square Output from TABLE Using the Restaurant Data

```
MTB > table 'owner' 'size';
SUBC>    chisquare 2.

 ROWS: OWNER      COLUMNS: SIZE

              1         2         3       ALL

   1         83        18         2       103
           54.85     26.05     22.10    103.00

   2         16         6         4        26
           13.85      6.57      5.58     26.00

   3         40        42        50       132
           70.30     33.38     28.32    132.00

 ALL        139        66        56       261
          139.00     66.00     56.00    261.00

 CHI-SQUARE =    67.917    WITH D.F. =    4

   CELL CONTENTS --
                     COUNT
                     EXP FREQ
```

TABLE C...C

CHISQUARE [K]

TABLE prints a table of counts. The subcommand CHISQUARE tells Minitab to also do a chi-square test. If K = 2 is specified, both the Observed and Expected counts are printed in each cell. If K is omitted, or set equal to 1, only the Observed counts are printed.

More than two variables may be listed on the TABLE command. A separate chi-square test is then done for each two-way table printed.

Stat > Tables > Cross Tabulation, check Chisquare analysis. Option K = 2 is not available in the dialog box in Releases 8 and 9. It is available in Release 10.

To see how size and ownership are related, we can compare the Observed and Expected counts in Exhibit 12.3. For sole proprietorships (OWNER=1), there are more small restaurants (SIZE=1) than expected and fewer large ones (SIZE=3) than expected. For corporations (OWNER=3), the reverse is true—there are fewer small restaurants than expected and more large ones. The overall conclusion is not surprising: Corporate-owned restaurants tend to be larger than those owned by a single person.

Exercises

12-9   In the Restaurant survey (described in Appendix A), is there a relationship between the type of food sold and the size of the restaurant? Do an appropriate chi-square test.

12-10 In the Restaurant survey (described in Appendix A), is there a relationship between the type of ownership and the overall outlook of the owner? Do an appropriate test. Describe this relationship, using appropriate row and/ or column percents. (Recall that TABLE has subcommands to calculate these.)

12-11 Using the Pulse data (described in Appendix A), test to see whether there is a relationship between
(a)   sex and activity;
(b)   sex and smokers;
(c)   smokers and activity.

12-12 Using the Cartoon data (described in Appendix A), see whether there is a relationship between education and whether or not a person took the delayed test. You will have to create a variable for the second classification. The following command uses CODE (described in Appendix B) to create a column, C20, in which those who took the test are assigned 1 and those who did not (those who have * for their score) are assigned 0.

```
CODE ('*')0 (0:10)1 'CARTOON2' C20
```

(a)   Do an appropriate test.
(b)   Calculate appropriate percents and describe the relationship. Can you give some possible reasons for this relationship?

12.4 Tables with Small Expected Counts

A survey was done in an introductory statistics class. Each student was asked to give his or her year in college and political preference. This gave

Table 12.3 Data from a Survey of Political Preference

	Freshman	Sophomore	Junior	Senior
Democrat	4	16	16	6
Republican	1	7	7	3
Other	3	4	12	5

the contingency table shown as Table 12.3. Exhibit 12.4 uses the CHI-SQUARE command. Notice that the output contains the message

```
5 cells with expected frequencies less than 5.0
```

This indicates that the chi-square analysis done on this table may not be quite appropriate. As with many statistical tests, the chi-square test is an approximate test. The approximation becomes better and better as the expected cell frequencies increase. Consequently, if too many cells have expected frequencies that are small, a chi-square analysis may not be appropriate.

A good rule of thumb is that not more than 20% of the cells should have expected cell frequencies of less than 5, and no cell should have an expected frequency of less than 1. The table in Exhibit 12.4 has 12 cells, so no more than $.2 \times 12 = 2.4$ cells should have expected frequencies of less than 5. But, as the message says, and as we can see ourselves from the table, there are five such cells.

There are several procedures we can follow with tables having too many small expected frequencies. We can try to combine cells so that we reduce the number of rows and/or columns. For example, if we combine columns 1 and 2 into one level, called "lower class," and combine columns 3 and 4 into another level, called "upper class," we get the following 3×2 table:

	Lower Class	Upper Class
Democrat	20	22
Republican	8	10
Other	7	17

Now if no more than one cell in this smaller table has expected frequency less than 5, we can comfortably do the chi-square analysis. The smaller table was read into the worksheet and a second analysis was done. In this case, no cell had expected frequency less than 5, so the chi-square approximation

Exhibit 12.4   Chi-Square Output for Political-Preference Data

```
MTB > read c1-c4
DATA>        4              16              16              6
DATA>        1               7               7              3
DATA>        3               4              12              5
DATA> end
         3 rows read.
MTB > chisquare c1-c4

Expected counts are printed below observed counts

               c1        c2        c3        c4     Total
    1           4        16        16         6        42
             4.00     13.50     17.50      7.00

    2           1         7         7         3        18
             1.71      5.79      7.50      3.00

    3           3         4        12         5        24
             2.29      7.71     10.00      4.00

 Total          8        27        35        14        84

ChiSq =   0.000 +   0.463 +   0.129 +   0.143 +
          0.298 +   0.255 +   0.033 +   0.000 +
          0.223 +   1.788 +   0.400 +   0.250 = 3.982
df = 6
5 cells with expected counts less than 5.0
```

is probably good enough. Of course, we have changed our analysis slightly. We are now testing a different null hypothesis: that there is no relationship between a student's political preference and whether the student is in the upper or the lower class. Before, we were testing the null hypothesis that there is no relationship between a student's political preference and whether the student is a freshman, sophomore, junior, or senior. We do not have quite enough data here to analyze the larger 3×4 table, so if we want to get something out of the data, we can try looking at a table with fewer levels.

Another way to handle the problem is to remove one or more levels of a classification. In the political preference table, we could omit the first column, corresponding to freshmen. This leads to the following table:

	Sophomore	Junior	Senior
Democrat	16	16	6
Republican	7	7	3
Other	4	12	5

This table provides an analysis relevant to students above the freshman level. Here, however, the omission doesn't help much. There are still too many cells with small expected frequencies.

13

Nonparametric Statistics

The tests and confidence intervals described in Chapters 8, 9, and 10 assume we have random samples from normal populations. The methods they use are for population means. There are several nonparametric methods that also do tests and confidence intervals, but do not assume normality. These methods are based on medians rather than means.

There are several reasons why it may be more appropriate to use nonparametric procedures instead of the normal-theory (parametric) procedures. Among them are the following:

1. *Some populations are not normal.* To be strictly valid, the procedures in Chapters 8, 9, and 10 require that we have random samples from normal populations. If the samples do not come from normal populations, these procedures may give misleading results. For example, if we construct a 95% *t* confidence interval from nonnormal data, the real confidence may be only 91%, not 95% as we had planned. Another, and usually more serious, problem is loss of efficiency. Loosely speaking, a more efficient procedure makes better use of the data and enables us to get a better estimate or test with a smaller sample size. Some nonparametric procedures are more efficient than normal theory procedures if we are sampling from a nonnormal population.

2. *Occasional outliers can distort results.* Normal-theory methods are quite sensitive to even a few extreme, or outlying, observations. As a simple example, consider the five numbers 3, 7, 9, 10, and 11. Normal theory uses their mean, whereas many of the methods presented in this chapter use the median. The median of these numbers is 9 and the mean

is 8. But what would happen if one of the observations had been measured or recorded incorrectly? Suppose the 11 had been recorded as a 61 and we failed to notice this. Then the median would remain unchanged but the mean would more than double, to 18. The median, and the other procedures described in this chapter, are more *resistant* than normal-theory methods to distortion by a few gross errors.

3. *The median is a more informative measure of the center of the population.* If a population is very skewed, the mean can be much larger (or smaller) than most of the observations. A very common example is income in a small town where most people have moderate incomes but a few people are very wealthy. Those few can pull the town mean far upward, so the median would be a more informative measure of town income.

13.1 Sign Procedures

Exhibit 13.1 summarizes some of the data from the Wisconsin restaurant survey (described in Appendix A). These are the market values of the fast-food restaurants owned by one person. In the original data set, there are two observations of zero and three asterisks in the data set. We omitted these when we entered the data in Exhibit 13.1. Notice that the stem-and-leaf display is not symmetric, but is skewed toward high values: 22 observations are on the top three lines and the remaining 15 observations are spread over the next eight lines.

Sign Tests

Suppose we want to do a hypothesis test using these values. We might not want to assume we have a sample from a normal distribution. A sign test does not make any assumptions about the shape of the population in our hypothesis: It could be very skewed; it could have "fat tails," like the t-distribution; it could even be bimodal (i.e., have two peaks).

As an example, suppose the median market value for all fast-food restaurants in Michigan that are owned by one person is $105,000. Are fast-food restaurants in Wisconsin worth more than those in Michigan? Let η be the (unknown) population median for Wisconsin. Then we test $H_0: \eta = 105,000$ versus $H_1: \eta > 105,000$.

Minitab's STEST command does the calculations for a sign test. It first counts the number of values that are above $\eta = 105,000$ and the number that are below. If η were 105,000 we would expect, on the average, about half

Exhibit 13.1   Market Value of 37 Fast-Food Restaurants Owned by One Person
(*Note:* Market value was recorded in thousands. Thus 275 represents $275,000.)

```
MTB > set c50
DATA>    275 225 100   25   40 275 250 365 400 100 125
DATA>     42 300   75 225 135   70   90   70 200 175 140
DATA>     40 175   65 120   30   45 125 125 130 150 150
DATA>     75 160   80 500
DATA> end
MTB > stem-and-leaf c50

Stem-and-leaf of C50          N  = 37
Leaf Unit = 10

    5      0 33444
   13      0 57778889
   (9)     1 002333344
   15      1 55688
   10      2 033
    7      2 588
    4      3 0
    3      3 7
    2      4 0
    1      4
    1      5 0
```

our observations to be above 105,000 and half to be below. In the restaurant example, we have 22 above and only 17 below. Is this surprising? The output from STEST in Exhibit 13.2 tells us that it is not. The *p*-value is 0.162. That is, more than 16% of the time we will get 22 or more observations above the median even if $\eta = 105,000$.

How Minitab Does a Sign Test. The sign test is based on the binomial distribution. Suppose H_0 is true and $\eta = 105,000$. If we pick one fast-food restaurant in Wisconsin at random, there is a 50-50 chance that its market value is above $105,000. If we pick a second fast-food restaurant, again there is a 50-50 chance that its market value is above $105,000, and so on, for all 37 observations. On each trial (sampling a fast-food restaurant at random), the probability of a success (value over $105,000) is 1/2. Let X be the total number of successes. Then X has a binomial distribution with $p = 1/2$ and $n = 37$.

Sometimes there are several values equal to the hypothesized median value. This would be the case if our hypothesis were $\eta = 100,000$. There are

Exhibit 13.2   Sign Test of H_0: η = \$105,000 Versus H_1: η > \$105,000 on VALUE

```
MTB > stest median = 105 'Value';
SUBC>    alt = 1.

SIGN TEST OF MEDIAN = 105.0 VERSUS  G.T.   105.0

               N   BELOW   EQUAL   ABOVE    P-VALUE     MEDIAN
Value         37     15      0      22      0.1620      125.0
```

two market values of \$100,000. In such cases, it is conventional practice to set aside the ties and apply the sign test to the remaining data. Then X would have a binomial distribution with $p = 1/2$ and $n = 35$.

Suppose we test H_0: $\eta = 105,000$ versus H_1: $\eta > 105,000$. In our sample, $X = 22$. The probability of getting this many, or even more, observations over 105,000 is .1620 (you can use Minitab's PDF command to get this value). This is the p-value given by STEST in Exhibit 13.2.

Suppose we test H_0: $\eta = 105,000$ versus H_1: $\eta < 105,000$. Then we let X be the number of observations below 105,000. If H_0 is true, X again has a binomial distribution with $p = 1/2$ and $n = 37$. In our sample, $X = 15$. The probability of getting this many, or even more, observations under 105,000 is .838, certainly not significant.

Suppose we do a two-sided test—for example, H_0: $\eta = 105,000$ versus H_1: $\eta \neq 105,000$. Here we count the number of observations below η and the number above η, and let X be the larger of the two counts. (If they are equal, either will do.) Here there are 22 observations below 105,000 and 15 above. The number below is the larger, so we calculate $P(X \geq 22) = .1620$. The p-value of the two-sided test is twice this value, or $2(.1620) = .3240$. Thus, over 32% of the time we can expect to get a split this extreme or more so—that is, with 22 or more observations on one side or the other.

Sign Confidence Intervals

In Exhibit 13.3, we used the SINTERVAL command to get a 95% confidence interval for the median market value for fast-food restaurants owned by one person. First, notice that SINTERVAL prints out three confidence intervals. The top and bottom intervals are exact; that is, they have the achieved confidence given in the output. The middle interval is an approximate interval calculated by a nonlinear interpolation, abbreviated NLI, and

Exhibit 13.3   Sign Confidence Interval for the Median Market Value

```
MTB > sint c50

SIGN CONFIDENCE INTERVAL FOR MEDIAN

                         ACHIEVED
           N    MEDIAN   CONFIDENCE   CONFIDENCE INTERVAL   POSITION
  C50     37    125.0      0.9011     (    100.0,   150.0)        14
                           0.9500     (     91.0,   159.0)       NLI
                           0.9530     (     90.0,   160.0)        13
```

corresponds to the confidence specified on SINTERVAL. Here it is the default confidence of 95%. With nonparametric methods, it is rare that we can get the exact confidence we want.

Let's look at the 90.1% interval. The 14 under POSITION tells us that this interval goes from the fourteenth smallest observation to the fourteenth largest observation. If we order the 37 observations and count 14 observations up from the bottom, we get 100,000; if we count 14 observations down from the top, we get 150,000. This gives the interval ($100,000, $150,000), shown on the output. Details on the way Minitab calculates a sign confidence interval are given in Section 13.5.

Paired Data

One natural use of sign tests and confidence intervals is for paired data. In this case, the null hypothesis is often that the median difference is zero. To test this, we first compute the differences, then use STEST. Exhibit 13.4 gives an example using the cholesterol data (described in Appendix A). We were interested in the change from the second to the fourth day. Therefore, we first calculated those changes and put them in C5. Then we tested H_0: $\eta = 0$ versus H_1: $\eta \neq 0$, where η is the median of the population of all changes. We also found a 90% confidence interval for η.

Exercises

13-1   Consider the Cartoon experiment (described in Appendix A). The national median of all OTIS scores is 100. Do the OTIS scores for the participants in this study differ significantly from the national median? Do a sign test.

Exhibit 13.4   Sign Procedures for Paired Data, Using the Cholesterol Study

```
MTB > let c5 = '2-DAY' - '4-DAY'
MTB > name c5 'Differ'
MTB > stem 'Differ'

Stem-and-leaf of Differ     N  = 28
Leaf Unit = 10

     1    -0 4
     3    -0 32
     9    -0 000000
    14     0 01111
    14     0 2333
    10     0 4455
     6     0 666
     3     0 8
     2     1 01

MTB > stest 'Differ'

SIGN TEST OF MEDIAN = 0.00000 VERSUS  N.E.   0.00000

               N  BELOW  EQUAL  ABOVE   P-VALUE      MEDIAN
Differ        28      9      0     19    0.0872       19.00

MTB > sinterval 90 'Differ'

SIGN CONFIDENCE INTERVAL FOR MEDIAN

                             ACHIEVED
           N    MEDIAN     CONFIDENCE    CONFIDENCE INTERVAL    POSITION
Differ    28     19.00       0.8151    (     8.00,    30.00)         11
                             0.9000    (     3.28,    36.29)        NLI
                             0.9128    (     2.00,    38.00)         10
```

STEST [median = **K**] for data in **C...C**

ALTERNATIVE = **K**

Does sign tests. For each column, STEST tests the null hypothesis, H_0: $\eta =$ K, where η is the population median. The default alternative hypothesis is H_1: $\eta \neq$ K. If η is not specified on STEST, H_0: $\eta = 0$ is tested.

The subcommand ALTERNATIVE allows you to do a one-sided test. Use K = −1 for H_1: $\eta <$ K and K = +1 for H_1: $\eta >$ K.

Stat > Nonparametrics > 1-Sample Sign, choose option Test median

---

SINTERVAL [K% confidence] for data in **C...C**

Calculates sign confidence intervals for the population median. If the confidence level is not specified, 95% is used.

Specifically, SINTERVAL calculates three intervals for each column listed. The first interval corresponds to the achievable confidence just below K; the third, to that just above K. Each interval goes from the dth smallest to the dth largest observation, where d is the value printed under POSITION on the output. The middle interval, labeled NLI on the output, is calculated by a nonlinear interpolation method and corresponds to the confidence specified on SINTERVAL.

Stat > Nonparametrics > 1-Sample Sign, choose option Confidence interval

---

13-2   Suppose we wanted to test H_0: $\eta = \$105,000$ versus H_1: $\eta \neq \$105,000$, using the VALUE data in Exhibit 13.1. Use the output in Exhibit 13.2 to find the corresponding p-value.

13-3   The data below, collected in a chemistry class, are the results of a titration to determine the acidity of a solution. They are in C1 of the file ACID.

.123	.109	.110	.109	.112	.109	.110	.110
.110	.112	.110	.101	.110	.110	.110	.110
.106	.115	.111	.110	.107	.111	.110	.113
.109	.108	.109	.111	.104	.114	.110	.110
.110	.113	.114	.110	.110	.110	.110	.110
.090	.109	.111	.098	.109	.109	.109	.109
.111	.109	.108	.110	.112	.111	.110	.111
.111	.107	.111	.112	.105	.109	.109	.110
.110	.109	.110	.104	.111	.110	.111	.109
.110	.111	.112	.123	.110	.109	.110	.109
.110	.109	.110	.110	.111	.111	.109	.107
.120	.133	.107	.103	.111	.110	.122	.109
.108	.109	.109	.114	.107	.104	.110	.114
.107	.101	.111	.109	.110	.111	.110	.126
.110	.109	.114	.110	.110	.110	.110	.110
.111	.107	.110	.107				

(a)   The instructor knew the correct value for this solution was .110. Do a two-sided sign test of the null hypothesis H_0: $\eta = .110$. (This is a check to see whether the class is "biased"—that is, to see whether it tends to be systematically too high or too low.)

(b)   Make a histogram of the data. Do you think the population is symmetric? If it is, then $\eta = \mu$.

(c)   A distribution is called *heavy-tailed* if it has a higher probability of very extreme values than in a normal distribution. Does the histogram for part (b) give any indication that the distribution of titration results from this class is heavy-tailed? (It may help to compare your histogram with those of normal distributions.)

13-4   The data from a second titration experiment are given below. These data are in C2 of the worksheet ACID.

.109	.111	.110	.110	.105
.110	.111	.110	.110	.111
.109	.111	.109	.112	.109
.109	.111	.110	.112	.112
.109	.110	.110	.109	.113
.108	.105	.110	.109	.109
.110	.110	.110	.104 ·	.109
.110	.111			

(a)   Make a histogram. Do you think $\mu = \eta$? It will be if the population is symmetric. Exercise 13-3 gives us some more information about the shape of the distribution for titration data. What did you conclude there?

(b)   Find (as closely as possible) a 95% sign confidence interval for η.

(c)   Find a 95% confidence interval for η using normal theory methods and compare it to the sign confidence interval in part (a).

(d)   Repeat parts (a) and (b), using 90% confidence intervals.

13-5   The following is a sample (which we got by simulation) from a normal distribution:

$$62, 60, 65, 70, 60, 67, 61, 66, 64, 64, 62, 63$$

(a)   Calculate an approximately 95% sign confidence interval for η (the closest confidence will be 96.1).

(b)   Find a 96.1% t confidence interval using the TINTERVAL command. Compare this with the sign interval of part (a).

(c)   Now suppose a mistake was made in recording the last observation and the number 36 was recorded instead of 63. Repeat parts (a) and (b).

(d)   Repeat part (c), only now suppose the last observation was mistak-
      enly recorded as 630.

13-6   We can use simulation to get a feel for how well sign confidence intervals
       work in various cases. If your data came from a normal population, a t
       confidence interval is the best way to estimate $\mu = \eta$. But how much worse
       might a sign confidence interval be? The following commands simulate 10
       samples, each containing 18 observations, from a normal distribution with
       $\mu = 50$ and $\sigma = 8$, and then get a 90% sign confidence interval and a 90% t
       confidence interval for each sample. Compare the two confidence intervals
       for each sample. Do both cover μ? Which is narrower?

```
RANDOM 18 C1-C10;
  NORMAL 50 8.
SINT 90 C1-C10
TINT 90 C1-C10
```

13-7   (a)   Repeat Exercise 13-6, but simulate data from a Cauchy distribution.
             Use:

```
RANDOM 18 C1-C6;
  CAUCHY 0 1.
```

       This gives an example of data from a very "heavy-tailed" distribution.

       (b)   Repeat Exercise 13-6, but simulate data from a uniform distribution.
             Use:

```
RANDOM 18 C1-C6;
  UNIFORM 0 1.
```

       A uniform distribution is an example of a distribution that has very
       "skinny tails."

13-8   The job of President of the United States is a very demanding, high-
       pressure job. This might cause premature deaths of presidents. On the
       other hand, only vigorous people would run for president, so presidents
       might tend to live longer than other people. We list below the "modern"
       (since Lincoln) presidents who died before December 1993, the number
       of years they lived after inauguration, and the life expectancy of a man
       whose age is that of the president at the time of his first inauguration.
       These data are in the worksheet PRES.

	Life Expectancy After First Inauguration	Actual Years Lived After First Inauguration
Andrew Johnson	17.2	10.3
Ulysses S. Grant	22.8	16.4
Rutherford B. Hayes	18.0	15.9
James A. Garfield	21.2	.5
Chester A. Arthur	20.1	5.2
Grover Cleveland	22.1	23.3
Benjamin Harrison	17.2	12.0
William McKinley	18.2	4.5
Theodore Roosevelt	26.1	17.3
William H. Taft	20.3	21.2
Woodrow Wilson	17.1	10.9
Warren G. Harding	18.1	2.4
Calvin Coolidge	21.4	9.4
Herbert C. Hoover	19.0	35.6
Franklin D. Roosevelt	21.7	12.1
Harry S. Truman	15.3	27.7
Dwight D. Eisenhower	14.7	16.2
John F. Kennedy	28.5	2.8
Lyndon B. Johnson	19.3	9.2

(a) Do a paired sign test of the null hypothesis that being president has no effect on length of life.

(b) Find an approximately 95% sign confidence interval for the median difference between the expected and attained life spans of presidents.

(c) If our main interest is the effect of stress on length of life, then perhaps we should not include the presidents who were assassinated (Garfield, McKinley, and Kennedy). Carry out the analysis of (a) and (b) without these three presidents.

(d) We have, perhaps, stretched the use of statistics rather far here, as we often must in real problems. Comment on this statement, paying particular attention to the assumptions needed for a sign test.

13.2 Wilcoxon Signed Rank Procedures

With sign procedures, the population can have any shape; with Student's t procedures, the population should be approximately normal. Wilcoxon procedures are in between: the population should be approximately symmetric but it need not be normal or have any other specific shape. Recall that for any symmetric population, the mean and median are equal. The Wilcoxon test can then be viewed as a test for either the population mean μ or the population median η. Similarly, a Wilcoxon confidence interval can be viewed as an interval for either μ or η.

Wilcoxon Signed Rank Test

The Wilcoxon test, like the sign test, can be used to do a test on the median of one sample, but in most studies these methods are used to do paired tests. Therefore, we will explain the Wilcoxon method using a paired example, the same example used in Exhibit 13.4. We want to see if there is a significant change in cholesterol from the second to the fourth day. Thus we test H_0: $\eta_1 = \eta_2$ versus H_1: $\eta_1 > \eta_2$, or equivalently we test H_0: $\eta = 0$ versus H_1: $\eta > 0$, where $\eta = \eta_1 - \eta_2$.

Exhibit 13.5 shows how Minitab calculates the test statistic, W. As usual, we begin by calculating the differences, 2-DAY − 4-DAY. Then we take the absolute value of each difference. Next, we assign ranks. The smallest absolute value is 2; it gets rank 1. The second smallest absolute value is 4, and there are three of these. Their ranks would be 2, 3, and 4. When there are ties, we average the ranks and assign the average rank to all of them. Thus they all get rank 3. The next rank is 5. The fifth smallest value is the last observation, so it gets rank 5. After assigning all ranks, we compute W. This is the sum of all the ranks that correspond to differences that are positive. Here W is 324.

If the null hypothesis is true, W has mean = $n(n + 1)/4$ and variance = $n(n + 1)(2n + 1)/24$, where n is the number of observations in our sample. There are 28 observations in our example, so the mean of $W = 28(29)/4 = 203$ and the variance $28(29)(57)/24 = 1928.5$. The standard deviation, then, is $\sqrt{1928.5} = 43.9$.

The exact distribution of W is given in special tables (see M. Hollander and D. Wolfe, *Nonparametric Statistical Methods*. New York: Wiley, 1973). If we don't have such tables, we can use the fact that W has approximately a normal distribution. Minitab uses the normal approximation (with a continuity correction). In our example, $W = 324$, which is more than two standard deviations above its mean.

Suppose we test H_0: $\eta = 0$ versus H_1: $\eta > 0$. The p-value for this test is $P(W \geq 324)$. Output from the command WTEST, shown in Exhibit 13.6, gives us the p-value. It is 0.003. Thus, there is little chance of W being as large as 324 by chance alone. Therefore, there is a statistically significant decrease in cholesterol from the second to the fourth day.

Now suppose we test H_0: $\eta = 0$ versus the alternative H_1: $\eta \neq 0$. Since this is a two-sided alternative hypothesis, we need two probabilities—one

Exhibit 13.5   Example of Calculating the Wilcoxon Statistic, *W*

2-Day	4-Day	Differences	Absolute Values	Ranks	Ranks of Positive Differences
270	218	52	52	21	21
236	234	2	2	1	1
210	214	−4	4	3	
142	116	26	26	13.5	13.5
280	200	80	80	2	26
272	276	−4	4	3	
160	146	14	14	10.5	10.5
220	182	38	38	17.5	17.5
226	238	−12	12	8.5	
242	288	−46	46	20	
186	190	−4	4	3	
266	236	30	30	15.5	15.5
206	244	−38	38	17.5	
318	258	60	60	24	24
294	240	54	54	22	22
282	294	−12	12	8.5	
234	220	14	14	10.5	10.5
224	200	24	24	12	12
276	220	56	56	23	23
282	186	96	96	27	27
360	352	8	8	6	6
310	202	108	108	28	28
280	218	62	62	25	25
278	248	30	30	15.5	15.5
288	278	10	10	7	7
288	248	40	40	19	19
244	270	−26	26	13.5	
236	242	−6	6	5	
					Sum = 324

Exhibit 13.6   Wilcoxon Test for Paired Data, Using the Cholesterol Data

```
MTB > wtest 'Differ';
SUBC>    alt 1.

TEST OF MEDIAN = 0.000000 VERSUS MEDIAN G.T. 0.000000

                    N FOR   WILCOXON            ESTIMATED
              N     TEST  STATISTIC  P-VALUE      MEDIAN
Differ       28      28      324.0    0.003       22.00
```

Exhibit 13.7   Confidence Interval Associated with Wilcoxon Signed Rank Test

```
MTB > wint 'Differ'

                  ESTIMATED   ACHIEVED
              N    MEDIAN   CONFIDENCE   CONFIDENCE INTERVAL
Differ       28     22.0        95.1  (    7.0,     38.0)
```

for the lower tail and one for the upper tail. The distribution of the test statistic is symmetric, so all we need to do is double the one-sided p-value, giving $2(.003) = .006$ as the two-sided p-value.

Wilcoxon Confidence Interval

There is also a confidence interval associated with the Wilcoxon test. This is calculated by Minitab's WINTERVAL command. The confidence interval is essentially the set of all values d for which the test $H_0: \eta = d$ versus $H_1: \eta \neq d$ is not rejected, using $a = 1 - $ (specified confidence)/100. The exact details of how to calculate the confidence interval and the estimated median printed on the output are given in Section 13.5.

As with most nonparametric procedures, it is seldom possible to achieve the specified confidence, so the closest value is printed in the output. Exhibit 13.7 gives a confidence interval for the change in cholesterol scores from the second to the fourth day.

WTEST [median = **K**] for data in **C...C**

ALTERNATIVE = **K**

Does Wilcoxon signed rank tests. For each column, WTEST tests the null hypothesis, H_0: η = K, where η is the population median. The default alternative hypothesis is H_1: $\eta \neq$ K. If η is not specified on WTEST, H_0: $\eta = 0$ is tested.

The subcommand ALTERNATIVE allows you to do a one-sided test. Use K = −1 for H_1: $\eta <$ K and K = +1 for H_1: $\eta >$ K.

Stat > Nonparametrics > 1-Sample Wilcoxon, choose option Test median

---

WINTERVAL [**K**% confidence] for data in **C...C**

Calculates confidence intervals for the population medians, using a Wilcoxon signed rank procedure. The achievable confidence closest to the specified confidence, K, is printed. If K is not given, 95% is used.

Stat > Nonparametrics > 1-Sample Wilcoxon, choose option Confidence interval

Exercises

13-9   Do Exercise 13-1 using a Wilcoxon test in place of the sign test. Do you think a Wilcoxon test is appropriate for these data?

13-10 Do Exercise 13-3 using a Wilcoxon test. Do you think a Wilcoxon test is appropriate for these data?

13-11 (a)   Do Exercise 13-4 using a Wilcoxon confidence interval.
       (b)   Compare this interval to the sign interval and to the *t* interval of Exercise 13-4.

13-12 Do Exercise 13-5 using Wilcoxon confidence intervals in place of sign confidence intervals.

13-13 Do Exercise 13-6 using Wilcoxon confidence intervals in addition to sign confidence intervals. (Thus use SINT, TINT, and WINT for each sample.) Compare all three intervals for each sample.

13.3 Two-Sample Rank Procedures

In this section, we describe a nonparametric procedure for comparing two populations. We assume that (1) we have a random sample from each population, (2) the samples were taken independently of each other, and (3) the populations have approximately the same shape (this means the variances must be approximately equal). We use η_1 to represent the median of the first population and η_2 for that of the second population.

Our null hypothesis is that the medians of the two populations are equal. Our alternative is that one population is shifted from the other; that is, it has a different median. Since we assume the two populations have the same shape, this procedure is analogous to the pooled-t procedures discussed in Section 8.2.

This two-sample rank test was introduced by Wilcoxon, and is often called the Wilcoxon rank sum test. It was further developed by Mann and Whitney. To avoid confusion with the Wilcoxon test described in Section 13.2, we chose the name MANN-WHITNEY for the Minitab command.

An Example

In Chapter 9, we used two-sample t procedures to analyze data for patients with Parkinson's disease. Recall that Parkinson's disease, among other things, affects a person's ability to speak. Eight of the people in this study had received an operation to treat the disease. This operation seemed to improve the patients' condition overall, but it may also have affected their ability to speak. Each patient was given a test for speaking ability. The results were shown in Table 9.3. We will now use nonparametric methods to analyze these data.

Exhibit 13.8 shows how to do the two-sample rank test by hand. The first two columns give the original data. The second two columns show each sample ordered. Next, we combine the two ordered samples and assign ranks. The smallest observation in the combined sample is 1.2, and we give it rank 1. The next three observations are all equal to 1.3. Whenever we have two or more observations that are tied, we assign the average rank to each. The three observations of 1.3 are tied for ranks 2, 3, and 4. We therefore give each the average rank of $(2 + 3 + 4)/3 = 3$. It is not until we look at the eighth rank that we get an observation from the first sample. When we have assigned all ranks, we sum the ranks corresponding to the observations in the first sample. This sum is usually denoted by W. Here $W = 126.5$.

Exhibit 13.8   Two-Sample Rank Test, Done by Hand

Original Data		Ordered Data		Ranks of Ordered Data	
Operation	*No Operation*	*Operation*	*No Operation*	*Operation*	*No Operation*
2.6	1.2	1.7	1.2	8	1
2.0	1.8	2.0	1.3	11.5	3
1.7	1.8	2.5	1.3	15.5	3
2.7	2.3	2.5	1.3	15.5	3
2.5	1.3	2.6	1.5	17.5	5.5
2.6	3.0	2.6	1.5	17.5	5.5
2.5	2.2	2.7	1.6	19.5	7
3.0	1.3	3.0	1.8	21.5	9.5
	1.5		1.8		9.5
	1.6		2.0		11.5
	1.3		2.2		13
	1.5		2.3		14
	2.7		2.7		19.5
	2.0		3.0		21.5
				Sum = 126.5	

The value of W reflects the relative locations of the two samples. If the values in the first sample tend to be larger than those in the second sample, W will be large; if the values in the first sample tend to be smaller, W will be small. Minitab's MANN-WHITNEY command does all this work for us. See Exhibit 13.9.

The output gives the value of $W = 126.5$ and tells us that the attained significance of the two-sided test is 0.0203. This means that the chance of observing two samples as separated as these, when in fact the two populations have the same median, is only 0.0203. We therefore have statistical evidence that the two populations differ.

The output also contains a confidence interval and a point estimate for $(\eta_1 - \eta_2)$. Both of these are calculated using procedures developed from the two-sample rank test. The specific calculations are discussed in Section 13.5.

Exhibit 13.9   Mann-Whitney Test and Confidence Interval

```
MTB > set c1
DATA>   2.6     2.0     1.7     2.7     2.5     2.6     2.5     3.0
DATA> end
MTB > set c2
DATA>   1.2 1.8 1.8 2.3 1.3 3.0 2.2 1.3 1.5 1.6 1.3 1.5 2.7 2.0
DATA> end
MTB > name c1='Op' c2='No Op'
MTB > mann-whitney 'Op' 'No Op'

Mann-Whitney Confidence Interval and Test

Op          N =   8      Median =        2.5500
No Op       N =  14      Median =        1.7000
Point estimate for ETA1-ETA2 is        0.7000
95.6 Percent C.I. for ETA1-ETA2 is (0.2002,1.2001)
W = 126.5
Test of ETA1 = ETA2  vs.  ETA1 ~= ETA2 is significant at 0.0203
The test is significant at 0.0199 (adjusted for ties)
```

MANN-WHITNEY [K% confidence] sample in **C** and **C**

ALTERNATIVE = K

MANN-WHITNEY gives the results of both a test and confidence interval
to compare two independent samples. The test is also called the Wilcoxon
rank sum test.

If the confidence level is not specified, 95% is used.

The default hypothesis for the test is H_0: $\eta_1 = \eta_2$ versus H_1: $\eta_1 \neq \eta_2$. The
subcommand ALTERNATIVE allows you to do a one-sided test. Use K =
−1 for H_1: $\eta_1 < \eta_2$ and K = +1 for H_1: $\eta_1 > \eta_2$.

Stat > Nonparametrics > Mann-Whitney

Exercises

13-14 (a) By hand, compute the W statistic of a two-sample rank test, using the
following data:

Sample A	10	7	6	12	14
Sample B	8	4	6	11	

(b) Use Minitab's MANN-WHITNEY command to check your answer.

13-15 Do migratory birds store, then gradually use up, a layer of fat as they migrate? To investigate this question, two samples of migratory song spar-
rows were caught, one sample on April 5 and one sample on April 6. The
amount of stored fat on each bird was subjectively estimated by an expert.
The higher the fat class, the more fat on the bird.

	Number of Birds in Class	
Fat Class	Found on April 5	Found on April 6
1	0	3
2	0	1
3	0	11
4	2	6
5	0	1
6	10	9
7	9	2
8	7	2
9	6	1
10	4	0
11	1	0
12	2	0

(a) Use DOTPLOT to compare these two groups visually.
(b) Do an appropriate test to see whether there is any evidence that birds
use up a layer of fat as they migrate.

13-16 A study was done at Penn State to see how much one type of air pollution,
ozone, damages azalea plants. Eleven varieties of azaleas were included in
the study. We will look at data from just two varieties.

During week 1, 10 plants of each variety were fumigated with ozone.
A short time later, each plant was measured for leaf damage. The procedure
was repeated four more times, each time using new plants. These data are
in C1–C5 (variety A) and C6–C10 (variety B) of the worksheet AZALEA.

Leaf Damage for Variety A				
Week 1	Week 2	Week 3	Week 4	Week 5
1.58	1.09	0.00	2.22	0.20
1.62	1.03	0.00	2.40	0.40
2.04	0.00	0.07	2.47	0.34
1.28	0.46	0.18	1.85	0.00
1.43	0.46	0.40	2.50	0.00
1.93	0.85	0.20	1.20	0.00
2.20	0.30	0.63	1.33	0.10
1.96	0.90	0.63	2.40	0.06
2.23	0.00	0.56	2.23	0.17
1.54	0.00	0.26	2.57	0.25

Leaf Damage for Variety B				
Week 1	Week 2	Week 3	Week 4	Week 5
1.29	0.00	0.78	0.40	0.00
0.70	0.20	0.64	0.00	0.40
1.93	0.00	1.00	0.20	0.47
0.98	0.98	0.42	0.40	0.00
0.94	0.00	0.97	0.14	0.00
1.06	0.62	2.43	0.44	0.00
0.94	0.67	0.65	1.23	0.00
1.65	0.00	0.00	0.35	0.00
0.70	0.00	0.30	0.17	0.00
0.35	0.00	0.00	0.20	0.00

(a) Compare the first week's data for the two varieties, using a two-sample rank test. Does there appear to be any difference between the two varieties' susceptibilities to ozone damage?

(b) Repeat the test of part (a) for each of the other four weeks. Overall, how do the two varieties compare?

(c) Susceptibility to ozone varies with weather conditions, and weather conditions vary from week to week. Does this seem to show up in the data? One simple way to look for a "week effect" is to calculate the median leaf damage for all 20 plants sprayed during week 1, the median for week 2, and so on. Do these five medians seem to be very different?

(d) Another way we could test for a "week effect" would be to use a contingency table analysis (discussed in Section 12.2). First form a contingency table as follows: For each week, count the number of damaged plants and the number of undamaged plants (a plant is undamaged if its leaf damage is .00); form a table that has two rows and five columns. Then do a chi-square test for association between damage and week. Do the results agree with what was indicated in part (c)?

13.4 Kruskal-Wallis Test

The Kruskal-Wallis test is a generalization of the procedure used by MANN-WHITNEY, and offers a nonparametric alternative to one-way analysis of variance for comparing several populations.

We assume that (1) we have a random sample from each population, (2) the samples were taken independently of each other, and (3) the populations have approximately the same shape (this means the variances must be approximately equal). We use η_1 to represent the median of the first population, η_2 for the second population, η_3 for the third population, and so on.

Our null hypothesis is that the medians of all the populations are equal. Our alternative is that they are not.

In Chapter 10, we used analysis of variance to analyze data from a study of fabric flammability. Recall that there were 11 observations from each of five labs. Thus there are five populations.

Exhibit 13.10 shows how to do the Kruskal-Wallis test by hand. The five columns headed "Data" show the original data from Table 10.1, with the numbers for each lab ordered from smallest to largest. The next step is to combine all the data and rank all 55 observations. If two or more observations are tied, we assign the average rank to each. The columns in Exhibit 13.10 headed "Ranks" give these ranks. Underneath each column of ranks, we show the average of the ranks in that column. The test statistic is

$$H = \frac{12 \Sigma n_i [\overline{R}_i - \overline{R}]^2}{N(N+1)}$$

where n_i is the number of observations in group i, N is the total number of observations, $\overline{R}_i$ is the average of the ranks in group i, and $\overline{R}$ is the average of all the ranks. Using the fabric data, we get

$$H = \frac{12}{55(55+1)} [11(26.1 - 28)^2 + 11(35.7 - 28)^2 + 11(26.1 - 28)^2$$
$$+ 11(14.7 - 28)^2 + 11(37.4 - 28)^2] = 14.9$$

Exhibit 13.10   Kruskal-Wallis Test, Done by Hand

	Laboratory								
	1		2		3		4		5
Data	Ranks	Data	Ranks	Data	Ranks	Data	Ranks	Data	Ranks
2.9	11.5	2.7	3.0	2.8	7.0	2.6	1.0	2.8	7.0
2.9	11.5	3.2	21.5	2.8	7.0	2.7	3.0	3.1	17.0
3.0	14.0	3.3	26.0	2.8	7.0	2.7	3.0	3.5	34.5
3.1	17.0	3.4	30.0	3.2	21.5	2.8	7.0	3.5	34.5
3.1	17.0	3.4	30.0	3.3	26.0	2.9	11.5	3.5	34.5
3.1	17.0	3.6	38.0	3.3	26.0	2.9	11.5	3.7	40.5
3.1	17.0	3.8	44.5	3.5	34.5	3.2	21.5	3.7	40.5
3.7	40.5	3.8	44.5	3.5	34.5	3.2	21.5	3.9	47.5
3.7	40.5	4.0	49.0	3.5	34.5	3.3	26.0	4.1	51.0
3.9	47.5	4.1	51.0	3.8	44.5	3.3	26.0	4.1	51.0
4.2	53.5	4.3	55.0	3.8	44.5	3.4	30.0	4.2	53.5
Average Rank	26.1		35.7		26.1		14.7		37.4

Large values of H suggest that there are some differences among the populations. Under the null hypothesis, the distribution of H can be approximated by a chi-square distribution with $k - 1$ degrees of freedom, where k = the number of populations. Minitab's KRUSKAL-WALLIS command does all this work for us.

First we enter the data into two columns as we did for ONEWAY, in Section 10.1. All 55 observations go into one column called "Charred," and integers for the labs go into a second column, called "Labs." Next, we use KRUSKAL-WALLIS to get the output shown in Exhibit 13.11.The output gives the p-value for the test, 0.007. This means that the chance of observing five samples as separated as these, when in fact the populations have the same median, is only 0.007. We therefore have statistical evidence that the two populations differ.

Some authors (e.g., see E. L. Lehman, *Nonparametrics: Statistical Methods Based on Ranks.* New York: Wiley, 1975), suggest adjusting H when there are ties in the data, so that the chi-square approximation is better. Minitab prints this adjusted value. Minitab also prints a z-value, z_i, for each group i. Under the null hypothesis, each z_i is approximately normal, with mean 0 and variance 1. The value of z_i indicates how the mean rank, $\overline{R}_i$, for group i differs from the mean rank, $\overline{R}$, for all N observations.

Exhibit 13.11   Kruskal-Wallis Test

```
MTB > KRUSKAL-WALLIS 'Charred' 'Lab'

LEVEL      NOBS      MEDIAN   AVE. RANK    Z VALUE
   1        11       3.100       26.1       -0.44
   2        11       3.600       35.7        1.78
   3        11       3.300       26.1       -0.44
   4        11       2.900       14.7       -3.07
   5        11       3.700       37.4        2.18
OVERALL     55                   28.0

H = 14.19   d.f. = 4   p = 0.007
H = 14.26   d.f. = 4   p = 0.007 (adjusted for ties)
```

KRUSKAL-WALLIS on data in C levels in C

Does a Kruskal-Wallis test. The first column contains all the data; the second says what level (group or treatment) each observation belongs to. The number for levels must be integers.

Stat > Nonparametrics > Kruskal-Wallis

Exercises

13-17 (a)   Using the azalea data for variety *A*, do a Kruskal-Wallis test to see whether ozone damage varied over the five weeks.

    (b)   Do the same test using the data for variety *B*. How do the results compare to those for variety *A*?

13-18 A study was done to compare Pinot Noir wine made in three different regions. Wine samples from each region were taken and scored by a panel of judges on a number of characteristics. The scores for three characteristics are given below. The higher the number, the more favorable the rating.

| Region 1 | | | Region 2 | | | Region 3 | | |
Aroma	Body	Flavor	Aroma	Body	Flavor	Aroma	Body	Flavor
3.3	2.8	3.1	3.3	5.4	4.3	4.3	4.3	3.9
4.4	4.9	3.5	3.4	5.0	3.4	5.1	4.3	4.5
3.9	5.3	4.8	4.7	4.1	5.0	5.9	5.7	7.0
3.9	2.6	3.1	4.1	4.0	4.1	7.7	6.6	6.7
5.6	5.1	5.5	4.3	4.6	4.7	7.1	4.4	5.8
4.6	4.7	5.0	5.1	4.9	5.0	5.5	5.6	5.6
4.8	4.8	4.8	3.9	4.4	5.0	6.3	5.4	4.8
5.3	4.5	4.3	4.5	3.7	2.9	5.0	5.5	5.5
4.3	3.9	4.7	5.2	4.3	5.0	6.4	5.4	6.6
4.6	4.1	4.3				5.5	5.3	5.3
3.9	4.0	5.1				6.0	5.4	5.7
4.2	3.8	3.0				6.8	5.0	6.0
3.3	3.5	4.3						
5.0	5.7	5.5						
3.5	4.7	4.2						
4.3	5.5	3.5						
5.2	4.8	5.7						

(a)   Use appropriate displays to compare the aromas of wines made in the different regions. Also do a Kruskal-Wallis test.
(b)   Repeat part (a) using the other two characteristics, body and flavor.
(c)   Compare the results for all three characteristics.

13.5 Technical Details (Optional)

Sign Confidence Interval

We continue with the restaurant example from Section 13.1. Suppose we form the confidence interval that goes from the fourteenth smallest to the fourteenth largest observation. Let X be the number of observations that are less than η. If $X = 0, 1, \ldots$, or 13, the fourteenth smallest observation must be above η; if $X = 25, 26, \ldots$, or 37, the fourteenth largest observation must be below η. In both cases, η is not in the confidence interval. On the other hand, if $X = 14, 15, \ldots$, or 24, then η is in the confidence interval. Therefore $P(\eta$ is in the confidence interval) $= P(X = 14, 15, \ldots, 24) = 1 - [P(X = 0, 1, \ldots, 13) + P(X = 25, 26, \ldots, 37)]$. For a binomial with $n = 37$ and $p = 1/2$, $P(X = 0, 1, \ldots, 13) = P(X = 25, 26, \ldots, 37) = .0494$. Therefore, $P(\eta$ is in the confidence interval) $= 1 - 2(.0494) = .9012$. (Note that the output in Exhibit 13.3 shows .9011. The difference is due to round-off error.)

In general, the confidence of the interval that goes from the dth smallest to the dth largest observation is given by the formula $1 - 2P(X < d)$. The third interval in Exhibit 13.3 goes from the thirteenth smallest to the thirteenth largest observation. Therefore, it has confidence $1 - 2P(X < 13)$ $= 1 - 2P(X < 12) = 1 - 2(.0235) = .9530$.

The problem of ties, which we discussed for the sign test, does not arise in constructing a confidence interval. Therefore, we always use all the data for the interval.

Wilcoxon Signed-Rank Test and Confidence Interval

In order to explain the methods used by Minitab, we will use a very simple example and the one-sample case. Exhibit 13.12 contains a sample of eight observations. We will test H_0: $\eta = 10$. The first line contains the eight observations. If there are any observations that are equal to $\eta = 10$, we omit these and do the Wilcoxon test using the remaining observations. There are no "ties" in our example. Next, we subtract the null hypothesis value of η from each observation. Then we take absolute values of these differences. Next, we assign ranks. Finally we compute W as the sum of all the ranks corresponding to the observations that were above 10 (i.e., had positive differences).

The way we calculate the p-value for the test depends on the alternative hypothesis. Suppose we use H_1: $\eta > 10$. The p-value is then $P(W >$ the value of W in our sample$) = P(W \geq 32)$.

Suppose we test H_0: $\eta = 10$ versus H_1: $\eta \neq 10$. Since this is a two-sided test, we need two probabilities: one for small values of W and one for large values of W. In our sample, $W = 32$. Since this value is above the mean of W, the probability for large values of W is $P(W \geq 32)$. The probability for small values is then given by $P(W \leq R - 32)$, where R is the sum of all the ranks. The value of R is given by the formula $R = n(n + 1)/2$. For $n = 8$,

Exhibit 13.12 Calculating the Wilcoxon Statistic W for H_0: $\eta = 10$

X	20	18	23	5	14	8	18	22
$X - 10$	10.	8	13	−5	4	−2	8	12
$\lvert X - 10 \rvert$	10	8	13	5	4	2	8	12
Ranks	6	4.5	8	3	2	1	4.5	7

$W = 6 + 4.5 + 8 + 2 + 4.5 + 7 = 32$

$R = 8(9)/2 = 36$. Thus, $P(W \leq R - 32) = P(W \leq 4)$. Since the distribution of W is symmetric, $P(W \geq 32) = P(W \leq 4)$. Thus, the p-value $= P(W \leq 4) + P(W \leq 32) = 2P(W \geq 32)$.

Suppose we test H_0: $\eta = 20$ versus H_1: $\eta < 20$ using the data in Exhibit 13.12. Here the p-value is $P(W \leq$ the value of W in our sample). Notice that there is one observation equal to 20. We set aside this observation when we do the test. This gives $n = 7$ and $W = 6$. The p-value is then $P(W \leq 6)$.

There is a second way to calculate the value of W; this is the method that we will use to find a point estimate of η and get a confidence interval. Exhibit 13.13 shows the calculations for testing H_0: $\eta = 10$ using the data in Exhibit 13.12. We first write the observations across the top and down the left side of the table. Each entry in the table is the average (often called the Walsh average) of the observations in the corresponding row and column. The entries in the bottom part of the table are not used because they are just repeats of the entries in the top part of the table. Therefore, we did not write them in. Underneath this table are the ordered Walsh averages. We can compute the Wilcoxon statistic W from the ordered Walsh averages: $W = $ (number of Walsh averages above 10) $+ 1/2$(number of Walsh averages equal to 10) $= 32 + 1/2(0) = 32$, which is the same value we got in Exhibit 13.12.

Since we are working with a population that we believe to be approximately symmetric, there are several ways to estimate η. We might use the sample mean or the sample median. Or we could use the median of all the Walsh averages, a method that involves both averaging and finding a median. For the data in Exhibit 13.13, there are 36 Walsh averages, so the median of the Walsh averages is the average of the eighteenth and nineteenth ordered averages. These are circled in Exhibit 13.13. The average of these two values is $(16 + 17)/2 = 16.5$. This is the estimate of η that WTEST and WINTERVAL print.

The Walsh averages can also be used to form a confidence interval for η. The method is very similar to the method we used for sign confidence intervals. If our interval goes from the dth smallest to the dth largest Walsh average, then its confidence is given by the formula $1 - 2P(W < d)$. Minitab uses a normal approximation with a continuity correction to calculate $P(W < d)$.

Mann-Whitney Test and Confidence Interval

In Section 13.3, we calculated W from ranks. There is a second way to calculate the statistic W; this is the method we will use to find a point estimate and confidence interval for $\eta_1 - \eta_2$. Exhibit 13.14 shows the calculations that are used when this technique is done by hand. The first sample is

Exhibit 13.13   Walsh Averages for Use in Wilcoxon Procedure

		Data							
		20	*18*	*23*	*5*	*14*	*8*	*18*	*22*
	20	20	19	21.5	12.5	17	14	19	21
	18		18	20.5	11.5	16	13	18	20
	23			23	14	18.5	15.5	20.5	22.5
Data	5				5	9.5	6.5	11.5	13.5
	14					14	11	16	18
	8						8	13	15
	18							18	20
	22								22

Walsh Averages in Order from Smallest to Largest

5	6.5	8	9.5	11	11.5	11.5	12.5	13	13
13.5	14	14	14	15	15.5	16	(16)	(17)	18
18	18	18	18.5	19	19	20	20	20	20.5
20.5	21	21.5	22	22.5	23				

written down the left side of the table; the second sample, across the top. Each entry inside the table is $X - Y$, where X is the observation in the corresponding row and Y is the observation in the corresponding column. This gives a total of $8 \times 14 = 112$ differences. We can calculate the Mann-Whitney statistic, W, using the formula $W = $ (number of positive differences) + $(1/2)$(number of zeroes) + $n_1(n_1 + 1)/2$, where n_1 is the number of observations in the first sample. In the example, $W = 89 + (1/2)(3) + 8(9)/2 = 126.5$, the value we got in Exhibit 13.8.

We estimate the difference in the population medians $(\eta_1 - \eta_2)$ by the median of these differences. Since there are 112 differences, the median is the average of the 56th and 57th observations. These are circled in Exhibit 13.14. The average of these two values is $(.7 + .7)/2 = .7$. The confidence interval is also based on these differences. It goes from the dth smallest difference to the dth largest difference, where the value of d depends on the confidence you specify. Minitab does not print the value of d, but just gives the confidence interval. Here the confidence interval, given in Exhibit 13.9, goes from .2 to 1.2.

If there is no difference between the two populations, the mean of W is $n_1(n_1 + n_2 + 1)/2$, where $n_1 = $ number of observations in the first sample and $n_2 = $ number of observations in the second sample. Here this average is $8(8 + 14 + 1)/2 = 92$. If W is much larger than 92, many numbers in the first

Exhibit 13.14   Calculations Used by MANN-WHITNEY

First Sample						Second Sample								
	1.2	1.8	1.8	2.3	1.3	3.0	2.2	1.3	1.5	1.6	1.3	1.5	2.7	2.0
2.6	1.4	.8	.8	.3	1.3	−.4	.4	1.3	1.1	1.0	1.3	1.1	−.1	.6
2.0	.8	.2	.2	−.3	.7	−1.0	−.2	.7	.5	.4	.7	.5	.7	0.0
1.7	.5	−.1	−.1	−.6	.4	−1.3	−.5	.4	.2	.1	.4	.2	−1.0	−.3
2.7	1.5	.9	.9	.4	1.4	−.3	.5	1.4	1.2	1.1	1.4	1.2	0.0	.7
2.5	1.3	.7	.7	.2	1.2	−.5	.3	1.2	1.0	.9	1.2	1.0	−.2	.5
2.6	1.4	.8	.8	.3	1.3	−.4	.4	1.3	1.1	1.0	1.3	1.1	−.1	.6
2.5	1.3	.7	.7	.2	1.2	−.5	.3	1.2	1.0	.9	1.2	1.0	−.2	.5
3.0	1.8	1.2	1.2	.7	1.7	0.0	.8	1.7	1.5	1.4	1.7	1.5	.3	1.0

The differences in order:

−1.3	−1.0	−1.0	−.7	.6	−.5	−.5	−.5	−.4	−.4
−.3	−.3	−.3	−.2	−.2	−.2	−.1	−.1	−.1	−.1
0.0	0.0	0.0	.1	.2	.2	.2	.2	.2	.2
.3	.3	.3	.3	.3	.4	.4	.4	.4	.4
.4	.4	.5	.5	.5	.5	.5	.5	.6	.6
.7	.7	.7	.7	.7	(.7)	(.7)	.7	.7	.8
.8	.8	.8	.8	.8	.9	.9	.9	.9	1.0
1.0	1.0	1.0	1.0	1.0	1.0	1.0	1.1	1.1	1.1
1.1	1.2	1.2	1.2	1.2	1.2	1.2	1.2	1.2	1.2
1.2	1.3	1.3	1.3	1.3	1.3	1.3	1.3	1.3	1.4
1.4	1.4	1.4	1.4	1.4	1.5	1.5	1.5	1.7	1.7
1.7	1.8								

sample must have been large. In that case, we might suspect that the first population has a larger median. If W is much smaller than 92, many numbers in the first sample must have been small, so we might suspect that the first population has a smaller median than the second.

The exact distribution of W is given in special tables (see M. Hollander and D. Wolfe, *Nonparametric Statistical Methods*. New York: Wiley, 1973). If we do not have such tables, we can use the fact that W has an approximately normal distribution, with variance $n_1 n_2 (n_1 + n_2 + 1)/13$. (A slightly more complicated version of this formula is often used when there are ties.) In our example, the variance is $(8)(14)(8 + 14 + 1)/12 = 214.67$ and, therefore, the standard deviation is $\sqrt{214.67} = 14.65$. If the two populations were the same, W would be near its mean of 92. But, in fact, W is 126.5, more than 2.3 standard deviations above 92.

PART THREE
Additional Topics

14

Control Charts

14.1 Control Charts and Process Variation

The concept of statistical process control was developed by Walter Shewhart in the 1920s to help managers at Western Electric (part of the Bell Telephone System) understand and detect changes in manufacturing processes. His ideas were used only on a limited basis in the United States. Then, after World War II, Japan adopted them and started the quality revolution that eventually enabled Japan to produce products of very high quality and at reasonable cost. Now manufacturing and service industries worldwide are using process-control ideas to improve quality and reduce waste. Control charts, discussed in this chapter, are one of the techniques used to study a process.

A process is simply a sequence of steps that results in an outcome. A receptionist answers the phone and directs the call to the appropriate person. A laboratory measures the amount of cholesterol in a sample of blood. A machine fills bottles with 12 ounces of soda. In each case, the specific activity is repeated again and again over time. Ideally each activity would be done "perfectly." The receptionist would answer each call on the first ring and direct the call to the correct person. The laboratory would determine the exact amount of cholesterol in each sample of blood. The machine would fill each bottle with precisely 12 ounces of soda. But this is the real world, and processes vary.

There are two basic types of process variation. Common-cause variation results from chance or the inherent variability in the system. The only way to reduce this type of variation is to change the process itself. For example, in the case of the receptionist, we might install a different type of phone system, create a chart so that the receptionist can more quickly and accurately determine to whom the calling party should be referred, or reduce distractions and noise in the office so the receptionist can work more efficiently.

The second type of variation, special-cause variation, results from special causes. In the receptionist example, perhaps the company placed some extra ads one month, causing more calls than the receptionist could handle quickly, so response time went up. A process is considered to be in control if there are no special causes affecting it, only common causes.

Control charts are used to study a process over time. There are two basic types: charts for measurement data and charts for attribute data (count data). If we were to record the time from when a call first comes into a company until the calling party reaches the appropriate person within the company, we would have measurement data. If we were to record for each day, the total number of calls and the number that were incorrectly referred, we would have attribute data.

14.2 Control Charts for Measurement Data

Charts for Means

One part of the process of making ignition keys for automobiles consists of cutting grooves into raw key blanks. Some of the groove dimensions are critical to the proper functioning of the keys. Suppose we wish to study one critical groove. Keys are produced at a high volume, so we take a sample of five keys every twenty minutes and carefully measure the dimension of this groove. Each sample is often called a subgroup. Data are shown in Exhibit 14.1. Here there are $k = 20$ subgroups and each subgroup has $n = 5$ observations. We use x_{ij} to denote observation j in subgroup i. We calculated three quantities for each subgroup: the mean (denoted $\bar{x}_i$), the standard deviation (denoted s_i), and the range (denoted r_i). We will use these to create control charts.

A control chart for means first plots the sample means versus the sample numbers. This gives us an overall picture of how the sample means vary over time.

If the process is in control, the individual observations will all have come from the same population, and thus will all have the same population mean, μ, and standard deviation, σ. The sample means will then have mean μ and standard deviation $\sigma/\sqrt{n}$. In a normal distribution, almost all observations (99.7%) are within three standard deviations of the mean. We add

Exhibit 14.1   Data for Ignition-Key Study (measurements are in thousandths of an inch)

Sample Number			Samples				Standard	
	x1	x2	x3	x4	x5	Means	Deviations	Ranges
1	6.1	8.4	7.6	7.6	4.4	6.82	1.588	4.0
2	8.8	8.3	7.6	7.4	5.9	7.60	1.102	2.9
3	8.0	8.0	9.4	7.5	7.0	7.98	0.896	2.4
4	6.7	7.6	6.4	7.1	8.8	7.32	0.942	2.4
5	8.7	8.4	8.8	9.4	8.6	8.78	0.377	1.0
6	7.1	5.2	7.2	8.8	5.2	6.70	1.526	3.6
7	7.8	8.9	8.7	6.5	6.8	7.74	1.083	2.4
8	8.7	9.4	8.6	7.3	7.1	8.22	0.983	2.3
9	7.4	8.1	8.6	8.3	8.7	8.22	0.517	1.3
10	8.1	6.5	7.5	8.9	9.7	8.14	1.236	3.2
11	7.8	9.8	8.1	6.2	8.4	8.06	1.292	3.6
12	8.9	9.0	7.9	8.7	9.0	8.70	0.464	1.1
13	8.7	7.5	8.9	7.6	8.1	8.16	0.631	1.4
14	8.4	8.3	7.2	10.0	6.9	8.16	1.222	3.1
15	7.4	9.1	8.3	7.8	7.7	8.06	0.666	1.7
16	6.9	9.3	6.4	6.0	6.4	7.00	1.325	3.3
17	7.7	8.9	9.1	6.8	9.4	8.38	1.094	2.6
18	8.9	8.1	7.3	9.1	7.9	8.26	0.740	1.8
19	8.1	9.0	8.6	8.7	8.0	8.48	0.421	1.0
20	7.4	8.4	9.2	7.4	10.3	8.54	1.240	2.9

three lines to the plot to help us judge the variability in the data: The center line is at the mean; the upper control limit (UCL) is three standard deviations above the mean; and the lower control limit (LCL) is three standard deviations below the mean.

In order to draw these lines, we must estimate μ and σ. We estimate μ by $\bar{\bar{x}}$, often called the grand mean:

$$\text{center line} = \bar{\bar{x}} = \frac{1}{k}\Sigma_i \bar{x}_i = \frac{1}{kn}\Sigma_i\Sigma_j x_{ij}$$

Several different methods are commonly used to estimate σ. Minitab gives you a choice of two. By default Minitab uses the pooled standard deviation, s_p, defined by

$$s_p = \sqrt{\Sigma_i \Sigma_j (x_{ij} - \bar{x}_i)^2 / kn}$$

Then the control limits are

$$\text{UCL} = \bar{\bar{x}} + 3s_p / \sqrt{n}$$

$$\text{LCL} = \bar{\bar{x}} - 3s_p / \sqrt{n}$$

The subcommand RBAR says to estimate σ with s_r, defined by

$s_r = \bar{r}/d_n$, where $\bar{r} = \Sigma r_i / k$, and d_n is a value from special table[1]

Then the control limits are

$$\text{UCL} = \bar{\bar{x}} + 3s_r / \sqrt{n}$$

$$\text{LCL} = \bar{\bar{x}} - 3s_r / \sqrt{n}$$

Exhibit 14.2 shows an $\bar{x}$-chart for the ignition-key data. We first enter all the data, in order, into one column, C1; then we use XBARCHART to create the plot. The first argument on XBARCHART, C1, is the column of data; the second, 5, is the subgroup size. The subcommand TEST asks Minitab to do tests for special causes, which we will discuss later.

The overall mean, $\bar{\bar{x}}$, is equal to 7.966. The control limits are straight lines at 6.578 and 9.354. All the sample means are within these limits.

Finding Problems. A process is considered to be in control if no special causes are influencing the variability in the process, only common causes. Only when a process is in control can we hope to make improvements in that process.

A process can be out of control for many different reasons. One sample could be unusually large or small. The mean level could change so that from some point on, the mean is larger (or smaller) than it was in the past. There could be a trend such that the mean increases (or decreases) as time goes on. There could be cycles, in which the mean decreases for, say, five samples, then returns to its former level, then decreases for another five samples, and so on.

Control charts help us find unusual patterns. The next step is to determine what caused the patterns, by carefully examining all aspects of the process. For example, we might discover that one sample was unusually

---

[1]See standard books on control charts, such as H. M. Wadsworth, K. S. Stephens, and A. B. Godfrey, *Modern Methods for Quality Control and Improvement*. New York: Wiley, 1986.

Exhibit 14.2   An $\overline{X}$-Chart of the Ignition-Key Data

```
MTB > name c1 'Groove'
MTB > set c1
DATA>    6.1      8.4     7.6     7.6     4.4
DATA>    8.8      8.3     7.6     7.4     5.9
DATA>    8.0      8.0     9.4     7.5     7.0
DATA>    6.7      7.6     6.4     7.1     8.8
DATA>    8.7      8.4     8.8     9.4     8.6
DATA>    7.1      5.2     7.2     8.8     5.2
DATA>    7.8      8.9     8.7     6.5     6.8
DATA>    8.7      9.4     8.6     7.3     7.1
DATA>    7.4      8.1     8.6     8.3     8.7
DATA>    8.1      6.5     7.5     8.9     9.7
DATA>    7.8      9.8     8.1     6.2     8.4
DATA>    8.9      9.0     7.9     8.7     9.0
DATA>    8.7      7.5     8.9     7.6     8.1
DATA>    8.4      8.3     7.2    10.0     6.9
DATA>    7.4      9.1     8.3     7.8     7.7
DATA>    6.9      9.3     6.4     6.0     6.4
DATA>    7.7      8.9     9.1     6.8     9.4
DATA>    8.9      8.1     7.3     9.1     7.9
DATA>    8.1      9.0     8.6     8.7     8.0
DATA>    7.4      8.4     9.2     7.4    10.3
DATA> end
MTB > xbarchart 'Groove' 5;
SUBC>    test 1:8.
                      X-bar Chart for Groove
      -
      -------------------------------------------------UCL=9.354
S     -
a  9.00+
m     -              +
p     -                                +                    +
l     -                                             +    +
e     -                    + + +        + +          +      =
   8.00+------+------------------+--------+----------X=7.966
M     -               +
e     -     +
a     -         +
n     -
   7.00+                                   +
   - +
      ------------+----------------------------------LCL=6.578
      -
      +----------+----------+----------+----------+
      0          5         10         15         20
                      Sample Number
```

low because the raw material used to produce that sample was mistakenly taken from an old batch. There could be a downward trend in the mean because a chemical used to produce the samples gradually deteriorated over time owing to improper storage conditions. We attempt to remove as many of the special causes of variation as we can in order to bring the process into control.

The TEST subcommand of XBARCHART will do up to eight tests for unusual patterns. In all cases, these are patterns that are extremely unlikely to occur if the process is under control. The patterns are summarized in Exhibit 14.3. List the tests you want, by number, in TEST. We requested all eight in Exhibit 14.2. Nothing was indicated, so the chart passed all the tests. Therefore, as far as we can tell, this process is in control.

The fact that a process is in control, however, does not mean the process is acceptable. For example, suppose that in the ignition-key study, the value of $\bar{\bar{x}}$ was 10.00, not 7.966. Suppose also that all points were within the control limits and no unusual patterns were detected. From a statistical standpoint, the process is in control. From a practical standpoint, keys produced by this company are probably not very good.

Charts for Dispersion

Two charts are used to study the spread, or dispersion, of a process. An s-chart plots the standard deviation of each sample; an r-chart plots the range of each sample. These values were calculated in Exhibit 14.1.

Exhibit 14.4 shows an s-chart for the data from Exhibit 14.1. As with an $\bar{x}$ chart, there is a center line, a UCL, and an LCL. To draw these lines, we need to know the mean and the standard deviation of s. This requires some mathematical derivations and special formulas. Fortunately, Minitab does the work for you. Formulas for the three lines are given below. The value of $c_4(n)$ and $c_5(n)$ are taken from special tables and depend on the sample size.[2]

$$\text{center line} = c_4(n)\sigma$$
$$\text{UCL} = c_4(n)\sigma + 3c_5(n)\sigma$$
$$\text{LCL} = c_4(n)\sigma - 3c_5(n)\sigma \quad \text{(if LCL} < 0 \text{ then set LCL} = 0)$$

---

[2]See standard books on control charts, such as H. M. Wadsworth, K. S. Stephens, and A. B. Godfrey, *Modern Methods for Quality Control and Improvement*. New York: Wiley, 1986.

Exhibit 14.3   Eight Tests for Special Causes

Test 1	Test 2	Test 3	Test 4
One point beyond zone A	Nine points in a row on same side of center line	Six points in a row all increasing or all decreasing	Fourteen points in a row alternating up and down

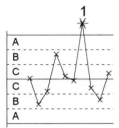
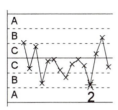
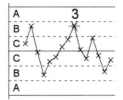
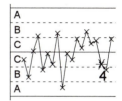

Test 5	Test 6	Test 7	Test 8
Two out of three points in a row in zone A or beyond (one side of center line)	Four out of five points in a row in zone B or beyond (one side of center line)	Fifteen points in a row in zones C (both sides of center line)	Eight points in a row beyond zones C (both sides of center line)

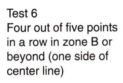
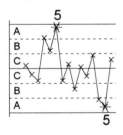
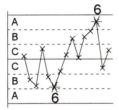
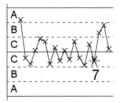
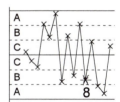

Note: The solid line in the center is the center line. The dotted lines are at one sigma limit and two sigma limits beyond the center line. The solid lines at the top and bottom are at three sigma limits beyond the center line.

XBARCHART for measurements in **C** and sample sizes in **E**

> **RBAR**
> **SLIMITS** are **K...K**
> **TESTS K...K**
>
> **ESTIMATE** using only samples **K...K**
> **MU**   = **K**
> **SIGMA** = **K**
>
> **INCREMENT** = **K**
> **START** at **K** [end at **K**]
> **TITLE**   'text'
> **FOOTNOTE** 'text'
> **XLABEL**   'text'
> **YLABEL**   'text'

Prints an $\bar{x}$-chart, a chart for the sample means. If all subgroups have the same size, you can use a constant for E.

By default, Minitab uses the pooled standard deviation to estimate the process standard deviation. RBAR tells Minitab to use an estimate based on the sample ranges.

By default, Minitab calculates the UCL and LCL based on three standard deviations. SLIMITS allows you to specify the number of standard deviations you want. You may list one or more values.

Minitab can do eight different tests, numbered 1, 2, . . . , 8, to determine whether there are any problems. List the numbers of the tests you want in TEST. You can use a colon to abbreviate a consecutive list.

By default, Minitab uses all the samples to calculate the center line and control limits. ESTIMATE tells Minitab to use only certain samples. List the numbers of the samples you want. You can use a colon to indicate a range. For example, ESTIMATE 1:20 35:50 would use only samples 1 through 20 and 35 through 50.

If you have historical values for μ and σ, you can specify these on MU and SIGMA. Minitab will use these values to calculate the center line and control limits.

You may specify the scale on the y-axis. INCREMENT is the distance between tick marks (the + symbols). START specifies the first and, optionally, the last point plotted on the y-axis. Points outside are omitted from the plot.

You may add titles, footnotes, and axis labels with subcommands. Text must be enclosed in single quotes.

Release 8: **Stat > SPC Charts > Xbar**
Windows: Character control charts are not in the menu. You must type the commands.

Exhibit 14.4 *S*-Chart for the Ignition-Key Data

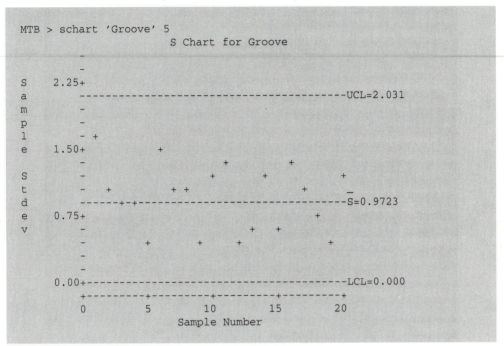

```
MTB > schart 'Groove' 5
                          S Chart for Groove
```

SCHART for measurements in **C** and sample sizes in **E**

RBAR
SLIMITS are **K...K**

ESTIMATE using only samples **K...K**
SIGMA = K

Prints an *s*-chart, a chart for standard deviations. If all subgroups have the same size, you can use a constant for E.

SCHART can also use the subcommands INCREMENT, START, TITLE, FOOTNOTE, XLABEL, and YLABEL. The subcommands for SCHART are all the same as for XBARCHART.

Release 8: **Stat > SPC Charts > S**
Windows: Character control charts are not in the menu. You must type the commands.

In order to calculate these three lines from data, we must estimate σ. Recall that Minitab gives you two choices, s_p and s_r, defined in the section on $\bar{x}$-charts. The formulas for the UCL and LCL are approximations. Therefore, it is possible for the LCL to be less than zero. If this happens, Minitab sets the LCL equal to zero.

Exhibit 14.5 shows an r-chart of the data from Exhibit 14.1. As in an s-chart, there is a center line, a UCL, and an LCL. These three lines are based on the formulas given below. The value of $d_2(n)$ and $d_3(n)$ are taken from special tables and depend on the sample size.[3]

$$\text{center line} = d_2(n)\sigma$$
$$\text{UCL} = d_2(n)\sigma + 3d_3(n)\sigma$$
$$\text{LCL} = d_2(n)\sigma - 3d_3(n)\sigma \quad (\text{if LCL} < 0 \text{ then set LCL} = 0)$$

In order to calculate these three lines from data, we must estimate σ. Minitab offers two choices, s_p and s_r, defined in the section on $\bar{x}$-charts. The formulas for the UCL and LCL are approximations. Therefore, it is possible for the LCL to be less than zero. If this happens, Minitab sets the LCL equal to zero.

Finding Problems. If a process is in control, the values plotted in an s-chart (or an r-chart) should vary about the center line, be within the control limits, and show no special patterns. What problems would we typically detect in a chart for spread? One value of s (or r) could be unusually large. The variability could change so that from some point on, s (or r) is larger than it was in the past. It is also possible for spread to decrease, but this is usually not a problem but a desirable goal. We would still investigate what was going on, but in the hope of using what we might find to improve the process.

As we mentioned for the $\bar{x}$-chart, a process that is in control is not necessarily a process that is acceptable. For example, suppose that in the ignition-key study the center line was at 7.966, all points were within the control limits, and no unusual patterns were detected. Thus, from a statistical standpoint, the process is in control. Suppose, however, the groove dimension for individual keys varied from 3 to 13. This much variability may not be acceptable.

[3]See standard books on control charts, such as H. M. Wadsworth, K. S. Stephens, and A. B. Godfrey, *Modern Methods for Quality Control and Improvement*. New York: Wiley, 1986.

Exhibit 14.5   *R*-Chart for the Ignition-Key Data

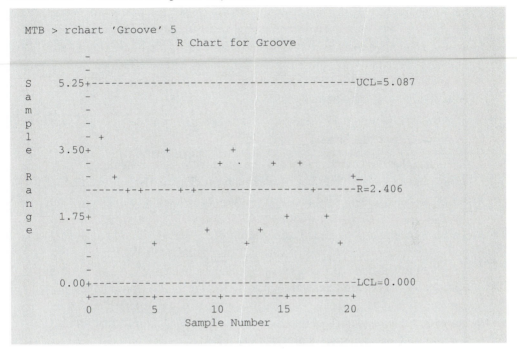

```
MTB > rchart 'Groove' 5
                            R Chart for Groove
         -
         -
  S    5.25+-------------------------------------------------UCL=5.087
  a      -
  m      -
  p      -
  l      -  +
  e    3.50+              +              +
         -                       +   .      +     +
  R      -      +                                  +_
  a      -------+-+------+-+-------------------+------R=2.406
  n      -
  g    1.75+                            +        +
  e      -
         -          +          +          +
         -
         -
       0.00+-------------------------------------------------LCL=0.000
         +---------+---------+---------+---------+
         0         5        10        15        20
                      Sample Number
```

RCHART for measurements in **C** and sample sizes in **E**

> **RBAR**
> **SLIMITS** are **K...K**
>
> **ESTIMATE** using only samples **K...K**
> **SIGMA = K**

Prints an *r*-chart, a chart for ranges. If all subgroups have the same size, you can use a constant for E.

RCHART can also use the subcommands INCREMENT, START, TITLE, FOOTNOTE, XLABEL, and YLABEL. The subcommands for RCHART are the same as for XBARCHART.

Release 8: **Stat > SPC Charts > R**
Windows: Character control charts are not in the menu. You must type the commands.

When we first study a process, we usually do a chart for dispersion before an $\bar{x}$-chart. If the chart for dispersion shows the process is not in control, we cannot properly interpret an $\bar{x}$-chart. In order to draw the UCL and LCL on an $\bar{x}$-chart, we need an estimate of s. If the dispersion of the process is not in control, the limits on the $\bar{x}$-chart will be affected.

SCHART and RCHART do not have the subcommand TEST to do automatic tests for special causes.

Unequal Subgroup Sizes

The examples we have discussed so far are for the case in which all subgroups have the same number of observations. This is the most common situation. Minitab will also produce charts when the subgroups are not all equal. In this case, you must supply a column containing integers to tell Minitab which subgroup each observation is in. This is similar to the way you enter data in the analysis of variance command, ONEWAY.

Exhibit 14.6 does an $\bar{x}$-chart using the data from Exhibit 14.1 with some observations removed. Three subgroups, 10, 11, and 12, now have only three observations. Since the subgroups are not all equal, we must tell Minitab which subgroup each observation is in. We enter this information into C2 and use C2 as the second argument in XBARCHART.

The control limits for the $\bar{x}$-chart are no longer straight lines. The limits are wider for the subgroups with only three observations. This is not surprising: when a sample mean is based on fewer observations, the standard deviation of that sample mean is larger.

Below, we give the formulas for the center line and control limits when s_p is used to estimate σ. We use n_i for the number of observations in subgroup i. Consult the Minitab manual for the formulas using s_p.

$$\text{center line} = \bar{\bar{x}} = \frac{i}{N}\Sigma_i\Sigma_j x_{ij}, \text{ where } N = \Sigma_i n_i$$

$$s_p = \sqrt{[\Sigma_i\Sigma_j(x_{ij} - \bar{x}_i)^2]/[\Sigma_i(n_i - 1)]}$$

$$\text{UCL} = \bar{\bar{x}} + 3s_p/\sqrt{n_i}$$

$$\text{LCL} = \bar{\bar{x}} - 3s_p/\sqrt{n_i}$$

Exhibit 14.6   Data for the Ignition-Key Study, with Several Observations Removed

```
MTB > name c1 'Groove' c2 'Count'
MTB > set 'Groove'
DATA>    6.1     8.4     7.6     7.6     4.4
DATA>    8.8     8.3     7.6     7.4     5.9
DATA>    8.0     8.0     9.4     7.5     7.0
DATA>    6.7     7.6     6.4     7.1     8.8
DATA>    8.7     8.4     8.8     9.4     8.6
DATA>    7.1     5.2     7.2     8.8     5.2
DATA>    7.8     8.9     8.7     6.5     6.8
DATA>    8.7     9.4     8.6     7.3     7.1
DATA>    7.4     8.1     8.6     8.3     8.7
DATA>    8.1     6.5     7.5
DATA>    7.8     9.8     8.1
DATA>    8.9     9.0     7.9
DATA>    8.7     7.5     8.9     7.6     8.1
DATA>    8.4     8.3     7.2    10.0     6.9
DATA>    7.4     9.1     8.3     7.8     7.7
DATA>    6.9     9.3     6.4     6.0     6.4
DATA>    7.7     8.9     9.1     6.8     9.4
DATA>    8.9     8.1     7.3     9.1     7.9
DATA>    8.1     9.0     8.6     8.7     8.0
DATA>    7.4     8.4     9.2     7.4    10.3
DATA> end
MTB > set 'Count'
DATA>    (1:9)5 (10:12)3 (13:20)5
DATA> end
MTB > xbarchart 'Groove' Count'
                        X-bar Chart for Groove
            -
    10.00+
            -                       ------
  S         -
  a         -------------------           ----------------UCL
  m         -
  p   8.75+              +
  l         -                     +  +           +     +  +
  e         -             +  +          +  +        +     =
            ------+-----------------------------+----------X=7.933
  M         -             +
  e   7.50+    +
  a         -       +             +
  n         -                                    +
            -  +          +
            -------------------           ----------------LCL
    6.25+                      ------
            -
            +---------+---------+---------+---------+
            0         5        10        15        20
                        Sample Number
```

Exercises

14-1   Procter & Gamble was involved in studying a process that tightens the caps
on containers of a hair conditioner. Cap torque, a measure of how tightly
the caps are screwed onto the containers, was recorded for 17 samples, each
of size 4. Data are given below.

Sample Number	Samples			
	1	2	3	4
1	24	14	18	27
2	17	32	31	27
3	21	27	24	21
4	24	26	31	34
5	28	32	24	16
6	22	37	36	21
7	16	17	22	34
8	20	19	16	16
9	18	30	21	16
10	14	15	14	14
11	25	15	16	15
12	19	15	15	19
13	19	30	24	10
14	15	17	17	21
15	34	22	17	15
16	17	20	17	20
17	15	17	24	20

(a)   Do an r-chart to check the dispersion of cap torque. Does it appear to
be in control?

(b)   Do an $\bar{x}$-chart to check the mean level. Does it appear to be in control?

(c)   Use the TEST subcommand to see whether Minitab flags any points
as possible problems. How do these results compare to the results of
your visual inspection in part (b)?

(d)   Display a histogram and a normal probability plot of the individual
observations. If the process is in control and data come from a normal
distribution, then your displays should look "normal." Do they?

14-2   The TEST subcommand helps you find unusual patterns in a control chart.
There is a problem, however—the problem you always have when you do
many tests. You may find a problem when nothing is actually wrong.

(a)   Test 1 detects points that are more than 3 sigma limits from the mean.
Suppose a process is generated from a normal distribution with a

constant mean and standard deviation, and all observations are independent. That is, suppose there are no problems with the process. What is the probability that a sample will fail test 1? Does this depend on the process mean, standard deviation, or sample size?

(b) Suppose you were to chart a process like the one described in part (a). Each day you collect 50 samples, each of size 5, and do an $\bar{x}$-chart. You create a separate chart each day for 10 days. Do you think it is likely that at least one chart will fail test 1? On the average, how many of the samples you collect over the 10 days would fail test 1?

(c) Use simulation to create the 10 charts described in part (b). Use the TEST subcommand for all eight tests. Did any chart fail test 1? How many charts failed one or more of the eight tests for special causes?

14-3 (a) Control charts are very useful displays. However, they do not show all the important patterns that may occur in a data set. What other Minitab displays might be useful for investigating the control chart data?

(b) Below, we give a small data set. Use control charts and the displays you listed in part (a) to investigate these data. What are some of the important features?

Sample	Samples				
Number	1	2	3	4	5
1	701	630	687	691	681
2	699	691	676	689	677
3	708	707	694	690	672
4	708	696	692	682	687
5	719	699	681	700	659
6	711	692	673	690	661
7	715	699	679	682	669
8	715	683	687	679	679
9	698	674	675	671	684
10	691	690	679	664	679

14.3 Control Charts for Attribute Data

We use a chart for attribute data when the variable of interest is counts: for example, the number of purchase orders per day that contain one or more errors, or the number of incoming phone calls per day that are misdirected.

Let n_i denote the number of observations in the ith subgroup, and y_i the number of defectives in the ith subgroup. Then $p_i = y_i/n_i$ is the proportion of defectives in the ith subgroup. There are two charts you can do. A p-chart plots the proportions, p_i, and an np-chart plots the counts, n_i. We will discuss the p-chart.

If the process is in control, then the probability of getting a defective will be the same for all samples. We denote this value by p. We estimate p by $\bar{p}$, using the formula

$$\bar{p} = \frac{\text{total number of defectives}}{\text{total number of observations}} = \frac{\Sigma y_i}{\Sigma n_i}$$

The distribution of each y_i is binomial with mean $n_i p$ and variance $n_i p(1 - p)$. If the sample sizes, n_i, are not too small, we can approximate the distribution of each y_i by a normal with the same mean and variance. It then follows, using some basic probability theory, that we can approximate the distribution of each p_i by a normal, with mean p and variance $p(1 - p)/n_i$. Now we can calculate a center line and control limits for the chart.

center line $= \bar{p}$

$$\text{UCL}_i = \bar{p} + 3\sqrt{\bar{p} + \bar{p}(1 - \bar{p})/n_i}$$

$$\text{LCL}_i = \bar{p} - 3\sqrt{\bar{p} + \bar{p}(1 - \bar{p})/n_i} \quad \text{(if LCL} < 0, \text{ set LCL} = 0)$$

Because we are using a normal approximation, it is possible for the LCL to be less than zero. If this happens, Minitab sets the LCL equal to zero.

Exhibit 14.7 shows data from a manufacturing process. Thirty samples, each containing 50 assembled parts, were taken. For each sample, the number of parts that had a plating defect was recorded. We put these counts into C1, then do a p-chart with sample size 50. We also use TEST to do tests for special causes. Only the first four tests from Exhibit 14.3 are used by PCHART. Points 20 and 21 failed test 1. They are both beyond the upper control limit. These samples had 15 and 12 defective parts, respectively. The next step would be to try to determine why these two samples had so many defective parts.

You might wish to redo the p-chart, calculating the center line and control limits with samples 20 and 21 omitted. This would give a better picture of the process. In Exhibit 14.8, we use the subcommand ESTIMATE to do this. Now the center line is slightly lower and the control limits are closer together. All points, except points 20 and 21, are within the limits.

Exhibit 14.7 *P*-Chart for Plating Defects

```
MTB > set c1
DATA>   1   6   5   5   4   3   2   2   4   6   2   1   3   1   4
DATA>   5   4   1   6  15  12   6   3   4   3   3   2   5   7   4
DATA> end
MTB > name c1 'Plating'
MTB > pchart 'Plating' 50;
SUBC>    test 1:4.

                    P Chart for Plating
             -
             -                   1
     0.300+                      *
  P          -
  r          -                  1
  o          -                  *
  p          -
  o   0.200+------------------------------UCL=0.2049
  r          -
  t          -
  i          -                            +
  o          - +         +          +   +
  n   0.100+  ++              +          +  _
           -----+---+-----+-+------+-----+P=0.08600
             -       +      +         + ++
             -       ++  +              +
             -+          + +     +
     0.000+------------------------------LCL=0.000
             +---------+---------+---------+
             0        10        20        30
                    Sample Number

TEST 1. One point beyond zone A.
Test Failed at points: 20 21
```

Exhibit 14.8   *P*-Chart with Two Samples Removed from Calculations

```
MTB > pchart c1 50;
SUBC>    estimate 1:19 22:30.

                    P Chart for Plating
            -
            -
   0.300+                      +
P          -
r          -
o          -                        +
p          -
o   0.200+
r          -------------------------------UCL=0.1831
t          -
i          -                              +
o          - +        +          + +
n   0.100+  ++               +        +   _
           -----+---+-----+-+-------+-----+P=0.07286
           -      +        +          + ++
           -      ++   +                  +
           -+            + +      +
   0.000+-------------------------------LCL=0.000
           +---------+---------+---------+
           0        10        20        30
                  Sample Number
```

NPCHART number of defectives in **C** sample sizes in **E**

TESTS K...K
SLIMITS are **K...K**
ESTIMATE using only samples **K...K**
P = K

Prints an *np*-chart, a chart for the number of defectives. If all subgroups have the same size, you can use a constant for E.

The subcommands are the same as for PCHART.

Release 8: **Stat > SPC Charts > NP**
Windows: Character control charts are not in the menu. You must type the commands.

PCHART number of defectives in **C** sample sizes in **E**

> **TESTS K...K**
> **SLIMITS** are **K...K**
> **ESTIMATE** using only samples **K...K**
> **P = K**

Prints a *p*-chart, a chart for the proportion of defectives. If all subgroups have the same size, you can use a constant for E.

Minitab can do four different tests, numbered 1, 2, 3, and 4, to determine whether there are any problems. List the numbers of the tests you want in the subcommand TEST. You can use a colon to abbreviate a consecutive list.

If you have a historical value for p, you can specify it in the subcommand P. Minitab will use this value to calculate the center line and control limits.

PCHART can also use the subcommands INCREMENT, START, TITLE, FOOTNOTE, XLABEL, and YLABEL. These subcommands as well as SLIMITS and ESTIMATE, are all the same as for XBARCHART.

Release 8: **Stat > SPC Charts > P**
Windows: Character control charts are not in the menu. You must type the commands.

Exercises

14-4 Documents used in a manufacturing system were monitored for correctness. Each day, a sample of 100 documents was taken. The number of documents that contained one or more error was recorded.

	Number of Documents with Errors									
Days 1–10	10	12	10	11	6	7	12	10	6	11
Days 11–20	9	14	16	21	20	12	11	6	10	10
Days 21–25	11	11	11	6	9					

(a) What percentage of the documents (overall) have errors?

(b) Calculate, by hand, the center line and the upper and lower 3-sigma control limits for a *p*-chart.

(c) Use Minitab to display a *p*-chart. Compare your answers in part (b) to the values that Minitab prints on the chart.

(d) Does the process appear to be in control?

14.4 Optional Material on High-Resoution Graphs

Release 7 and 8

The commands and subcommands for control charts are the same for the character and high-resolution versions, with the usual exception: the high-resolution commands start with the letter G. Exhibit 14.9 shows the high-resolution version of the *p*-chart shown in Exhibit 14.7. Further information about the tests is given with the printed output. This chart was done on a Macintosh. The DOS chart would look similar, but may have slightly different scales.

Release 9, Windows

The high-resolution control charts are in **Stat > Control Charts**. Exhibit 14.10 shows the dialog box for a *p*-chart, filled in to produce a high-resolution version of the character chart shown in Exhibit 14.7. Exhibit 14.11 shows the output.

Release 9 VAX/VMS and UNIX Workstations

Standard Version. The graphics capability in the standard version of Release 9 is the same as Release 7. Refer to the material on Release 7, above.

Enhanced Version. Type GPRO, and Minitab will interpret all the commands you type for control charts as commands for high-resolution graphs. The syntax of high-resolution charts is the same as for character charts, with one exception: You can give only one data column on a high-resolution control chart; that is, you can produce just one chart at a time.

The high-resolution charts have the additional subcommands, AXIS, TICK, MINIMUM, and MAXIMUM, that were described in Section 3.8.

Exhibit 14.11 shows the high-resolution version of the character *p*-chart in Exhibit 14.7.

Exhibit 14.9   High-Resolution *P*-Chart

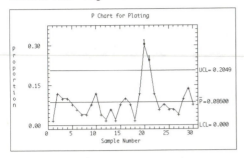

Exhibit 14.10   Dialog Box for a *P*-Chart

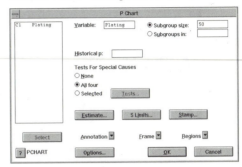

Exhibit 14.11   High-Resolution *P*-Chart

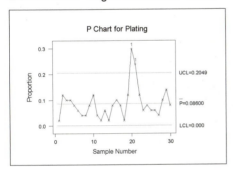

15

Additional Topics in Regression

15.1 Variable Selection in Regression

The Peru data described in Appendix A were collected to study the long-term effects of a change in environment on blood pressure. Some questions we might ask are: "What happens to the Indians' blood pressure as they live longer periods in the new environment?" "What factors are related to blood pressure?" In this section, we will show how variable selection procedures can be used to help answer these questions.

Both systolic and diastolic blood pressure were measured. Since systolic blood pressure is often a more sensitive indicator, we will use it as our dependent variable. There are eight predictors: age, number of years since migration, weight, height, three measures of body fat (chin skin fold, forearm skin fold, and calf skin fold), and pulse rate. To these we will add one derived predictor, fraction of life since migration. This is calculated with

```
NAME C11 'FRACTION'
LET 'FRACTION' = 'AGE' / 'YEARS'
```

This new variable might be a better measure of "how long" the Indians have lived in their new environment than just the number of years since migration. Younger people adapt to new surroundings more quickly than older people. A 25-year-old might be able to adapt as well in one year as a 50-year-old could in two years. There may be other derived variables that could be useful predictors, but we will stop here for now.

Now the problem is how to find a "good" regression equation, one that is parsimonious (i.e., has few predictors), but also has good predictive ability. With nine predictors, there are many possible subset models—$2^9 = 512$, to be exact.

Minitab has two "automatic" methods that will help you in this search: stepwise regression and best subsets regression. Each sorts through subsets of the variables, using numerical criteria to choose good subsets. Automatic procedures such as these can be a valuable tool, particularly in the early stages of building a model and when you have many predictors. At the same time, they present certain dangers. Since the procedures automatically "snoop" through many models, the model selected may fit the data "too well." That is, the procedure may select variables that by pure chance fit well. In addition, automatic procedures cannot take into account special knowledge the experimenter may have about the data, how important a variable is from a scientific standpoint, how difficult a variable is to measure, and so on. Therefore, the model may not be the best from a practical point of view.

Stepwise Regression

In its simplest form, STEPWISE starts by fitting all models with just one variable. For each model, it checks the F-value associated with that variable. The variable with the largest F-value is added to the equation, provided its F-value is larger than FENTER. If no variable has an F-value larger than FENTER, STEPWISE terminates. Let's assume variable X_3 entered the equation on step 1.

Next, STEPWISE fits all two-variable models in which one of the variables is X_3. The variable with the largest F-value is added to the equation, provided its F-value is larger than FENTER. If no variable has an F-value larger than FENTER, STEPWISE terminates. Let's assume variable X_5 entered the equation on step 2.

Note that the F-value Minitab uses to enter variables is just the square of the t-ratio that is printed by REGRESS. If you used REGRESS to fit all models containing two predictors, X_3 and X_i, then chose the variable X_i with the largest t-ratio, this would be the variable STEPWISE chooses. Traditionally, most people use the F-statistic to guide STEPWISE, and Minitab does the same.

Now STEPWISE looks to see if any of the other variables in the equation (other than the one that was just entered) can be removed. At this stage there is just one possibility, X_5. Later, when the equation contains more variables, there will be many candidates. The variable with the smallest F-value is removed, provided its F-value is smaller than FREMOVE. If no

STEPWISE y in **C,** predictors in **C...C**

    FENTER   = **K**
    FREMOVE = **K**

    FORCE C...C
    ENTER C...C

Does stepwise regression.

The value of FENTER controls which variables enter the equations; FRE-MOVE controls which variables are removed. The default for both is 4.0. You must have FENTER ≥ FREMOVE.

Variables listed on FORCE are included on the first step and never removed. Variables listed on ENTER are included on the first step but may be removed later.

Stat > Regression > Stepwise

variable has an F-value smaller than FREMOVE, STEPWISE goes to the next step, where it tries to enter another variable.

    Note that the F-value Minitab uses to remove variables is just the square of the t-ratio printed by REGRESS. Thus, if you were to examine all the t-ratios printed by REGRESS, and choose the variable with the smallest t-ratio, this would be the variable STEPWISE chooses to remove.

    STEPWISE continues in this manner, first adding the variable with the largest F-value (provided its value is larger that FENTER), then removing the variable with the smallest F-value (provided its value is less than FRE-MOVE). STEPWISE terminates when it can no longer add or remove any variables.

    At each step, Minitab prints the coefficient and t-ratio for each variable in the equation, and s (the square root of MSE) and R^2 for the equation.

    STEPWISE has several advantages: you can have many predictors, you can have more predictors than you have observations, and you can have predictors that are highly correlated. STEPWISE can still look for good subsets. However, STEPWISE uses a heuristic to add and remove variables, and it may not find models with the highest R^2 value.

    In Exhibit 15.1 we used STEPWISE on the nine predictors in the Peru data set. We used the default value of 4.0 for both FENTER and FREMOVE. STEPWISE added WEIGHT and then FRACTION, and then terminated. Therefore, no other variable had an F-value greater than 4.0.

Exhibit 15.1   Using STEPWISE with the Peru Data

```
MTB > Stepwise 'SYSTOL' 'AGE' 'YEARS' 'WEIGHT' 'HEIGHT' 'CHIN' 'FOREARM'  &
CONT>      'CALF' 'PULSE' 'FRACTION';

 Stepwise regression of  SYSTOL  on  9 predictors, with N =    39

    STEP          1        2
CONSTANT     66.60    60.90

WEIGHT        0.96     1.22
T-RATIO       3.72     5.21

FRACTION               -26.8
T-RATIO                -3.71

S            11.3      9.78
R-SQ        27.18     47.31
 More? (Yes, No, Subcommand, or Help)
SUBC> y
 No variables entered or removed
 More? (Yes, No, Subcommand, or Help)
SUBC> n
```

Notice the end of the output. Minitab asked "More? (Yes, No, Subcommand, or Help)". STEPWISE has a feature that allows you to intervene in the algorithm at various points by typing certain subcommands. We will not discuss this feature here. If you are interested, read the Minitab manual or consult the Help facility. We answered "yes" because we wanted to see more steps, but Minitab told us no variables could be entered or removed (using the current values for FENTER and FREMOVE). We then answered "no", because we did not wish to intervene.

Note that like REGRESS, STEPWISE does not print the value of R^2 if you fit a model without a constant.

There are two special cases of stepwise regression, forward selection and backwards elimination.

Forward Selection. In forward selection, variables are entered into the equation as in STEPWISE, but are never removed. The procedure terminates when no variable that is not in the equation has an F-value greater than FENTER. To do forward selection using Minitab's STEPWISE command, just set FREMOVE = 0. Then no variables will ever be removed.

Backwards Elimination. Backwards elimination starts with all variables in the equation, then removes them, one by one, using the same rule that STEPWISE uses. No variables are ever added back in. The procedure terminates when no variable in the equation has an F-value that is less than FREMOVE. To do backwards elimination using STEPWISE, first put all the variables into the equation using the subcommand ENTER. Then set FENTER = 100000 (or any other large number), so that no variable that is removed can re-enter the equation.

Note that since backwards elimination starts with all variables in the model, Minitab must be able to fit this model. Therefore, data sets with highly correlated variables cannot be used.

Best Subsets Regression

In Exhibit 15.2, we used BREG on the same data we used in STEPWISE. BREG first looked at all one-predictor models and selected the model with the largest R^2 value, the model with WEIGHT. Information on this model and on the next-best one-predictor model was printed. Then BREG looked at all two-predictor models, found the one with the largest R^2, and printed information on it and on the next-best one. And so on, until all nine predictors were used. Our task now is to determine which of these models are "good," and are thus candidates for further study.

To aid us, BREG prints four statistics for each model: R^2, adj R^2, C_p, and s. On Minitab's output, R^2 and adj R^2 are both multiplied by 100 to convert them to percentages.

R^2 was discussed in Chapter 11. It is calculated as

$$R^2 = \frac{\text{Regression SS}}{\text{Total SS}} = 1 - \frac{\text{Error SS}}{\text{Total SS}}$$

First, notice that maximizing R^2 is the same as minimizing SSE. Also notice that R^2 always increases as the number of predictors increases, because SSE always decreases as the number of predictors increases. Thus R^2 is not always useful when you are comparing models with different numbers of predictors. In general, we look for increasing values of R^2, until the rate of increase tapers off. Then we stop adding predictors into our model. Here the number of predictors is probably about four or five.

Adjusted R^2 takes into account the number of predictors in the model. Suppose there are n observations in the data set and p parameters. Thus,

Exhibit 15.2   Using BREG with the Peru Data

```
MTB > breg 'SYSTOL' 'AGE'-'PULSE' 'FRACTION'

Best Subsets Regression of SYSTOL
```

The variable-selection columns (read from the vertical labels in the original output) are, in order: AGE, YEARS, WEIGHT, HEIGHT, CHIN, FOREARM, CALF, PULSE, FRACTION.

Vars	R-sq	Adj. R-sq	C-p	s	AGE	YEARS	WEIGHT	HEIGHT	CHIN	FOREARM	CALF	PULSE	FRACTION
1	27.2	25.2	28.5	11.338			X						
1	7.6	5.1	45.5	12.770									X
2	47.3	44.4	12.9	9.7772			X						X
2	42.1	38.9	17.5	10.251	X	X							
3	50.3	46.1	12.3	9.6273			X			X			X
3	49.0	44.7	13.4	9.7509	X	X							X
4	59.7	55.0	6.1	8.7946	X	X	X						X
4	52.5	46.9	12.4	9.5502	X	X			X				X
5	63.9	58.4	4.5	8.4571	X	X	X	X					X
5	63.1	57.6	5.1	8.5417	X	X	X			X			X
6	64.9	58.3	5.6	8.4663	X	X	X		X	X			X
6	64.9	58.3	5.6	8.4681	X	X	X			X		X	X
7	66.1	58.4	6.6	8.4556	X	X	X	X	X	X			X
7	65.5	57.7	7.1	8.5220	X	X	X			X	X	X	X
8	66.6	57.7	8.1	8.5228	X	X	X	X	X	X		X	X
8	66.2	57.2	8.5	8.5760	X	X	X	X	X	X	X		X
9	66.7	56.4	10.0	8.6554	X	X	X	X	X	X	X	X	X

p = (number of predictors) $+1$ if the equation has a constant. Then R^2-adjusted is defined by

$$R^2\text{-adjusted} = 1 - \frac{(\text{Error SS})/(n-p)}{(\text{Total SS})/(n-1)}$$

It is an approximately unbiased estimate of the population value of R^2. R^2-adjusted is directly related to MSE. As R^2-adjusted increases, MSE decreases. The fourth statistic, s, is the square root of MSE, and thus is also equivalent to R^2-adjusted.

In Exhibit 15.2, both R^2 and R^2-adjusted increased greatly when we went from a one-predictor model to a two-predictor model. Then they each increased more slowly. Once we reached five predictors, R^2-adjusted no longer increased at all, and in fact decreased a bit.

BREG y in **C**, predictors in **C...C**

   BEST K models

   INCLUDE C...C in all models
   NVARS from **K** [up to **K**]

BREG does best subsets regression. This method is also called all possible regressions.

The value of BEST says how many models of each size to print. By default, Minitab prints the best two models of each size.

Variables listed on INCLUDE are included in all models. Only columns that were listed on BREG can be listed here.

By default, BREG prints the best one-predictor models, the best two-predictor models, the best three-predictor models, on up to the model with all predictors. If you type NVARS 5 12, then only the models with 5, 6, 7, . . ., 12 predictors are printed. Note that NVARS does not count the variables listed on INCLUDE. Thus, INCLUDE C1–C3 with NVARS 2 6 prints models with 3+2=5 to 3+6=9 predictors.

Stat > Regression > Stepwise

   Note that if you fit models without a constant, BREG does not print R^2 or R^2-adjusted.

   Let's look at the remaining statistic, C_p. Again, p is the number of parameters in the candidate model and n is the number of observations in the data set. MSE_m is calculated from the full model, the model with all the predictors in it. Then C_p is given by

$$C_p = \frac{SSE_p}{MSE_m} - (n - 2p)$$

This criterion attempts to measure both random error and any bias from fitting a model with too few of the predictors. In general, we look for models in which C_p is small and is also close to p. If a model has little or no bias, C_p should be close to p. If a model has considerable bias, C_p will tend to be larger than p. If C_p is small, the model is relatively precise (has a small variance) in estimating the true regression coefficients. Adding more predictors will not improve this precision much.

In Exhibit 15.2, the models with four and five predictors have small values of C_p (6.1 and 4.5 respectively), and values that are close to the number of parameters (5 and 6 respectively).

15.2 Diagnostics in Regression

Diagnostics for Individual Observations

REGRESS has subcommands—HI, RESIDUALS, SRESIDUALS, TRESIDUALS, DFITS, and COOKD—that help you detect unusual properties or problems with individual observations. RESIDUALS and SRESIDUALS were described in Chapter 11. The other subcommands will be described here.

We assume the data set has n observations and the regression model has p parameters, including the constant term if it is in the equation. We use the following notation: Y_i is the response value for the ith observation; $\hat{Y}_i$ is the fitted value for the ith observation; $\mathbf{X}$ is the $n \times p$ design matrix.

HI. HI stores the leverages, also called the diagonal of the hat matrix. The $n \times n$ projection, or hat matrix, is defined as $\mathbf{H} = \mathbf{X}(\mathbf{X'X})^{-1}\mathbf{X'}$. It has the property that $\mathbf{H}Y = \hat{Y}$. The leverage of the ith observation is the ith diagonal, h_i (also called h_{ii}) of $\mathbf{H}$. Note that h_i depends only on the predictors; it does not involve the response Y. If h_i is large, the ith observation has unusual predictors. That is, it has predictor values that are far from the center of the data, using Mahalanobis distance, a metric that adjusts for the fact that the predictors may have different variances. This observation will have a large influence in determining the regression coefficients.

Many people consider h_i to be large enough to merit checking if it is more than $2p/n$ or $3p/n$. If an observation has $h_i > 3p/n$, Minitab lists it as an unusual value and labels it with an X.

TRESIDUALS. These are called Studentized residuals or Studentized deleted residuals. To calculate the Studentized residual for the ith observation, first remove the ith observation from the data set. Then calculate the regression equation using this smaller data set. Let $\hat{Y}_{(i)}$ = the fitted or predicted value for the deleted observation, and let its residual $e_{(i)} = Y_i - \hat{Y}_{(i)}$. Then the Studentized residual for the ith observation is $e_{(i)}/\text{stdev}(e_{(i)})$. Thus TRESIDUALS are similar to SRESIDUALS, but the ith observation is omitted. Because TRESIDUALS estimates the equation with the ith observation deleted, the ith observation cannot influence the estimates. Therefore, unusual Y values stand out more clearly.

Each Studentized residual has a Student's t distribution with $(n - 1 - p)$ degrees of freedom. That is why we chose the name TRESIDUAL for this subcommand.

COOKD. Recall that leverages, stored by HI, tell us if an observation has unusual predictors. Standardized residuals, stored by SRESIDUALS, tell us if an observation has an unusual response. Cook's distance, stored by COOKD, combines these two into one overall measure of how unusual an observation is. Specifically, COOKD for the ith observation is

$$\text{COOKD}_i = \frac{1}{p} \frac{h_i}{1 - h_i} (\text{standardized residual}_i)^2$$

Generally it is a good idea to check observations where COOKD > F(0.5, $p, n - p$). Here F is the F-distribution, printed by Minitab's command CDF.

DFITS. These are also called DFFITS. They combine leverage, stored by HI, and Studentized residual, stored by TRESIDUALS, into one overall measure of how unusual an observation is. Specifically, DFITS for the ith observation is

$$\text{DFITS}_i = \sqrt{h_i / (1 - h_i)} \; (\text{Studentized residual}_i)$$

Generally, you should check observations where DFITS > $2\sqrt{p/n}$.

Pure Error Lack-of-Fit Test

PURE. This subcommand helps you assess how well your model fits the data. To use it, you must have replicates. A replicate is one setting of the predictors for which you have two or more responses. In this case, it is possible to calculate an estimate of σ^2, the variance of Y_i, without assuming any form for the model. The estimate is MSPE, defined below.

Suppose X_j is one setting of the predictors. Suppose there are n_j responses for this setting. Let Y_{ji} be the ith response corresponding to X_j and $\bar{Y}_j$ = the average of the n_j responses at X_j. Then the sum of squares for pure error is

$$\text{SSPE} = \Sigma_j \Sigma_i (Y_{ji} - \bar{Y}_j)^2$$

If SSE is the sum of squares for fitting a specific model, then the sum of squares for lack-of-fit is

$$\text{SSLOF} = \text{SSE} - \text{SSPE}$$

REGRESS C on K predictors C...C

HI	**C**
RESIDUALS	**C**
SRESIDUALS	**C**
TRESIDUALS	**C**
DFITS	**C**
COOKD	**C**
PURE	
DW	

These subcommands of REGRESS help you discover problems in your data and in your regression model. The first six subcommands store the corresponding diagnostic in a column. The last two print their results.

Stat > Regression > Regression, use Options button for DW and PURE

As usual, each sum of squares has degrees of freedom associated with it. The degrees of freedom for SSPE is $\Sigma(n_j - 1)$. The degrees of freedom for SSLOF = df(SSE) − df(SSPE). We then use these to form mean squares and an F-statistic as follows:

$$F = \frac{\text{MSLOF}}{\text{MSPE}} = \frac{\text{SSLOF}/\text{df(SSLOF)}}{\text{SSPE}/\text{df(SSPE)}}$$

If you use the subcommand PURE, Minitab prints this F-statistic and its p-value. If the p-value is small, your model has significant lack-of-fit. For example, if you fit a line and find significant lack-of-fit, a line is not an adequate model. Perhaps you need a quadratic or a transformation.

Durbin-Watson Statistic

DW. This subcommand prints the Durbin-Watson statistic, used to test for autocorrelation in the data. Regression assumes that observations are independent. However, observations collected over time are often correlated, usually positively.

The Durbin-Watson statistic, D, is calculated from the residuals, e_i, as follows:

$$D = \frac{\Sigma(e_i - e_{i-1})^2}{\Sigma e_i^2}$$

Minitab prints the value of D. To finish the test, you need to use special tables.[1]

There are also graphical ways to check for autocorrelation. Suppose you've done a regression and stored the residuals in 'Resid'. You can plot each residual, e_i, versus its lagged value, e_{i-1}. Minitab's command LAG will shift a column down to give you the lagged values, e_{i-1}. If this plot shows a tilt, the observations are autocorrelated. The following commands will do the plot:

```
NAME c10 'LagRes'
LAG 'Resid' 'LagRes'
PLOT 'Resid' 'LagRes'
```

15.3 Additional Features of REGRESS

Storage of Some Results

Both MSE and XPXINV are usually stored because you want them for additional calculations, which Minitab does not do directly. For example, the product MSE $(\mathbf{X'X})^{-1}$ is the variance-covariance matrix of the coefficients in the regression equation.

Minitab stores the diagonal of the hat matrix (see the subcommand HI in Section 15.2), but it does not store the full matrix. The full hat matrix $\mathbf{H} = \mathbf{X(X'X)}^{-1}\mathbf{X'}$. You can calculate it using the matrix stored by XPXINV along with the matrix commands TRANSPOSE and MULTIPLY, described in Section 17.6.

REGRESS C on K predictors C...C

 MSE into **K**
 XPXINV into **M**

MSE stores the mean square error in a constant. XPXINV stores the matrix $(\mathbf{X'X})^{-1}$ in a Minitab matrix. (Matrices are described in Section 17.6.)

Stat > Regression > Regression

---

[1]See most any book on regression, for example J. Neter, W. Wasserman, and M. Kutner, *Applied Linear Regression Models*. Homewood, IL: Irwin, 1989.

REGRESS C on K predictors C...C

WEIGHTS C

WEIGHTS tells Minitab to do a weighted regression using the weights in the specified column.

Stat > Regression > Regression, use Options button

Weighted Regression

REGRESS assumes that for each x, the amount of variation in the population of ys is approximately the same. This variance is usually called the variance of y about the regression line, and is denoted by σ^2. Suppose the variances are not equal. Suppose the variance associated with the ith observation is σ_i^2. If this is the case, we will usually detect it when we plot residuals versus a predictor or versus the fitted values. When variances are not equal, we often use a weighted regression analysis. This is what the subcommand WEIGHTS does.

The weight associated with the ith observation is $w_i = 1/\sigma_i^2$. Thus, observations with a high variance are given a low weight. Actually, you can use $w_i = k/\sigma_i^2$, where k is any constant of proportionality. Of course, in practice you do not usually know σ_i^2, so these values must be estimated from the data. There is no exact way to do this.

In many cases, the variances are related to the level of a predictor, X, or to the level of the response, Y. Plotting the residuals versus each predictor and versus Y can often give you an idea as to what that relationship is. Here are some common situations and appropriate weights to use for each:

$$\sigma_i^2 = \sigma^2 X_i \qquad w_i = 1/X_i \qquad\qquad \sigma_i^2 = \sigma^2 Y_i \qquad w_i = 1/Y_i$$
$$\sigma_i^2 = \sigma^2 X_i^2 \qquad w_i = 1/X_i^2 \qquad\qquad \sigma_i^2 = \sigma^2 Y_i^2 \qquad w_i = 1/Y_i^2$$
$$\sigma_i^2 = \sigma^2 \sqrt{X_i} \qquad w_i = 1/\sqrt{X_i} \qquad\qquad \sigma_i^2 = \sigma^2 \sqrt{Y_i} \qquad w_i = 1/\sqrt{Y_i}$$

Ill-Conditioned Data

Predictor variables can have several types of problems. These can cause both statistical and computational difficulties.

Multicollinearity. If one predictor is highly correlated with other predictors, we say the data are multicollinear. If this correlation is moderately high, Minitab prints a warning message. If this correlation is very high, Minitab removes the predictor from the equation, prints a message, and fits the smaller model. In all cases, calculations are done with a high degree of numerical accuracy.

Minitab uses the following rules: A predictor X_j is regressed on the remaining predictors. If R^2 for this regression is greater than 99%, a warning is printed. If R^2 is greater than 99.99%, X_j is removed from the equation.

Multicollinearity does not affect the fits and residuals, but it does affect the coefficients. Their standard deviations can be very large. Thus the fitted coefficients can vary greatly from one sample to another. There are cases in which no coefficient in the equation is statistically significant, but R^2 for the equation as a whole is very high.

You can investigate the correlation structure of the predictors using the CORRELATION command, and by regressing each predictor on the remaining predictors.

Some possible solutions to the problem of multicollinearity are: (1) Eliminate predictors from the equations, especially if this has little effect on R^2. (2) Change predictors by taking linear combinations of them. (3) If you are fitting polynomials, subtract a value near the mean of the predictor before squaring it.

Small Coefficient of Variation. The coefficient of variation of a predictor, X_j, is

$$\text{(standard deviation of } X_j)/(\text{mean } X_j)$$

If a predictor has a small coefficient of variation, the variability in the predictor is very small relative to the magnitude of the predictor. This means the predictor is essentially constant. For example, the variable YEAR with values from 1970 to 1975 has a small coefficient of variation. All the information in YEAR is in the fourth digit. This can cause numerical problems.

If the coefficient of variation is moderately small, Minitab prints a warning that tells you the predictor is nearly constant. If the coefficient of variation is very small, Minitab removes it from the equation. The solution is simple: Subtract a constant from the data. For example, replace YEAR by YEAR − 1970, which has values from 0 to 5.

Exercises

15-1   Use STEPWISE with the Peru data, but now set both F-values to 2 instead of 4. Any changes?

15-2   We will look at two of the predictors in the Peru data.
    (a)   Use REGRESS to fit the equation with just the predictor FRACTION. Does fraction of life since migration appear to be a useful predictor of systolic blood pressure?
    (b)   Add WEIGHT to the model in part (a). Compare this two-predictor model to the one in part (a). Is FRACTION a useful predictor? Try to explain what is going on.

15-3   Use STEPWISE to fit models to the Peru data. Use the original predictors plus derived variables that may be useful. Some possibilities are chin+fore-arm+calf as a composite measure of body fat, or wt/ht as a measure of body mass. Can you think of others? Can you improve the model in Exhibit 15.1 in a meaningful and useful way?

15-4   Use backwards elimination to fit models to the Peru data. Any changes from the results in Exhibit 15.1?

15-5   Use REGRESS to further investigate the 4- and 5-predictor models suggested by BREG.

15-6   In Exhibit 15.1 STEPWISE stopped with a 2-predictor model. Why didn't it go further when BREG, in Exhibit 15.2, seems to indicate that a 4- or 5-predictor model fits the data better?

16

Additional Topics in Analysis of Variance

Minitab has a fairly extensive analysis of variance capability. It includes analysis of balanced and unbalanced designs, crossed and nested factors, analysis of covariance, random and mixed models for balanced designs, multiple comparisons for one-way designs, and, in Release 9, MANOVA.

This chapter gives a brief introduction to a few features. To learn more, consult the Minitab manual and books on analysis of variance.

16.1 Multiple Comparisons with ONEWAY

Note: The subcommands for multiple comparisons that are discussed in this section are available in Releases 8 and 9, but not in Release 7.[1]

If a one-way analysis of variance results in a significant F-test, we conclude that the population means are not all equal, but we do not know what the differences are. The procedures discussed in this section help you study these differences.

There are many procedures for making multiple comparisons among means. We will discuss three of the more common ones: Fisher's, Tukey's, and Dunnett's. These methods are usually presented as a set of tests for pairwise comparisons of means. Minitab, however, presents the results as

---

[1]Actually, Release 7 does contain preliminary, undocumented versions of these subcommands, but they have a few bugs, primarily in the printing of output answers. Use these subcommands in Release 7 if you wish, but with caution and at your own risk.

a set of confidence intervals for pairwise differences, $\mu_i - \mu_j$. This allows you to assess the practical significance of differences between means, as well as the statistical significance. You can easily convert each confidence interval to a test if you wish. Reject H_0: $\mu_i = \mu_j$ if and only if the confidence interval for $\mu_i - \mu_j$ does not contain zero.

In all three methods, the confidence intervals for $\mu_i - \mu_j$ have the same general form. Suppose group i has n_i observations with sample mean $\bar{x}_i$, and group j has n_j observations with sample mean $\bar{x}_j$. Then the confidence interval for $\mu_i - \mu_j$ is

$$\bar{x}_i - \bar{x}_j \pm Ds \sqrt{\frac{1}{n_i} + \frac{1}{n_j}}$$

where $s = \sqrt{\mathrm{MSE}}$ and has υ degrees of freedom. Each method determines the value of the constant D in a different way. We discuss these below.

Minitab prints two error rates for each procedure: the individual and the familywise error rate. The individual error rate is the probability that a given confidence interval will not contain the true difference in group means. Multiple-comparison procedures, however, calculate not just one confidence interval but a set, or family, of confidence intervals. The familywise error rate is the probability that this family will contain at least one confidence interval that does not contain the true difference in group means.

Fisher's Least Significant Difference (LSD)

Fisher's method gives confidence intervals for all pairwise differences, $\mu_i - \mu_j$. You specify the individual error rate, a, in the subcommand FISHER. The value of D is the upper $a/2$ point of the t-distribution with υ degrees of freedom. Minitab prints the critical value, D, in the output.

Minitab calculates the familywise error rate. This error rate is exact if all groups contain the same number of observations. If the group sizes are different, Minitab uses the Tukey-Kramer method to calculate an approximate familywise error rate. The true error rate will be slightly smaller, resulting in conservative (slightly larger) confidence intervals.

Tukey's Honestly Significant Difference (HSD)

Tukey's method also provides confidence intervals for all pairwise difference between group means.

You specify the familywise error rate, a, in the subcommand TUKEY. The value of $D = Q/\sqrt{2}$, where Q is the upper a point of the Studentized range distribution with parameters $r =$ number of groups and $v =$ degrees of freedom for MSE. Minitab prints the critical value, Q, in the output.

Minitab calculates the individual error rate. This error rate is always exact. The familywise error rate is exact if all groups contain the same number of observations. If the group sizes are different, Minitab uses the Tukey-Kramer method. The true error rate will be slightly smaller, resulting in conservative (slightly larger) confidence intervals.

The methods used by FISHER and TUKEY are closely linked: The results are identical if the specified error rate for FISHER is equal to the individual error rate for TUKEY, or, equivalently, if the specified error rate for TUKEY is equal to the familywise error rate for FISHER. Which method you use depends on which error rate you want to specify.

Dunnett's Procedure

Dunnett's procedure is designed for the case in which you want to compare several experimental treatments with one control treatment. You specify the familywise error rate and the control group in the subcommand DUN-NETT. The value of D is calculated from special tables.[2] Minitab prints the critical value D on the output.

Minitab calculates the individual error rate. Both individual and familywise error rates are exact in all cases.

Example Using Fabric Data

Exhibit 16.1 uses the fabric data from Table 10.1. The analysis of variance test says the means are significantly different. We did all three multiple-comparison methods using the default error rate of .05 for each.

FISHER. The output from FISHER prints 10 confidence intervals. Each has an error rate of .05. Let's look at the first one. This is a confidence interval for $\mu_2 - \mu_1$ and goes from $-.6112$ to $.0840$. Because the interval contains zero, the corresponding null hypothesis, H_0: $\mu_2 = \mu_1$, cannot be rejected. There are just two intervals that exclude zero: the intervals for $\mu_4 - \mu_2$ and for $\mu_5 - \mu_4$. These are the only cases in which there is evidence that the means are significantly different.

The family error rate is .277. This means that when we use FISHER (with this number of groups, this number of observations, and $a = .05$), we

[2]See, for example, B. J. Winer, *Statistical Principles in Experimental Design*. New York: McGraw-Hill, 1971.

Exhibit 16.1   Multiple Comparisons Using the Fabric Data

```
MTB > oneway 'Charred' 'Lab';
SUBC>   Fisher;
SUBC>   Tukey;
SUBC>   Dunnett 5.

ANALYSIS OF VARIANCE ON Charred
SOURCE       DF        SS        MS        F        p
Lab           4     2.987     0.747     4.53     0.003
ERROR        50     8.233     0.165
TOTAL        54    11.219
                                  INDIVIDUAL 95% CI'S FOR MEAN
                                  BASED ON POOLED STDEV
  LEVEL      N      MEAN     STDEV  --+---------+---------+---------+----
      1     11    3.3364    0.4523              (------*------)
      2     11    3.6000    0.4604                     (------*------)
      3     11    3.3000    0.3715            (------*------)
      4     11    3.0000    0.2864    (------*------)
      5     11    3.6455    0.4321                   (------*------)
                                  --+---------+---------+---------+----
POOLED STDEV =    0.4058           2.80      3.15      3.50      3.85


Fisher's pairwise comparisons

     Family error rate = 0.277
Individual error rate = 0.0500

Critical value = 2.009

Intervals for (column level mean) - (row level mean)

                 1         2         3         4

    2    -0.6112
          0.0840

    3    -0.3112   -0.0476
          0.3840    0.6476

    4    -0.0112    0.2524   -0.0476
          0.6840    0.9476    0.6476

    5    -0.6567   -0.3931   -0.6931   -0.9931
          0.0385    0.3022    0.0022   -0.2978
```

(continued)

Exhibit 16.1   *(continued)*

```
Tukey's pairwise comparisons

    Family error rate = 0.0500
Individual error rate = 0.00671

Critical value = 4.00

Intervals for (column level mean) - (row level mean)

               1          2          3          4

    2    -0.7530
          0.2257

    3    -0.4530    -0.1894
          0.5257     0.7894

    4    -0.1530     0.1106    -0.1894
          0.8257     1.0894     0.7894

    5    -0.7985    -0.5348    -0.8348    -1.1348
          0.1803     0.4439     0.1439    -0.1561


Dunnett's intervals for treatment mean minus control mean

    Family error rate = 0.0500
Individual error rate = 0.0149

Critical value = 2.52

Control = level 5 of Lab

Level     Lower    Center    Upper  --------+---------+---------+---------+---
    1    -0.7451   -0.3091   0.1269          (-----------*----------)
    2    -0.4815   -0.0455   0.3906                   (-----------*----------)
    3    -0.7815   -0.3455   0.0906            (----------*----------)
    4    -1.0815   -0.6455  -0.2094  (----------*----------)
                                     --------+---------+---------+---------+---
                                        -0.80      -0.40      -0.00       0.40
```

have a .277 probability that at least one interval will not contain the corresponding true difference in means.

Minitab did not print all intervals. For example, there is no interval for $\mu_1 - \mu_2$. You can easily calculate it, though. If the interval for $\mu_2 - \mu_1$ is (a, b), the interval for $\mu_1 - \mu_2$ is $(-b, -a)$.

TUKEY. The output from TUKEY also prints 10 confidence intervals. However, this subcommand controls the familywise error rate. Here we took the default, .05. Therefore, we are 95% confident that all 10 intervals cover their true differences. Each individual interval has an error rate of .00671, much smaller than the .05 we used in FISHER. This means each interval printed by TUKEY is wider than the corresponding interval printed by FISHER.

DUNNETT. There is no natural control in the fabric data set. Therefore, in order to illustrate Dunnett's procedure, we will pretend that lab 5 uses special methods and is the standard of accuracy. We then test each of the other four labs against this control. This gives four confidence intervals. Minitab prints and plots these intervals. The interval for lab 4 does not contain zero. The other three do. Thus, only lab 4 is significantly different from the control. The confidence interval for $\mu_5 - \mu_4$ is $(-1.0815, -.2094)$. This tells us that lab 4 is below the control by at least .2094 and perhaps by as much as 1.0815.

Exercises

16-1 Use ONEWAY with the subcommand TUKEY to investigate differences due to temperature using the plywood data in Exercise 10-2. Compare the analysis of variance tables, the display of individual confidence intervals, and the results from Tukey's test.

16-2 Exercise 10-8 shows data from an experiment to study crop damage due to a beetle. There are two factors, attractant and disease, each with two levels. You can use ONEWAY to study the individual cells in a two-way design if you make each cell a separate level. Create a new factor, called treatmnt, which is 1 if attractant = yes and disease = yes, 2 if attractant = yes and disease = no, 3 if attractant = no and disease = yes, and 4 if attractant = no and disease = no.

(a) Use ANOVA to analyze the beetle data in Exercise 10-8, if you have not already done so.

ONEWAY data in **C**, levels in **C** [put residuals in **C** [fits in **C**]]

> **TUKEY**   [family error rate = **K**]
> **FISHER**   [individual error rate = **K**]
> **DUNNETT**   [family error rate = **K**] control level is **K**

ONEWAY has subcommands for multiple comparisons. TUKEY and FISHER provide confidence intervals for all pairwise differences between level means. DUNNETT provides a confidence interval for the difference between each treatment mean and a control mean.

Specified error rates must be between .5 and .001. If you do not specify an error rate, Minitab uses .05.

These subcommands are officially available in Releases 8 and 9.

Stat > ANOVA > Oneway, use button for Comparisons

(b)   Use ONEWAY, with the subcommand TUKEY, to analyze the beetle data. Use the factor treatmnt. Compare the results to those in part (a).

(c)   Level 4 of treatmnt could be considered the control since neither attractant nor disease pellets were used. Analyze the data using Dunnett's procedure. How do the results compare to those in parts (a) and (b)?

16-3   Exhibit 10-2 shows data from a two-way design used to study driving a car under different conditions.

(a)   Analyze the data as a two-way design if you have not already done so.

(b)   Use the technique described in Exercise 16-2 to convert the data to a one-way analysis. Use ONEWAY with the subcommand TUKEY. Interpret the results and compare them to those in part (a).

16.2 Multifactor Balanced Designs

In Chapter 10, we used ANOVA to analyze balanced two-factor designs. ANOVA can fit balanced designs with up to nine factors. In this section, we will discuss designs with three factors.

The model for the design is specified in the ANOVA command. List the response variable, an equal sign, then the factors and interactions in your model. Here are some examples using three factors:

```
ANOVA Y = A   B   C
ANOVA Y = A   B   C   A*C   B*C
ANOVA Y = A   B   C   A*B   A*C   B*C   A*B*C
```

The first model contains just main effects. The second model contains main effects and two interactions. The third is the full model, and contains all interactions.

Several special rules apply to ANOVA. You may omit the quotes around variable names. Because of this, any variable name used in ANOVA must start with a letter and must contain only letters and numbers. If you want to use special symbols in a variable name, you must enclose the name in single quotes, as in other Minitab commands. You can always use column numbers instead of column names. You may not put any extra text on the ANOVA line, except after the symbol #.

You can fit reduced models. For example, Y = A B C A*B is a three-factor model with just one two-way interaction. Models, however, must be hierarchical. For example, if the term A*B*C is in the model, then all of its subterms (A, B, C, A*B, A*C, and B*C) must also be in the model.

Because models can be quite long and tedious to type, Minitab provides two shortcuts. A vertical bar (or an exclamation point) indicates crossed factors, and a minus sign removes terms. Here are some examples:

```
ANOVA   Y = A|B|C
```

is equivalent to

```
ANOVA   Y = A   B   C   A*B   A*C   B*C   A*B*C
```

```
ANOVA   Y = A|B|C - A*B*C
```

is equivalent to

```
ANOVA   Y = A   B   C   A*B   A*C   B*C
```

Example Using the Potato Rot Data

The potato rot data set was described in Exercise 10-10, which analyzed the data using two-way analysis of variance and plots. Recall that this data set has three factors: (1) amount of bacteria injected into the potato (1 = low amount, 2 = medium amount, 3 = high amount); (2) temperature during storage (10° C, 16° C); (3) amount of oxygen during storage (2%, 6%, 10%). The response variable is the amount of rot in the potato after five days of storage.

In Exhibit 16.2, we use ANOVA to do a three-way analysis. There are seven terms in this model. As always, we start by checking the interactions. The three-factor interaction, BACTERIA*TEMP*OXYGEN, is not

Exhibit 16.2 Three-Way ANOVA

```
MTB > ANOVA ROT = BACTERIA|TEMP|OXYGEN

Factor      Type Levels Values
BACTERIA    fixed      3     1     2     3
TEMP        fixed      2     1     2
OXYGEN      fixed      3     1     2     3

Analysis of Variance for ROT

Source                      DF          SS         MS        F      P
BACTERIA                     2      651.81     325.91    13.91  0.000
TEMP                         1      848.07     848.07    36.20  0.000
OXYGEN                       2       97.81      48.91     2.09  0.139
BACTERIA*TEMP                2      152.93      76.46     3.26  0.050
BACTERIA*OXYGEN              4       30.07       7.52     0.32  0.862
TEMP*OXYGEN                  2        1.59       0.80     0.03  0.967
BACTERIA*TEMP*OXYGEN         4       81.41      20.35     0.87  0.492
Error                       36      843.33      23.43
Total                       53     2707.04
```

Exhibit 16.3 Three-Way ANOVA, Using a Reduced Model

```
MTB > ANOVA ROT =  BACTERIA  TEMP  OXYGEN  BACTERIA*TEMP

Factor      Type Levels Values
BACTERIA    fixed      3     1     2     3
TEMP        fixed      2     1     2
OXYGEN      fixed      3     1     2     3

Analysis of Variance for ROT

Source               DF          SS        MS        F      P
BACTERIA              2      651.81    325.91    15.68  0.000
TEMP                  1      848.07    848.07    40.79  0.000
OXYGEN                2       97.81     48.91     2.35  0.106
BACTERIA*TEMP         2      152.93     76.46     3.68  0.033
Error                46      956.41     20.79
Total                53     2707.04
```

ANOVA C = termlist

FITS   C
RESIDUALS C

Does balanced analysis of variance for up to nine factors.

For one-way analysis of variance, your design may be unbalanced. For two or more factors, your design must be balanced (all cells must have the same number of observations).

The subcommands FITS and RESIDUALS save the fitted values and residuals in columns.

Stat > ANOVA > Balanced ANOVA

---

significant, nor are the two-factor interactions, BACTERIA*OXYGEN and TEMP*OXYGEN. That leaves the three main effects and one two-way interaction. In Exhibit 16.3, we fit a model with just these terms. Because this model is balanced, the sums of squares for terms in the models do not change. The sums of squares for the terms we dropped go into SSE. Thus SSE increases, as does its degrees of freedom. MSE, however, decreases. Now two factors, bacteria and temperature, and the interaction between them are all significant. The remaining factor, oxygen, is marginally significant.

Exercises

16-4   In this exercise, we will do further analyses using the potato data, described in Exercise 10-10.

(a)   Use ANOVA to store the fits and residuals so you can analyze the residuals. Some possible displays are (1) a histogram of the residuals, (2) a plot of the residuals versus the fitted values, (3) plots of the residuals versus each factor. Do you see any problems?

(b)   Exhibit 16.2 showed that there was one significant interaction. Investigate this further. Use TABLE to get cell means. Use appropriate plots (either hand-drawn or done with Minitab) to display your results.

16-5   Plywood is made by cutting a thin, continuous layer of wood off logs as they are spun on their axis. Several of these thin layers then are glued together to make plywood sheets. Considerable force is required to turn a log hard enough for a sharp blade to cut off a layer. Chucks are inserted into the centers of both ends of the log to apply the torque necessary to turn the log.

A study was done to determine how various factors affect the amount of torque that can be applied to a log. There are three factors: the diameter of the test logs; the distance the chucks were inserted into the logs; and the temperature of the logs at the time they were tested. For each treatment combination, 10 trials were done. On each trial, the average torque that could be applied to the log before the chuck spun out was recorded. Then the torque for the 10 trials was averaged. This average is the response variable. The data are shown below and are stored in the file PLYWOOD.

Note: The levels of two factors, diameter and penetration, are not integers. ANOVA requires levels of a factor to be integers. One solution is to multiply diameter by 10 and penetration by 100.

Diameter (inches)	Penetration (inches)	Temperature (°F)	Torque	Diameter (inches)	Penetration (inches)	Temperature (°F)	Torque
4.5	1.00	60	17.30	7.5	1.00	60	29.55
4.5	1.50	60	18.05	7.5	1.50	60	31.50
4.5	2.25	60	17.40	7.5	2.25	60	36.75
4.5	3.25	60	17.40	7.5	3.25	60	41.20
4.5	1.00	120	16.70	7.5	1.00	120	23.20
4.5	1.50	120	17.95	7.5	1.50	120	25.90
4.5	2.25	120	18.60	7.5	2.25	120	35.65
4.5	3.25	120	18.55	7.5	3.25	120	37.60
4.5	1.00	150	15.75	7.5	1.00	150	22.55
4.5	1.50	150	16.65	7.5	1.50	150	22.90
4.5	2.25	150	15.25	7.5	2.25	150	28.90
4.5	3.25	150	15.85	7.5	3.25	150	35.20

(a)   Analyze the data using a three-way analysis of variance. Can you fit the full model?

(b)   Several interactions are significant. Investigate these further.

(c)   Sometimes it is easier to look at subsets of the data separately to see what is going on, especially when interactions are present. Analyze the data separately for 4.5-inch logs and then for 7.5-inch logs. How do the results compare?

(d)   Use appropriate displays to present your conclusions.

16-6   The data presented in Table 10.2 are part of a three-way design. The third factor was time of day. A total of 48 drivers were used; 24 drove during the day (this is a data set in Table 10.2), and 24 drove at night. The number of driving corrections made by each driver was recorded. The results are shown below.

Analyze these data. Use appropriate displays (graphs and tables) and tests.

	Day			Night		
	1st Class Road	2nd Class Road	Dirt Road	1st Class Road	2nd Class Road	Dirt Road
	4	23	16	21	25	32
Inexperienced	18	15	27	14	33	42
Driver	8	21	23	19	30	46
	10	13	14	26	20	40
	6	2	20	11	23	17
Experienced	4	6	15	7	14	16
Driver	13	8	8	6	13	25
	7	12	7	16	12	12

16.3 Unbalanced Designs

The command GLM will fit what is called the general linear model. This includes balanced and unbalanced analysis of variance, analysis of covariance, and more. In this section, we will use GLM in a simple form, to do an unbalanced two-way analysis.

As in ANOVA, you specify the model on the GLM command. You use the same rules for specifying the model as you do in ANOVA. Unfortunately, the analysis of unbalanced design is not as simple as in ANOVA.

In Chapter 10, we analyzed data from an experiment to study driving performance. The design was balanced, with four observations in each cell. Suppose the experiment did not go as planned. Suppose that on two of the drives the car broke down, so no result was obtained. Our carefully designed experiment is no longer balanced. Exhibit 16.4 shows the data and Exhibit 16.5 shows the output from GLM.

GLM prints two types of sums of squares: sequential sums of squares (Seq SS), and adjusted sums of squares (Adj SS). The mean squares, F-statistics, and p-values are all based on the adjusted sums of squares. In ANOVA, the sums of squares for the terms in the model add up to the total sum of squares. In GLM, the sequential sums of squares add up to the total sum of squares, but the adjusted ones do not. There are many ways to explain what this analysis means. Computationally, it is based on doing

Exhibit 16.4   Driving Performance Data from Table 10.2, with Two Observations Omitted

	Road Type		
	First Class	*Second Class*	*Dirt*
	4	23	16
Inexperienced	18	15	27
Driver	8	21	23
	10	13	14
	6	6	50
Experienced	13	8	15
Driver	7	12	8
			17

Exhibit 16.5   Using GLM to Analyze an Unbalanced Design

```
MTB > name c1='corrects' c2='expernc'  c3='road'
MTB > set 'corrects'
DATA>   4 18 8 10    23 15 21 13    16 27 23 14   6 13 7   6 8 12
DATA>  20 15 8 17
DATA> end
MTB > set 'expernc'
DATA>   (0)12   (1)10
DATA> end
MTB > set 'road'
DATA>   (1:3)4  1 1 1  2 2 2  3 3 3 3
DATA> end
MTB > glm corrects = expernc | road


Factor   Levels Values
expernc     2     0     1
road        3     1     2     3

Analysis of Variance for corrects

Source         DF     Seq SS     Adj SS     Adj MS      F      P
expernc         1     125.67     147.27     147.27    5.78   0.029
road            2     265.27     247.00     123.50    4.85   0.023
expernc*road    2      55.00      55.00      27.50    1.08   0.363
Error          16     407.33     407.33      25.46
Total          21     853.27
```

Exhibit 16.6 Population Means for the Driving
Performance Experiment

	Road Type		
	First Class	Second Class	Dirt
Inexperienced Driver	μ_{11}	μ_{12}	μ_{13}
Experienced Driver	μ_{21}	μ_{22}	μ_{23}

regression, and it is often explained using terminology from regression. We will discuss that approach later. Now we will look at what hypotheses are being tested.

This design has six cells. Each cell represents one population. Each population has a mean. These are shown in Exhibit 16.6. For example, μ_{11} is the mean number of corrections for the population of all inexperienced drivers driving on first-class roads. Similarly, μ_{23} is the mean number of corrections for the population of all experienced drivers driving on dirt roads.

Suppose we want to see whether experience makes a difference. If we use the adjusted sum of squares to do the test, as GLM does, then the specific hypothesis we are testing is

$$H_0: \frac{\mu_{11} + \mu_{12} + \mu_{13}}{3} = \frac{\mu_{21} + \mu_{22} + \mu_{23}}{3}$$

Thus, we are weighting the cell means in each row equally. In most cases, this is a reasonable thing to do. Similarly, if we use the adjusted sum of squares to see whether road type makes a difference, we are testing the hypothesis

$$H_0: \frac{\mu_{11} + \mu_{21}}{2} = \frac{\mu_{12} + \mu_{22}}{2} = \frac{\mu_{13} + \mu_{23}}{2}$$

Models in GLM

The sequential sums of squares can be used to do custom tests. We will not discuss these here. Note that sequential sums of squares depend on the order in which terms are listed on the GLM line. Thus, the sequential sums of squares for GLM Y = A B and GLM Y = B A will, in general, be different.

The adjusted sums of squares, however, do not depend on the order of the terms and will always be the same.

Although models can be unbalanced in GLM, they must be "full rank." Thus, there must be enough data to estimate all the terms in your model. For example, suppose you have a two-factor model with one empty cell. Then you can fit the model GLM Y = A B, but not GLM Y = A|B. Don't worry about figuring out whether or not your model is of full rank. Minitab will tell you whether it is not. In most cases, eliminating some of the high-order interactions (assuming, of course, that they are not important) will solve your problem.

Regression Approach to Analysis of Variance

When we discussed analysis of variance in Chapter 10, we did the calculations using a sum-of-squares approach. This works when the design is balanced. In unbalanced designs, we must use regression to do the calculations.

If a factor B has b levels, then it has $(b - 1)$ degrees of freedom. We form $(b - 1)$ predictors, called dummy variables. The first dummy variable, $B1$, is 1 when B is at its lowest level, -1 when B is at its highest level, and 0 otherwise. The second dummy variable, $B2$, is 1 when B is at its second level, 0 when B is at its highest level, and 0 otherwise. The third dummy variable is 1 when B is at its third level, 0 when B is at its highest level, and 0 otherwise. We continue in this way until we have $(b - 1)$ predictors.

We form one block of dummy variables for each factor. Suppose there is an interaction, say $A*B$, in the model, where A has a degrees of freedom and B has b degrees of freedom. Then $A*B$ has $(a - 1)(b - 1)$ degrees of freedom, so we need $(a - 1)(b - 1)$ dummy variables for $A*B$. To get these, we multiply each dummy variable for A by each dummy variable for B.

Exhibit 16.7 shows the driving performance data used in Exhibit 16.5. To save space, we use E for experience and R for road type. E has two levels and thus just one dummy variable, called E1. R has three levels and thus two dummy variables, R1 and R2. The interaction has two degrees of freedom and thus two dummy variables. ER11 is the product of columns E1 and R1; ER12 is the product of columns E1 and R2.

If you want to see the dummy variables that correspond to a given model, use the subcommand XMATRIX. This asks GLM to store the dummy variables in a matrix, which you can print out. The first column of this matrix is the dummy variable for the constant term. It is a column containing only 1s. The remaining columns are for the factors and interactions in your model. This matrix is often called a design matrix.

Exhibit 16.7   Dummy Variables for Data from Exhibit 16.4

Original Data			Dummy Variables				
Corrects	E	R	E1	R1	R2	ER11	ER12
4	0	1	1	1	0	1	0
18	0	1	1	1	0	1	0
8	0	1	1	1	0	1	0
10	0	1	1	1	0	1	0
23	0	2	1	0	1	0	1
15	0	2	1	0	1	0	1
21	0	2	1	0	1	0	1
13	0	2	1	0	1	0	1
16	0	3	1	−1	−1	−1	−1
27	0	3	1	−1	−1	−1	−1
23	0	3	1	−1	−1	−1	−1
14	0	3	1	−1	−1	−1	−1
6	1	1	−1	1	0	−1	0
13	1	1	−1	1	0	−1	0
7	1	1	−1	1	0	−1	0
6	1	2	−1	0	1	0	−1
8	1	2	−1	0	1	0	−1
12	1	2	−1	0	1	0	−1
20	1	3	−1	−1	−1	1	1
15	1	3	−1	−1	−1	1	1
8	1	3	−1	−1	−1	1	1
17	1	3	−1	−1	−1	1	1

To get the adjusted sums of squares that GLM prints, we must do one regression for each term in the model. In each case, we list the predictors for the term we want last on the regression line. The information we need is in the table of sequential sums of squares printed by REGRESS. Exhibit 16.8 shows the necessary output from the three regressions corresponding to the three terms in the driving performance model.

The first regression is for the main effect experience. There is just one dummy variable, E1. Its SEQ SS is 144.21, which is the Adj SS listed on

Exhibit 16.8   Regression Output Corresponding to GLM Calculations

Partial output from GLM:

```
MTB > glm corrects = expernc | road
Analysis of Variance for corrects
Source          DF      Seq SS      Adj SS      Adj MS       F       P
expernc          1      125.67      147.27      147.27     5.78   0.029
road             2      265.27      247.00      123.50     4.85   0.023
expernc*road     2       55.00       55.00       27.50     1.08   0.363
Error           16      407.33      407.33       25.46
Total           21      853.27
```

Partial output from REGRESS, fitting E1 last:

```
MTB > regress 'corrects 5 'ER11' 'ER12' 'R1' 'R2' 'E1'
corrects = 13.4 - 4.06 R1 - 0.06 R2 - 1.94 ER11 + 2.06 ER12 + 2.61 E1
SOURCE          DF      SEQ SS
R1               1      242.19
R2               1        1.37
ER11             1       16.02
ER12             1       39.09
E1               1      147.27
```

Partial output from REGRESS, fitting R1 and R2 last:

```
MTB > regress 'corrects 5 'E1' 'ER11' 'ER12' 'R1' 'R2'
corrects = 13.4 + 2.61 E1 - 1.94 ER11 + 2.06 ER12 - 4.06 R1 - 0.06 R2
SOURCE          DF      SEQ SS
E1               1      125.67
ER11             1       19.86
ER12             1       53.41
R1               1      246.96
R2               1        0.03
```

Partial output from REGRESS, fitting ER11 and ER12 last:

```
MTB > regress 'corrects 5 'E1' 'R1' 'R2' 'ER11' 'ER12'
corrects = 13.4 + 2.61 E1 - 4.06 R1 - 0.06 R2 - 1.94 ER11 + 2.06 ER12
SOURCE          DF      SEQ SS
E1               1      125.67
R1               1      264.69
R2               1        0.58
ER11             1       10.48
ER12             1       44.52
```

the GLM output. This sum of squares is the additional sum of squares for fitting the model with all five predictors in it over the model with just the predictors R1, R2, ER11, and ER12—that is, the model without E1.

The second regression is for road. There are two dummy variables, R1 with SEQ SS = 246.96, and R2 with SEQ SS = .03. The sum, 246.96 + .03 = 146.99, is the Adj SS for road (except for rounding error). This sum of squares is the additional sum of squares for fitting the model with all five predictors in it over the model with just the predictors E1, ER11, and ER12—that is, the model without R1 and R2.

The last regression is for the interaction. There are two dummy variables: ER11, with SEQ SS = 10.48, and ER12, with SEQ SS = 44.52. The sum, 10.48 + 44.52 = 55.00, is the Adj SS for the interaction. This sum of squares is the additional sum of squares for fitting the model with all five predictors in it over the model with just the predictors E1, R1, and R2.

GLM C = termlist

FITS	C
RESIDUALS	C
XMATRIX	M
COEFFICIENTS	C

GLM fits the general linear model. Using GLM, you can analyze balanced and unbalanced designs.

Output contains both the sequential sums of squares and the adjusted sums of squares, adjusted (i.e., each term is fitted after all other terms in the model). Automatic tests are done using the adjusted sums of squares. Observations that are considered unusual are printed out.

FITS and RESIDUALS store the fitted values and residuals, as they do in REGRESS.

GLM does its calculations using a regression approach. First a design matrix is formed from the factors and interactions. The design matrix will have as many columns as there are degrees of freedom in the model. The columns of the design matrix are used as predictors. XMATRIX stores the design matrix in the matrix M.

COEFFICIENTS stores the coefficients corresponding to the regression analysis in a column.

Stat > ANOVA > General Linear Model

Exercises

16-7   We will investigate two factors, COLOR and ED, from the Cartoon data (described in Appendix A).

(a)   Use the score on the cartoon test given immediately after presentation. Is there a difference between using color and black-and-white slides? Is the educational level of the participants important?

(b)   Repeat part (a) using the score on the realistic test given immediately after presentation. Compare your results to those in part (a).

(c)   Repeat part (a) using the delayed cartoon score. How do these results compare to those of part (a)?

(d)   Repeat part (a) using the delayed realistic score. How do these results compare to those of part (b)?

(e)   Many people did not take the delayed test. Do you think this affects your analysis and conclusions? Use appropriate tests and displays to investigate.

16-8   Researchers at Penn State University studied the feasibility of using a variety of fast-growing poplar trees as a renewable energy source. In an effort to determine how to maximize yield, the researchers designed an experiment to study three factors: site, irrigation, and fertilizer. Two sites were used: Site 1 had rich, moist soil, and site 2 had dry, sandy soil. Each site was divided into four areas. Area 1 received no irrigation and no fertilizer; area 2 received only fertilizer; area 3 received only irrigation; and area 4 received both fertilizer and irrigation. Trees were grown under each treatment combination and harvested. The dry weight of the wood in each tree was measured in kilograms. Data are shown below.

   Analyze the data as a three-way design using GLM. What conditions maximize yield?

Site	No Irrigation		Irrigation	
	No Fertilizer	Fertilizer	No Fertilizer	Fertilizer
	2.59	3.56	2.83	4.28
	1.89	5.69	1.70	2.36
	1.61	1.54	2.63	2.85
Rich	1.92	2.86	2.48	3.83
and	1.09	2.99	2.34	4.79
Moist	1.32	2.11	0.99	4.33
	0.64	2.73	0.39	2.12
	0.16	1.85	3.24	3.83
		0.66		
		1.93		
	1.24	3.82	2.26	3.37
	3.24	1.10	2.17	4.34
	2.51	4.16	2.84	2.90
Dry	2.68	4.92	1.68	5.21
and	1.71	4.38	0.35	5.12
Sandy	2.61	1.10	2.71	6.19
	0.25	4.49	0.99	4.80
	0.82	1.80	1.14	
	2.27		2.36	
			4.05	

16.4 Analysis of Covariance

In analysis of variance, all variables in the model are categorical. Often we want to study continuous variables, called covariates, as well. GLM analyzes models that have factors (categorical variables) as well as covariates. We will look at a simple case: one factor and one covariate.

Exhibit 16.9 analyzes data for the preprofessionals (ED = 1) in the Cartoon data (described in Appendix A). Here we use one-way analysis of variance to see whether the factor COLOR makes a difference. The F-test shows no evidence. If we look at the OTIS scores for the preprofessionals, we see that they vary greatly, from 78 to 129. This variation in ability might be masking a small but significant COLOR effect. Analysis of covariance allows us to "correct" for this variation.

Exhibit 16.10 uses GLM to do an analysis of covariance. The model is the same as the model for a one-way analysis with one extra term, the covariate. We used the subcommand COVARIATES to tell Minitab which variables in our model are covariates. We also used BRIEF 3 so that Mini-

Exhibit 16.9   One-Way ANOVA Using the Cartoon Data
(This example uses the subset of the Cartoon data where Ed = 1)

```
MTB >  GLM 'CARTOON1' = COLOR

Factor    Levels Values
COLOR          2    0    1

Analysis of Variance for CARTOON1

Source      DF     Seq SS      Adj SS      Adj MS       F      P
COLOR        1      7.664       7.664       7.664    1.00  0.322
Error       51    391.128     391.128       7.669
Total       52    398.792

Unusual Observations for CARTOON1

Obs. CARTOON1       Fit Stdev.Fit   Residual    St.Resid
 19   1.00000   6.53846   0.54311   -5.53846     -2.04R
 25   1.00000   6.53846   0.54311   -5.53846     -2.04R
 30   0.00000   5.77778   0.53296   -5.77778     -2.13R

R denotes an obs. with a large st. resid.
```

tab would print out information on the coefficients, information we will use later.

The analysis of variance table does a test for the covariate, OTIS. This test is very significant, as we would expect. A person's performance on the cartoon test depends on the person's ability, which is what the OTIS score measures. The test for color is also significant. Thus, once we condition on ability, we see a color effect.

Regression Approach to Analysis of Covariance

Suppose we use REGRESS to do the calculations that GLM did in Exhibit 16.10. First we form the dummy variables for the factors and their interactions, as we do in analysis of variance. Then we add the covariates as additional predictors. This example has one factor, COLOR, with one degree of freedom, so it has just one dummy variable. We will call it C. C = 1 if COLOR is at its low level of 0; and C = −1 if COLOR is at its high level of 1. The second predictor is OTIS, the covariate.

Exhibit 16.11 shows output from REGRESS. One regression equation is printed, but it represents two separate lines, one for each level of the

Exhibit 16.10   Using a Covariate in One-Way ANOVA
(This example uses the subset of the Cartoon data where Ed = 1)

```
MTB > glm 'CARTOON1' = OTIS COLOR;
SUBC>   covariates OTIS;
SUBC>   brief 3.

Factor    Levels Values
COLOR          2   0    1

Analysis of Variance for CARTOON1

Source     DF     Seq SS     Adj SS     Adj MS       F      P
OTIS        1    171.855    183.930    183.930   44.39  0.000
COLOR       1     19.740     19.740     19.740    4.76  0.034
Error      50    207.198    207.198      4.144
Total      52    398.792

Term        Coeff     Stdev   t-value       P
Constant   -8.168     2.169    -3.77   0.000
OTIS       0.13661   0.02050    6.66   0.000
COLOR
   0        0.6152    0.2819     2.18   0.034

Unusual Observations for CARTOON1

Obs. CARTOON1       Fit Stdev.Fit  Residual   St.Resid
 28  1.00000   5.56022   0.39312  -4.56022    -2.28R
 29  1.00000   6.24326   0.39795  -5.24326    -2.63R

R denotes an obs. with a large st. resid.
```

Exhibit 16.11   Regression Line for the Cartoon Analysis

```
MTB > regress 'CARTOON1' 2 'C' 'OTIS'

The regression equation is
CARTOON1 = - 8.17 + 0.615 C + 0.137 OTIS

Predictor     Coef      Stdev   t-ratio       p
Constant    -8.168      2.169     -3.77   0.000
C           0.6152      0.2819     2.18   0.034
OTIS        0.13661    0.02050     6.66   0.000
```

GLM C = termlist

COVARIATES C...C

The columns listed on COVARIATES are used as covariates. These columns must also be listed in termlist. If COVARIATE is used, it must be the first subcommand. This restriction is needed to allow proper error checking.

Stat > ANOVA > General Linear Model

---

factor. The lines have the same slope, 0.137, the coefficient of OTIS, but they have different constant terms.

When COLOR is at its low level, $C = 1$. Then the line is

$$CARTOON1 = -8.17 + 0.615(1) + 0.137(OTIS) = -7.555 + 0.137(OTIS)$$

When COLOR is at its high level, $C = -1$. Then the line is

$$CARTOON1 = -8.17 + 0.615(-1) + 0.137(OTIS)$$
$$= -8.785 + 0.137(OTIS)$$

Recall that the low level of COLOR is for the people who saw black-and-white slides. Thus, the first line above is for these people who saw black-and-white slides. The high level of COLOR is for the people who saw color slides. Thus, the second line is for these people who saw color slides. Exhibit 16.12 shows a plot of the data and the two regression lines. We used Release 9 to produce this plot.[3] A similar plot can be done with Release 8. With Release 7, you can use LPLOT to plot the data, then draw the lines by hand.

Notice that the line for black-and-white slides is above the line for color slides. Black-and-white slides seem to be a better teaching device than color slides, at least in this experiment. There is, however, quite a bit of variability, even if we look only at people with about the same OTIS scores.

Look at the output from GLM in Exhibit 16.10. GLM printed the same three coefficients that REGRESS printed. Therefore, we could have calculated the two lines we drew on the plot using information from GLM.

---

[3]First we used the PREDICT subcommand of REGRESS to calculate the endpoints of each line. Then we used the Lines option of PLOT twice, once to draw each line.

Exhibit 16.12   Plot for Analysis of Covariance

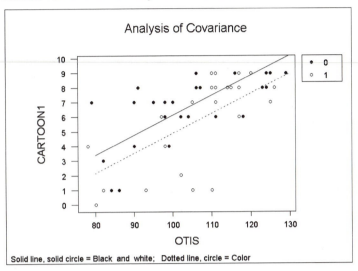

Solid line, solid circle = Black and white;   Dotted line, circle = Color

Exercises

16-9   In this exercise, we will repeat the analysis done in the text, using only data for the professionals.

(a)   Use a one-way analysis of variance to see whether there is a difference between color and black-and-white cartoon slides. Use only the data for the professionals.

(b)   Use the OTIS score as a covariate. Compare the results to those in part (a).

(c)   Calculate the regression line for each level of the factor COLOR.

16-10 (a)   Repeat the analysis of Exercise 16-9, using only the data for students.

(b)   Compare the results to those of Exercise 16-9. Investigate any differences.

16-11   Table 4.1, p. 114, contains data from the study of an energy-saving device for furnaces. Use analysis of covariance to see whether the device made a difference.

16.5 Random Effects and Mixed Models

Sometimes the levels of a factor are not of intrinsic interest in themselves, but are a sample from a larger population. Then the factor is called random.

Exhibit 16.13 A Two-Way Design to Illustrate
Mixed Models

	B1	B2	B3
	11	23	17
	16	12	12
A1	6	14	16
	7	13	25
	19	33	45
	25	21	42
A2	14	33	35
	20	27	47

A common example is subject. For example, suppose a large data-entry company is interested in comparing three keyboard designs for use with computers. The company selects 20 employees at random to test the keyboards. Each employee, or subject, tests each keyboard once. Keyboard design is called a fixed factor because we are interested in these three designs. Subject is called a random factor because we are not interested in these 20 subjects but want to generalize to the population of all subjects, all current and future employees who will use keyboards.

We have already discussed an example with a random factor, the randomized block design in Chapter 10. In a randomized block design, the blocking factor is usually random. Randomized block designs, however, are a very simple case. The analysis is the same whether the blocking factor is fixed or random, so we did not need to discuss random factors in Chapter 10. With more complicated designs, we must know which factors are random and which are fixed in order to do the correct analysis.

Minitab's ANOVA command will help you analyze a model with both fixed and random effects (called a mixed model). The sums of squares are the same in all cases. What changes is the way you do the tests; specifically, what term you use in the denominator of each F-test. In a model in which all factors are fixed, MSE is used as the denominator for all F-tests. In mixed models, different terms may use different denominators for F-tests.

We will use a formal example to illustrate ANOVA. Exhibit 16.13 shows data for a two-factor design. Exhibit 16.14 (pp. 422–423) shows Minitab output for three cases. There are two new subcommands. EMS asks Minitab to print a table of expected mean squares. This is the information used to do the proper tests. RESTRICT says to use the restricted model.

The literature contains two methods for analyzing mixed models. Most books use what is called the restricted model. RESTRICT requests this analysis. Minitab's default is the unrestricted model, which we will not discuss here.[4]

In the first ANOVA of Exhibit 16.14, both factors are fixed. In this case, each term is tested using MSE for the denominator of the F-statistic. This denominator is often called the error term for the test. The EMS table actually tells you the error for each term. Here it is term 4, Error. This is how the F-statistics shown in the Analysis of Variance table are calculated.

In the second ANOVA, A is fixed and B is random. Terms B and A*B are tested using MSE for error, but A is not. According to the EMS table, the proper error term is 3, the mean square for A*B or MSAB. This changes the F-statistic for A in the Analysis of Variance table. Now F = MSA/MSAB.

In the third ANOVA, A and B are both random. Terms A and B are both tested using term 3, the MS for A*B. Thus, the F-statistic for A is MSA/MSAB and the F-statistic for B is MSB/MSAB. The interaction term is still tested using MSE.

Mixed Models, Expected Mean Squares, and Variance Components

When you use ANOVA, you write a "computer" version of the analysis of variance model on the ANOVA command line. Let's look at the formal version of that model for the two-factor case. Suppose factor A has a levels, factor B has b levels, and each cell has n observations. In texts, this model is usually written as

$$Y = \mu + a_i + \beta_j + a\beta_{ij} + \varepsilon_{ijk}, \quad i = 1, \ldots, a; \quad j = 1, \ldots, b; \quad k = 1, \ldots, n$$

The a_i are the effects of factor A, the β_j are the effects of factor B, the $a\beta_{ij}$ are the effects of the interaction, and the ε_{ijk} are the random error. In all cases, ε_{ijk} are independent random variables, with each $\varepsilon_{ijk} \sim N(0, \sigma_\varepsilon^2)$.

There are additional conditions, depending on which terms are fixed and which are random. Here are the four cases:

1. If A and B are both fixed, then

$$\Sigma_i a_i = 0, \quad \Sigma_j \beta_j = 0, \quad \Sigma_i \Sigma_j a\beta_{ij} = 0$$

---

[4]See S. R. Searle, *Linear Models*. New York: Wiley, 1971. Chapter 9 includes a discussion of these two types of models.

Exhibit 16.14   Output from Three Different Models

```
Both factors are fixed:

MTB > ANOVA Y = A|B;
SUBC>   Restrict;
SUBC>   EMS.

Factor      Type Levels Values
A           fixed    2    1    2
B           fixed    3    1    2    3

Analysis of Variance for Y

Source        DF          SS          MS        F       P
A             1       1488.38     1488.38   56.91   0.000
B             2        915.58      457.79   17.50   0.000
A*B           2        255.25      127.62    4.88   0.020
Error        18        470.75       26.15
Total        23       3129.96

Source      Variance Error Expected Mean Square
            component  term (using restricted model)
 1 A                    4    (4) + 12Q[1]
 2 B                    4    (4) + 8Q[2]
 3 A*B                  4    (4) + 4Q[3]
 4 Error     26.15           (4)
```

Factor A is fixed and B is random:

```
MTB > ANOVA Y = A|B;
SUBC>   Random B;
SUBC>   Restrict;
SUBC>   EMS.

Factor      Type Levels Values
A           fixed    2    1    2
B           random   3    1    2    3

Analysis of Variance for Y

Source        DF          SS          MS        F       P
A             1       1488.38     1488.38   11.66   0.076
B             2        915.58      457.79   17.50   0.000
A*B           2        255.25      127.62    4.88   0.020
Error        18        470.75       26.15
Total        23       3129.96
```

Exhibit 16.14 *(continued)*

```
Source           Variance Error Expected Mean Square
                 component  term (using restricted model)
1 A                          3    (4) + 4(3) + 12Q[1]
2 B                53.95     4    (4) + 8(2)
3 A*B              25.37     4    (4) + 4(3)
4 Error            26.15          (4)
```

Both factors are random:

```
MTB > ANOVA Y = A|B;
SUBC>   Random A B;
SUBC>   EMS;
SUBC>   Restrict.


Factor      Type Levels Values
A           random   2   1    2
B           random   3   1    2    3

Analysis of Variance for Y

Source        DF        SS         MS        F      P
A              1    1488.38    1488.38    11.66  0.076
B              2     915.58     457.79     3.59  0.218
A*B            2     255.25     127.62     4.88  0.020
Error         18     470.75      26.15
Total         23    3129.96


Source           Variance Error Expected Mean Square
                 component  term (using restricted model)
1 A               113.40     3    (4) + 4(3) + 12(1)
2 B                41.27     3    (4) + 4(3) + 8(2)
3 A*B              25.37     4    (4) + 4(3)
4 Error            26.15          (4)
```

2. If A is fixed and B is random, then

$$\Sigma_i \alpha_i = 0, \quad \Sigma_i \alpha\beta_{ij} = 0 \text{ for each } j$$

β_j and $\alpha\beta_{ij}$ are random variables with each $\beta_j \sim N(0, \sigma_\beta^2)$ and each $\alpha\beta_{ij} \sim N(0, \sigma_{\alpha\beta}^2)$

3. If A is random and B is fixed, then

$$\Sigma_j \beta_j = 0, \quad \Sigma_i \alpha\beta_{ij} = 0 \text{ for each } i$$

α_i and $\alpha\beta_{ij}$ are random variables with each $\alpha_i \sim N(0, \sigma_\alpha^2)$, and each $\alpha\beta_{ij} \sim N(0, \sigma_{\alpha\beta}^2)$

4. If A and B are both random, then

a_i, β_j, and $a\beta_{ij}$ are random variables with each $a_i \sim N(0, \sigma_a^2)$, each $\beta_j \sim N(0, \sigma_\beta^2)$, and each $a\beta_{ij} \sim N(0, \sigma_{a\beta}^2)$

The terms σ_a^2, σ_β^2, $\sigma_{a\beta}^2$, and σ_ε^2 are called variance components. Minitab gives estimates of these in the EMS table.

If you are familiar with the formulas for expected mean squares, you may recognize them in the output shown in Exhibit 16.14. When we do a test, we want the expected mean square for the numerator and that for the denominator of the F-statistic to differ by just one term, the term being tested. Let's look at the second example in Exhibit 16.14, in which A is fixed and B is random. Shown below is Minitab's output, along with the expected mean squares as they are usually written in texts.

Source	Variance component	Error term	Expected Mean Square (using restricted model)	
1 A		3	(4) + 4(3) + 12Q[1]	$\sigma_\varepsilon^2 + 4\sigma_{a\beta}^2 + 12(\Sigma a_i^2)/(a-1)$
2 B	53.95	4	(4) + 8(2)	$\sigma_\varepsilon^2 + 8\sigma_\beta^2$
3 A*B	25.37	4	(4) + 4(3)	$\sigma_\varepsilon^2 + 4\sigma_{a\beta}^2$
4 Error	26.15		(4)	σ_ε^2

Minitab uses parentheses to represent a variance component. For example, (3) represents the third variance component, $\sigma_{a\beta}^2$. Minitab uses square brackets to represent a quadratic expression involving the constants for a fixed term. For example, Q[1] represents $(\Sigma a_i 2)/(a-1)$.

The hypotheses for the three tests can be expressed as follows:

Test for A H_0: $a_1 = a_2 = \ldots = a_a = 0$ or, equivalently, H_0: $\Sigma a_i 2 = 0$
Test for B H_0: $\sigma_\beta^2 = 0$
Test for A*B H_0: $\sigma_{a\beta}^2 = 0$

If H_0 for A is true, the EMS for A and for A*B are the same. Therefore, the F-statistic for A is MSA/MSAB. If H_0 for B is true, the EMS for B and Error are the same. Therefore, the F-statistic for B is MSB/MSE. If H_0 for A*B is true, the EMS for A*B and Error are the same. Therefore, the F-statistic for A is MSAB/MSE.

The variance component, 53.95, listed next to factor B on the output, is an estimate of σ_β^2. Similarly, 25.37 is an estimate of $\sigma_{a\beta}^2$, and 26.15 is an estimate of σ_ε^2.

The estimates of the variance components are the usual unbiased analysis of variance estimates. They are obtained by setting each calculated MS on the output equal to the formula for its EMS. This gives a system of linear equations in the unknown variance components. This system is then solved. Unfortunately, this method can result in negative estimates, which should be set to zero. Minitab, however, prints the negative estimates because they sometimes indicate that the model being fitted is inappropriate for the data. Variance components are not estimated for terms that are fixed.

Three-Factor Mixed Model

When you have more than two factors, there may be no exact F-test; that is, you may not be able to find a term to use for the denominator of the F-test. We will look at a simple example.

A company ran an experiment to see how several conditions would affect the thickness of a coating substance it manufactures. The manufacturing process was run at three settings: 35, 44, and 52. Three operators were chosen from a large pool of operators employed by the company. The experiment was run at two different times, selected at random. At each time, each operator made two determinations of thickness at each setting. Two factors, operator and time, are random; the other factor, setting, is fixed. The data are shown in Exhibit 16.15.

We can write the model as

$$Y_{ijkl} = \mu + T_i + O_j + S_k + TO_{ij} + TS_{ik} + OS_{jk} + TOS_{ijk} + \varepsilon_{ijkl}$$

where T = time, O = operator, S = setting, and Y is the response, thickness. Because operator and time are random, and because an interaction involving a random factor is random, and because error is always random, the following terms are random:

$$T_i \quad O_j \quad TO_{ij} \quad TS_{ik} \quad OS_{jk} \quad TOS_{ijk} \quad \varepsilon_{ijkl}$$

These terms are all assumed to be normally distributed random variables with mean zero and variances given by

$$\sigma_T^2 \quad \sigma_O^2 \quad \sigma_{TO}^2 \quad \sigma_{TS}^2 \quad \sigma_{OS}^2 \quad \sigma_{TOS}^2 \quad \sigma_\varepsilon^2$$

These variances are called variance components. The output from the subcommand EMS contains estimates of these variances.

In the restricted model, any term that contains one or more subscripts corresponding to fixed factors is required to sum to zero over each fixed subscript. In the example, there is one fixed factor: setting. Thus,

$$\Sigma_k S_k = 0 \quad \Sigma_k TS_{ik} = 0 \quad \Sigma_k OS_{jk} = 0 \quad \Sigma_k TOS_{ijk} = 0$$

Exhibit 16.15 Data for a Three-Factor ANOVA

	Time					
	1			*2*		
	Operator			*Operator*		
Setting	*1*	*2*	*3*	*1*	*2*	*3*
35	38	39	45	40	39	41
	40	42	40	40	43	40
44	63	72	78	68	77	85
	59	70	79	66	76	84
52	76	95	103	86	86	101
	78	96	106	82	85	98

Exhibit 16.16 shows the output of ANOVA using the restricted model. The F-tests for all terms except setting are given. In each case, the appropriate denominator is shown in the EMS table. The test for setting, however, is missing, and the note says that there is no exact F-test for this case. We must construct this test ourselves. The TEST subcommand helps us.

Let's look at the main effect for setting. The expected mean square for setting is

$$(8) + 2(7) + 4(6) + 6(5) + 12Q[3] = \sigma_\varepsilon^2 + 2\sigma_{TOS}^2 + 4\sigma_{OS}^2 + 6\sigma_{TS}^2 + 12(\Sigma S_k^2)/(3-1)$$

We remove the term for setting, $12(\Sigma S_k^2)/(3-1)$, to get the denominator for the test. This gives

$$(8) + 2(7) + 4(6) + 6(4) = \sigma_\varepsilon^2 + 2\sigma_{TOS}^2 + 4\sigma_{OS}^2 + 6\sigma_{TS}^2$$

There is no term in the model that has this as its expected mean square. Therefore, time has no exact F-test. We must "synthesize" the test by finding a linear combination of terms in the model that gives this denominator. The combination that does it is

$$\text{MSTS} + \text{MSOS} - \text{MSTOS} = [(8) + 2(7) + 6(5)] + [(8) + 2(7) + 4(6)] - [(8) + 2(7)]$$
$$= (8) + 2(7) + 6(5) + 4(6)$$

Thus,

$$F = \text{MSS}/[\text{MSTS} + \text{MSOS} - \text{MSTOS}]$$
$$= 7838.2/[57.3 + 107.1 - 24.0] = 140.4$$

Exhibit 16.16   ANOVA When There Is No Exact *F*-Test

```
MTB > anova thickness = T|O|S;
SUBC>    random T O;
SUBC>    restrict;
SUBC>    ems;
SUBC>    test S/ T*S + O*S - T*O*S.
```

```
Factor      Type Levels Values
T          random     2     1      2
O          random     3     1      2      3
S           fixed     3    35     44     52
```

Analysis of Variance for Thickness

Source	DF	SS	MS	F	P
T	1	9.0	9.0	0.29	0.644
O	2	1120.9	560.4	18.08	0.052
S	2	15676.4	7838.2	*	
T*O	2	62.0	31.0	9.15	0.002
T*S	2	114.5	57.3	2.39	0.208
O*S	4	428.4	107.1	4.46	0.088
T*O*S	4	96.0	24.0	7.08	0.001
Error	18	61.0	3.4		
Total	35	17568.2			

* No exact F-test can be calculated.

Source	Variance component	Error term	Expected Mean Square (using restricted model)
1 T	-1.222	4	(8) + 6(4) + 18(1)
2 O	44.120	4	(8) + 6(4) + 12(2)
3 S		*	(8) + 2(7) + 4(6) + 6(5) + 12Q[3]
4 T*O	4.602	8	(8) + 6(4)
5 T*S	5.542	7	(8) + 2(7) + 6(5)
6 O*S	20.778	7	(8) + 2(7) + 4(6)
7 T*O*S	10.306	8	(8) + 2(7)
8 Error	3.389		(8)

* No exact F-test can be calculated.

Approximate F-test with denominator: T*S + O*S - T*O*S
Denominator MS = 140.36 with   4 degrees of freedom

Numerator	DF	MS	F	P
S	2	7838	55.84	0.001

ANOVA C = termlist

> RANDOM   C...C
> EMS
> TEST   termlist / errorterm
> RESTRICT

Does analysis of variance for up to nine factors. The design must be balanced.

Factors may be fixed or random. List random factors in the subcommand RANDOM. ANOVA then calculates all exact F-tests, using the proper error terms.

EMS prints a table that contains expected mean squares, estimated variance components, and the error term (the denominator) used in each exact F-test. If there is no exact F-test for a term, the expected mean squares allow you to determine how to construct an approximate F-test using the subcommand TEST.

TEST allows you to specify your own F-tests. Termlist is a list of terms in the model. Each is used as the numerator in a test. Errorterm is a linear combination of terms in the model, to be used as the denominator in all the tests. If you want to use MSE in your errorterm, denote it by the reserved word ERROR. Minitab calculates the F-statistics and p-values for you. Here are two examples:

```
TEST  A / A*B + A*C + B*C - 2*A*B*C - 2*ERROR
TEST  A B / A*B
```

There are two methods used to do mixed model analysis of variance. One method requires the crossed, mixed terms to sum to zero over subscripts corresponding to fixed effects (we call this the restricted model); the other does not. Most textbooks use the restricted model. By default, ANOVA fits the unrestricted model. RESTRICT says you want the restricted model.

Stat > Balanced ANOVA, use Options button for TEST and EMS

This gives an approximate F-test. Now we need the degrees of freedom for the denominator. For a linear combination of the form $\Sigma a_i \, MS_i$, we use the approximation

$$df(\text{denominator}) = \frac{[\Sigma a_i MS_i]^2}{\Sigma[(a_i MS_i)^2/df_i]}$$

Using this formula to calculate the degrees of freedom for the test on setting, we find

$$\frac{[MSTS + MSOS - MSTOS]^2}{[(MSTS)^2/df_{TS}] + [(MSOS)^2/df_{OS}] + [(MSTOS)^2/df_{TOS}]}$$

$$= \frac{[57.3 + 107.1 - 24]^2}{[57.3^2/2] + [107.1^2/4] - [24^2/4]} = 4.52$$

To be conservative, we truncate this to 4. To use the TEST subcommand, we give the numerator for the test first, then a slash, then the denominator. Minitab does the remaining calculations. The subcommand for setting is

```
TEST  S / T*S + O*S - T*O*S
```

17

Additional Features of Minitab

This chapter describes some additional features of Minitab. Sections 17.1 through 17.5 discuss Minitab capabilities that help you manage data. You will not need these for most simple data analyses. Section 17.6 describes matrices and the commands that operate on them. Section 17.7 shows how to write very simple programs (also called macros or command files) using Minitab.

17.1 How to COPY and Subset Data

The COPY command has two functions: It allows you to copy data from one place in the worksheet to another, and it allows you to select a subset of data.

Copying Data from One Place to Another

You can copy a list of columns into another list of columns, copy a list of constants into another list of constants, copy one column into a list of constants, and copy a list of constants into a column. Suppose C1 and C2 each contain 10 numbers. Here are some examples:

```
COPY   C1-C2    C11-C12
COPY   C1       K1-K10
COPY   K1-K10   K21-K30
COPY   K1-K10   C5
```

COPY C...C into **C...C**
COPY K...K into **K...K**
COPY C   into **K...K**
COPY K...K into **C**

The COPY command allows you to make a second copy of data and to move data from columns to constants, and vice-versa.

Manip > Copy Columns. You cannot copy constants in a dialog box. Type the command to do this.

Selecting Subsets of Data

COPY has two subcommands, USE and OMIT, that allow you to create subsets of your data. We will use the same example we used in Chapter 1: columns C1–C3 contain the sex (coded 1 = male, 2 = female), height, and weight for seven people.

C1	C2	C3
SEX	HT	WT
2	66	130
1	70	155
2	64	125
2	65	115
2	63	108
1	66	145
1	69	160

To copy the data for just the women and store them in C12–C13, use

```
COPY C2 C3   C12 C13;
   USE 'SEX' = 2.
```

To copy rows that have any of the values 64, 65, or 66 for HT, and store them in C22–C23, use

```
COPY C2 C3   C22 C23;
   USE 'HT' = 64 65 66.
```

You can use a colon to indicate a range of values. For example, to copy the data for all people who are 64 inches or shorter, use

---

COPY C...C into C...C

USE rows numbered **K...K**
USE rows where **C** = **K...K**

OMIT rows numbered **K...K**
OMIT rows where **C** = **K...K**

The subcommands copy selected rows. USE says which rows to copy; OMIT says which rows not to copy. These rows can be specified by their row numbers or by the values in a column.

You can use any list of row numbers. Consecutive numbers can be abbreviated with a colon. Thus USE 2, 5:8 is the same as USE 2, 5, 6, 7, 8.

You can use any list of values for a selection column. An interval can be abbreviated with a colon. Thus USE C1 = 5:8, 20 says to copy all rows where C1 is a value in the range of 5 through 8, or is equal to 20.

Manip > Copy Columns

```
COPY C2 C3 C22 C23;
  USE 'HT' = 0:64.
```

You also can subset by giving row numbers. For example, to copy rows 1, 3, 4, and 5, use

```
COPY C2 C3 C22 C23;
  USE 1 3:5.
```

OMIT is the opposite of USE. USE says which rows to copy; OMIT says which rows not to copy. To copy data for the women, you can omit the men by using

```
COPY C2 C3 C22 C23;
  OMIT SEX = 1.
```

You also can use row numbers with OMIT. For example, to copy all but rows 2, 6, and 7, use

```
COPY C2 C3 C22 C23;
  OMIT 2 6 7.
```

17.2 How to STACK and UNSTACK Data

Exhibit 17.1 shows height and weight data for men and women, stored in two different forms. In the unstacked form, the data for the men are in one pair of columns, HtMen and WtMen, and the data for the women are in a second pair, HtWomen and WtWomen. In the stacked form, the data for men and women are stacked together, in columns Ht and Wt. The column Sex indicates which rows are for men (Sex = 1) and which rows are for women (Sex = 2).

Sometimes we need to change data from one form to the other. The STACK and UNSTACK commands make this relatively easy. For example, to stack the columns HtMen, WtMen on the columns HtWomen, WtWomen and store the stacked data in columns Ht and Wt, use

```
STACK ('HtMen' 'WtMen') ('HtWomen' 'WtWomen')  ('Ht' 'Wt');
  SUBSCRIPTS 'Sex'.
```

Similarly, to unstack the columns Ht and Wt into the columns HtMen, WtMen and HtWomen, WtWomen, use

```
UNSTACK ('Ht' 'Wt') ('HtMen' 'WtMen') ('HtWomen' 'WtWomen');
  SUBSCRIPTS 'Sex'.
```

STACK can also be used with numbers and stored constants. For example,

```
STACK 'HtMen' 61 64 'HtWomen' 66    C20
```

creates one column, which contains 10 numbers: 70, 66, 69, 61, 64, 66, 64, 65, 63, and 66. Note that we omitted the parentheses in this command. We can omit the parentheses if everything is stacked into one column.

Exhibit 17.1 Stacked and Unstacked Data

HtMen	WtMen	HtWomen	WtWomen	Ht	Wt	Sex
Unstacked Data				*Stacked Data*		
70	155	66	130	70	155	1
66	145	64	125	66	145	1
69	160	65	115	69	160	1
		63	108	66	130	2
				64	125	2
				65	115	2
				63	108	2

STACK (E...E) on ... on (E...E) store in (C...C)

SUBSCRIPTS into **C**

This command stacks blocks of columns and constants on top of each other. In general, each block must be enclosed in parentheses. However, if each block contains just one argument, the parentheses may be omitted.

SUBSCRIPTS creates a column of subscripts. All rows in the first block are given subscripts of 1, all rows in the second block are given subscripts of 2, and so on.

Release 8: **Calc > Stack.** You are limited to 5 blocks; use the STACK command for more.
Windows: **Manip > Stack.** You are limited to 5 blocks; use the STACK command for more.

UNSTACK (C...C) into (E...E) ... (E...E)

SUBSCRIPTS are in **C**

This command separates one block of columns into several blocks of columns and/or stored constants. In general, each block must be enclosed in parentheses. However, if each block contains just one argument, the parentheses may be omitted.

For most applications, the subcommand SUBSCRIPTS will be needed. The rows with the smallest subscript are stored in the first block, the rows with the second smallest subscript are stored in the second block, and so on. If you do not use SUBSCRIPTS, each row is stored in a separate block. The numbers in the subscript column must be integers between -10000 and $+10000$. They need not be in order, nor do they need to be consecutive integers.

Release 8: **Calc > Unstack.** You are limited to 5 blocks; type the UNSTACK command for more.
Windows: **Manip > Unstack.** You are limited to 5 blocks; type the UNSTACK command for more.

17.3　How to Change Values with CODE

The CODE command changes values in columns. Here is a simple example:

```
CODE   (-2 -1)0    C1-C2 C11-C12
```

C1	C2	C11	C12
8	16	8	16
-2	13	0	13
6	-2	6	0
0	-2	0	0
-1	14	0	14

CODE changes the values in C1–C2 that are equal to −2 or −1 into 0. All other values are left the same. The results are stored in C11 and C12. The original columns, C1–C2, are not changed.

You can make many changes at once, and you can specify a range with a colon. For example,

```
CODE (30:55)1 (56:80)2 (81:120)3 C1-C5, C11-C15
```

changes all values from 30 up through 55 into 1, values from 56 up through 80 into 2, and values from 81 up through 120 into 3. These changes are made in columns C1–C5 and the results are stored in C11–C15.

CODE (K...K) to K ... (K...K) to K for data in C...C, store in C...C

Changes the list of values in each set of parentheses to the number after the set. All other values in the columns are left as is. An interval can be abbreviated with a colon. For example, CODE (1:1.5, 2)0 changes all values in the range 1 through 1.5, and the value 2, into zeroes.

If several lists overlap, the last one is used. For example, CODE (10:20)1 (20:30)2 changes the value 20 into a 2. CODE can change values into the missing-value code, *. For example, to change all occurrences of −99 to missing values, and store the results back in the same columns, use

```
CODE (-99)'*' C1-C10    C1-C10
```

Release 8: **Calc > Code Data Values.** You are limited to 5 blocks; type CODE for more.
Windows: **Manip > Code Data Values.** You are limited to 5 blocks; type CODE for more.

17.4 How to SORT and RANK Data

The SORT command will sort any number of columns, by any number of columns, and any combination of alpha and numeric data. Here are two examples:

```
SORT C1-C4  C11-C14;
   BY C1.
SORT C1-C4  C21-C24;
   BY C1 C2.
```

C1	C2	C3	C4		C11	C12	C13	C14		C21	C22	C23	C24
10	0.2	31	131		10	0.2	31	131		10	0.1	35	210
10	0.1	35	210		10	0.1	35	210		10	0.2	31	121
12	0.1	37	176		10	0.4	31	140		10	0.4	31	140
12	0.1	36	190		11	0.2	29	180		11	0.1	33	182
10	0.4	31	140		11	0.1	33	182		11	0.2	29	180
12	0.1	30	110		12	0.1	37	176		12	0.1	37	176
11	0.2	29	180		12	0.1	36	190		12	0.1	36	190
11	0.1	33	182		12	0.1	30	110		12	0.1	30	110

In the first SORT, C1–C4 are sorted and are stored in C11–C14. The column listed on BY, C1, determines the order. The smallest value in C1 is 10, so all rows in which C1 = 10 are put first. The second smallest value in C1 is 11, so all rows in which C1 = 11 are put next, and so on.

In the second SORT, C1–C4 are sorted by two columns: first C1 and then, within C1, by C2. The sorted data are stored in C21–C24.

If you sort by just one column and do not wish to use the BY subcommand, just list that column first on SORT. Thus the following two commands are equivalent:

```
SORT C1-C4  C11-C14
```

```
SORT C1-C4  C11-C14;
   BY C1.
```

The RANK command does not reorder the data. It creates a new column containing the rank of each value in the original column. The smallest value is given rank 1; the second smallest, rank 2; the third smallest, rank 3; and so on. Here is an example:

```
RANK C1   C11
```

SORT C...C store in C...C

BY C...C
DESCENDING C...C

If you do not use the BY subcommand, the columns on SORT are sorted by the first column listed on SORT. If you use BY, the columns listed on SORT are sorted by the columns listed on BY.

By default, sorting is done in ascending order. DESCENDING requests descending order for the columns listed here. If you use BY, the columns listed on DESCENDING must also be listed on BY. If you do not use BY, the column listed on DESCENDING must be the first column listed on SORT.

Release 8: **Calc > Sort.** You are limited to 5 columns for BY; type the SORT command for more.

Windows: **Manip > Sort.** You are limited to 5 columns for BY; type SORT for more.

RANK C into C

Ranks data. The number 1 is the rank of the smallest value, the number 2 is the rank of the second smallest value, and so on. If several values are the same, their average rank is assigned to each of them.

Release 8: **Calc > Rank**
Windows: **Manip > Rank**

C1	C11
1.4	3
1.1	1
2.0	4.5
3.1	6
2.0	4.5
1.3	2

Ranks for C1 are put into C11. The smallest number in C1 is 1.1, so a 1 is put into C11. The second smallest number in C1 is 1.3, so a 2 is put into C11, and so on. If there are ties, the average rank is assigned to each value.

For example, the fourth and fifth values in C1 are both 2.0. They are both given the rank of $(4 + 5)/2 = 4.5$.

17.5 Formatted Input and Output of Data

Minitab has five commands that allow FORTRAN formats: READ, SET, INSERT, PRINT, and WRITE. In this section we will give only a brief introduction to the use of formats. You may need to consult a FORTRAN manual for more details.

Minitab can use all format specifications except I (for integers). These include F (for real numbers), E (for numbers using exponential notation), A (for alphabetic data), X (to skip over numbers), and / (to go to the next line).

Formatted Input with READ, SET, and INSERT

Suppose you have data that were typed without any blanks or commas between the numbers. READ by itself cannot enter these, but it can if you use a FORMAT subcommand. For example, suppose we have the data

```
231615.411821143   9
55431.8911941138128
```

and we want these numbers read into C1–C6 as follows:

C1	C2	C3	C4	C5	C6
231	61	5.4	1182	1143	9
554	31	.89	1194	1138	128

First we determine how wide each number is. In this case, the first number on each line uses three spaces, the second number uses two spaces, the third uses three spaces, the fourth and fifth both use four spaces, and the last uses three spaces. We use the following FORMAT subcommand with READ:

```
READ C1-C6;
   FORMAT (F3.0,F2.0,F3.0,2F4.0,F3.0).
```

This FORMAT describes one row of data. In F3.0, the 3 says the number to be read uses three spaces; F2.0 says the second number uses two spaces; the next F3.0 says the third number uses three spaces. Notice that in this instance, one of the spaces holds the decimal point. The next two numbers both use four spaces. To save typing, we can write 2F4.0 instead of F4.0, F4.0.

The last F3.0 says the last number uses three spaces. Notice that the number on the first line, 9, is just one space wide. The remaining two spaces are blank. When the field used by a number is wider than the number, the number must be right-justified; that is, all blanks must appear before the number.

You can skip over numbers with formats. For example, suppose we want the data read into the worksheet as follows:

C1	C2	C3	C4	C5	C6
231			1182		
554			1194		

We would use the command

```
READ C1 C4;
  FORMAT (F3.0,5X,F4.0).
```

The 5X says to skip over five spaces. Thus we read a number from the first three spaces, skip five spaces, then read a number from the next four spaces. The rest of the line is ignored because there are no more format specifications in the FORMAT subcommand.

Formats must be typed according to strict rules. You must use parentheses and you must put a comma between format specifications.

You can enter several data lines into one row of the worksheet by using a slash (/). For example, suppose a file contains data for 200 people: 25 measurements for each person, typed on three lines for each person. Here are the data for the first two people:

```
2 4 3 1 4 4 2 0 1 2
1 4 2 1 4 3 1 4 2 2
0 2 0 1 1
0 5 3 1 3 4 0 0 1 3
0 5 3 1 0 2 1 4 2 1
1 1 1 2 0
```

To enter these data into C1–C25, we can use

```
READ C1-C25;
  FORMAT(10F2.0/10F2.0/5F2.0).
```

The first 10F2.0 says to read 10 numbers from the first line. The slash says to go to the next line. The second 10F2.0 says to read 10 numbers from this line. The second slash says to go to another new line. The 5F2.0 says to read 5 numbers from this line. This procedure is repeated for each person, always using three lines of data. Notice that FORMAT describes one row of the worksheet.

You also can use slashes to skip over data lines. For example, suppose we use the following format on the data for 200 people:

```
READ C1-C3;
  FORMAT (3F2.0//).
```

This would take three numbers from the first line for each person and skip over the other two lines.

Formatted Output with PRINT and WRITE

Formats also can be used to output data. This is done if you want to control the spacing of the numbers, put data from one row of the worksheet onto several lines, or print the numbers to a different (usually larger) number of decimal places than Minitab would choose.

If you use FORMAT with PRINT, you must use a carriage control (see a FORTRAN manual for a complete discussion of carriage controls) at the beginning of the format statement. The most common is 1X, which gives single-spaced output.

As an example, consider the data that we read into C1–C6 in the first example of this section. The command

```
PRINT C1-C6;
  FORMAT (1X,F5.0,F4.0,F6.2,2F6.0,F5.0).
```

would print out the data as follows:

```
231 61 5.40 1182 1143    9
554 31 0.89 1194 1138 123
```

The first specification, 1X, is the carriage control. The next specification, F5.0, describes the first five spaces: the 5 says to use five spaces and the 0 says to print no digits after the decimal point. The number 231 is then right-justified in the field of five spaces. Similarly, the second number, 61, is right-justified in a field of four spaces. The next specification, F6.2, says to use six spaces and put two digits after the decimal point.

You can use the X format to skip spaces on the printed page. For example,

```
PRINT C1-C6;
  FORMAT (1X,F5.0,F4.0,12X,F6.2,2F6.0,F5.0).
```

would print two numbers, then skip twelve spaces, then print the last four numbers.

You can use slashes to print one row of the worksheet on several lines of output. You must put a carriage control at the beginning of each line. For example,

```
PRINT C1-C6;
  FORMAT (1X,F5.0,F4.0,F6.2/1X,2F6.0,F5.0).
```

puts three numbers on the first line and three on the second.

17.6 Matrix Calculations

In addition to columns and stored constants, Minitab has matrices, denoted M1, M2, M3, ... The total number of matrices available to you depends on your computer. In addition, columns can be used as column matrices.

There are two ways to enter data into a matrix: with READ and with COPY. Here is an example of each:

```
READ 3 4   M20
   1   4   6   3
   2   8   1   8
  16  12  14  10
END
COPY C2-C4   M21
```

C2	*C3*	*C4*
16	2	40
18	4	35
20	3	42

$$M20 = \begin{bmatrix} 1 & 4 & 6 & 3 \\ 2 & 8 & 1 & 8 \\ 16 & 12 & 14 & 10 \end{bmatrix} \qquad M21 = \begin{bmatrix} 16 & 2 & 40 \\ 18 & 4 & 35 \\ 20 & 3 & 42 \end{bmatrix}$$

READ the following data into a **K** by **K** matrix **M**

Data follow, one line for each row of the matrix. Note that you must specify the dimensions of the matrix you are creating.

Calc > Matrices > Read

PRINT M...M

Prints out the matrices.

Calc > Matrices > Print

Minitab has matrix commands to add, subtract, multiply, transpose, and invert matrices. Here are some simple examples:

```
COPY           C1-C3    M1
MULTIPLY       M1 C5    M2
TRANSPOSE      C4       M3
MULTIPLY       M3 M1    M4
MULTIPLY       M3 M2    K1
ADD            2 M1     M5
```

C1	*C2*	*C3*	*C4*	*C5*		*K1*
1	4	2	10	5		334
1	0	3	12	0		
				4		

$$M1 \qquad M2 \qquad\quad M3 \qquad\qquad M4 \qquad\qquad\quad M5$$

$$\begin{bmatrix} 1 & 4 & 2 \\ 1 & 0 & 3 \end{bmatrix} \quad \begin{bmatrix} 13 \\ 17 \end{bmatrix} \quad [\,10 \;\; 12\,] \quad [\,22 \;\; 40 \;\; 56\,] \quad \begin{bmatrix} 3 & 6 & 4 \\ 3 & 2 & 5 \end{bmatrix}$$

COPY C...C into M
COPY M into C...C

The first version of COPY forms a matrix out of columns. The second version puts a matrix into columns.

Calc > Matrices > Copy

DEFINE K into a K by K matrix M

Creates a matrix of the specified size, all of whose entries are the specified constant. For example,

```
DEFINE 1   4 3   M5
```

creates a matrix, M5, that has 4 rows and 3 columns. All entries of M5 are 1.

Calc > Matrices > Define

ADD   **M** to **M** put into **M**
SUBTRACT **M** from **M** put into **M**
MULTIPLY   **M** by **M** put into **M**

TRANSPOSE   **M** put into **M**
INVERT   **M** put into **M**

These five commands do the usual matrix arithmetic calculations. You can also do scalar addition, subtraction, and multiplication with ADD, SUB-TRACT, and MULTIPLY. For example:

```
ADD        K1 M1    M2
SUBTRACT   M1 K1    M3
MULTIPLY   M1 K1    M4
```

You can use a column as a column matrix and a constant as a 1-by-1 matrix when doing matrix arithmetic. For example, if M1 is a 4-by-3 matrix and C2 contains three numbers, we can calculate:

```
MULTIPLY     M1 C2    M2
TRANSPOSE    C2       M3
MULTIPLY     M3 C2    K1
TRANSPOSE    M1       M4
MULTIPLY     M4 M2    C1
```

Calc > Matrices > Arithmetic, for ADD, SUBTRACT, and MULTIPLY
Calc > Matrices > Transpose
Calc > Matrices > Invert

DIAGONAL is in **C** put into **M**
DIAGONAL of **M** put into **C**

In the first form of DIAGONAL, the matrix M has the entries in C on its diagonal and zeroes elsewhere. In the second form, the entries in C are the numbers on the diagonal of M.

Calc > Matrices > Diagonal

EIGEN analysis for **M** put values in **C** [vectors in **M**]

EIGEN calculates eigenvalues (also called characteristic values and latent roots) and eigenvectors for a symmetric matrix. The eigenvalues are stored in decreasing order of magnitude down the column. The eigenvectors are stored as columns of the matrix; the first column corresponds to the first eigenvalue (largest magnitude), the second column to the second eigenvalue, and so on.

Calc > Matrices > Eigen Analysis

17.7 Stored Command Files

You can store a block of Minitab commands in a computer file. Any time you want to use this block, you type one command, EXECUTE. An instructor might store commands to do a complicated exercise; then students would not have to type them. A researcher may have an analysis that she does each week, when she gets a new data set. If she stores the commands to do the analysis, she will not have to retype them each week.

As an example, we will use the commands in Exercise 8-7, which simulate fifty 90% confidence intervals.

```
STORE 'INTERVAL'
  RANDOM 50 C1-C5;
    NORMAL 30 2.
  RMEAN C1-C5 C10
  LET K1 = 1.645*2 / SQRT(5)
  LET C21 = C10 - K1
  LET C22 = C10 + K1
  SET C23
    1:50
  END
  MPLOT C21 C23  C22 C23
END
```

STORE says to put all the Minitab commands that follow into a computer file called INTERVAL. When you have typed all commands to be stored, type END to tell Minitab you are finished typing stored commands. Now, if at some time in the same session or in another session, you type

```
EXECUTE 'INTERVAL'
```

Minitab will execute the commands in the file. When Minitab is finished, C1–C5, C10, C21–C23, and K1 will contain the calculated results and an MPLOT will be printed.

You can EXECUTE a file several times just by specifying the number of times you wish to do so. For example,

```
EXECUTE 'INTERVAL' 3
```

would do the simulation three times, with different random data each time, and print three MPLOTS. C1–C5, C10, C21–C23, and K1 would contain the calculations from the last simulation.

As another example of using EXECUTE to loop through a block of commands, we will calculate a moving average of length 2 for data in C1. Here are the commands:

```
STORE 'MOVEAVG'
   # Calculates the moving average of length 2, using the
   # data stored in C1. The moving averages are stored
   # in C2.
   NOECHO
   LET C2(K1) = (C1(K1) + C1(K1+1)) / 2
   LET K1 = K1 + 1
   ECHO
END

LET K1 = 1
LET K2 = COUNT(C1) - 1
EXECUTE 'MOVEAVG' K2
```

There are two new commands in the stored file: NOECHO says do not echo (print) the commands that follow, just do the calculations and print output. ECHO returns to the usual state, in which commands in a STORE file are printed when the file is executed. Without NOECHO, the two LET commands would be printed every time the file was executed. But we do not need to see them. We also used # to add a comment to our program.

Suppose we use MOVEAVG on the data in C1, below. After EXE-CUTE, the worksheet is

C1	C2	K1	K2
4	3.0	6	5
2	2.5		
3	4.5		
6	5.5		
5	4.5		
4			

STORE 'filename'
END

EXECUTE 'filename' [**K** times]

Type STORE to start creating a file of Minitab commands. Then type the Minitab commands. Type END when you are done. The commands will be stored in the file specified.

Use EXECUTE to execute the Minitab commands that are stored in the specified file. If you use the repeat factor K, the commands in the file will be executed K times.

Release 8 and Windows: In these releases, it is generally easier to use STORE and EXEC through dialog boxes rather than typing them as commands.
File > Other Files > Start Recording Exec
File > Other Files > Stop Recording Exec
File > Other Files > Run an Exec

The calculations we did are very simple, although the Minitab commands may at first look somewhat complicated. C1 contains the original data. C2 contains the moving average of C1. The first entry in C2 is the average of the first and second entries in C1, the second entry in C2 is the average of the second and third entries in C1, and so on. Notice that since C1 contains six numbers, C2 contains $6 - 1 = 5$ numbers. Before we used EXECUTE, we set $K1 = 1$ and $K2 = 5$. We EXECUTED the file $K2 = 5$ times, once to calculate each value in C2. The first time, $K1 = 1$. We calculated

```
C2(1) = (C1(1) + C1(2)) / 2
```

and increased K1 to 2. The second time, $K1 = 2$. We calculated

```
C2(2) = (C1(2) + C1(3)) / 2
```

and increased K1 to 3. This continued until we had looped five times.

Creating STORE Files

If you use STORE to create a command file and make a typing error, there is, unfortunately, no way to correct the file in Minitab. You must start over. Thus STORE is most useful for short command files, in which you are not likely to make an error. If you want to store a long file, it is best to create the file outside of Minitab, using an editor. You also can create the file in

---

NOECHO
ECHO

These commands are used mostly in a STORED file; NOECHO is often the first command in the file, right after STORE, and ECHO is the last, just before END. When the file is EXECUTED, no commands are printed out; only output is printed. This is what you usually want.

There is one time when you may not want to use NOECHO. When you first create a file, especially one that is a little complicated, you should EXECUTE it to check for errors. Having the commands printed out will help you interpret any error messages. Once you are sure the commands in your file are correct, you can add NOECHO and ECHO to it.

Minitab, using STORE, then leave Minitab and use your editor to make corrections.

A Minitab file of stored commands must have an appropriate extension to its file name. Exactly how this is done varies with the type of computer. Usually the extension is .MTB. If you use STORE, Minitab automatically adds the appropriate extension. If you list the names of your files, you will see what extension Minitab uses on your computer. If you create a file of Minitab commands with your editor, make sure the name of the file includes the appropriate extension.

Looping Through Columns

There is a little device, called the CK capability, that allows you to loop through columns. As a simple example, suppose we want to plot each of C1 through C25 against C50. This would require 25 separate PLOT commands. With the CK capability, we can use

```
STORE 'PLOTS'
  PLOT CK1 C50
  LET K1 = K1+1
END

LET K1 = 1
EXECUTE 'PLOTS' 25
```

The file PLOTS is EXECUTED 25 times, once to produce each plot. The first time, K1 = 1, so Minitab substitutes a 1 for the K1 in the PLOT command and does PLOT C1 C50. Then K1 is increased to 2. The second time through the file, Minitab substitutes a 2 for K1 and does PLOT C2 C50, and so on.

Any time a column appears in a command, you may use CK1 (or CK2 or CK3, . . .) for the column. When Minitab executes the command, K1 is replaced by its current value.

Appendix A

Data Sets Used in This Handbook

Many of the data sets used in this handbook are available as saved worksheets (to be input with the command RETRIEVE). Table A.1 gives a complete list. The data sets that are starred have been distributed with Minitab software in Releases 7, 8, and 9. The data sets that are not starred are available from Duxbury Press.

Longer data sets that are used throughout this book are described in this appendix. These data sets are available with the Minitab software and are not displayed in this book.

Table A.1 Data Sets Available as Saved Worksheets

Name	*Description*	*Text Citation*
ACID	*Acid data set	Exercises 13-3 and 13-4
ALFALFA	*Alfalfa data	Exercise 10-11
AUTO	Automobile accident data	Exercise 4-12
AZALEA	*Azalea data	Exercise 13-16
BEETLE	Beetle data	Exercise 10-8
BILLIARD	Billiard ball data	Table 10.4
CAP	Cap torque data	Exercise 14-1
CARTOON	*Cartoon data	This appendix
CHOLEST	*Cholesterol data	Table 9.1
CITIES	*Temperatures for various cities	Exercise 4-10
DOCS	Documents data	Exercise 14-4
DRIVE	Driving performance data	Exercise 16-6
EMPLOY	*Employment	This appendix
FA	*Regression data from Frank Anscombe	Exercise 11-30
FABRIC	*Fabric data, organized as in Table 10.1	Table 10.1
FABRIC1	Fabric data, organized as used in Exhibit 10.1	
FALLS	Accidental deaths from falls data	Exercise 12-1
FURNACE	*Furnace data	This appendix
FURNACE1	Furnace data	Table 4.1
GRADES	*Full grades data set	This appendix
GA	*Sample A, first 100 observations	
GB	*Sample B, second 100 observations	
HCC	*Health care center data	Exercise 5-4
KEY	Ignition key data	Exhibit 14.1
LAKE	*Lake data set	This appendix
LEAF	Leaf-surface resistance data	Exercise 4-17
LIVER	Manganese in cow liver data	Exercise 10-5
MAPLE	*Maple trees data	Exercise 11-18
MARRIAGE	Marriage rate data	Exercise 4-5
MEATLOAF	*Meat loaf data	Exercise 10-12

OXYGEN	Oxygen in steel data	Exercise 10-6
PANCAKE	Pancake data	Table 10.3
PARKINSN	Parkinson's disease data	Table 9.3
PENDULUM	Pendulum data, as a two-way design	Exercises 10-3 and 10-9
PERU	*Peru data	This appendix
PLATING	Plating data	Exhibit 14.7
PLYWOOD	*Plywood data (full set)	Exercise 16-5
PLYWOOD1	Plywood data (short version)	Exercise 10-2
POPLARS	Poplar trees data	Exercise 16-8
POTATO	*Potato rot data	Exercise 10-10
PRES	*Presidents data set	Exercise 13-8
PULSE	*Pulse data	This appendix
PULSE1	Pulse data (portion of full data set)	Table 1.1
REACTOR	Penn State nuclear reactor data	Exercise 6-15
RESTRNT	*Restaurant data	This appendix
SAT	SAT scores data	Exercise 11-8
SCORES	Test scores data	Table 11.1
STREAM	Stream data	Exhibit 11.5, Exercises 11-24 and 11-26
TRACK15	Olympic track data, 1500 m	Exercise 11-9
TRACKM	Men's Olympic track data	Table 4.2
TRACKMW	Track records data for men and women	Exercise 4-6
TREES	*Trees data	This appendix
TUMOR	Tumor data	Exercise 11-32
TWAIN	*Mark Twain data	Exercise 12-8
WIND	Wind chill data	Exercise 4-11
WINE	Wine data	Exercise 13-18

Cartoon

When educators make an instructional film, they have two objectives: Will the people who watch the film learn the material as efficiently as possible? Will they retain what they have learned?

To help answer these questions, an experiment was conducted to evaluate the relative effectiveness of *cartoon* sketches and *realistic* photographs, using both *color* and *black-and-white* visual materials.

A short instructional slide presentation was developed. The topic chosen for the presentation was the behavior of people in a group situation and, in particular, the various roles or character types that group members often assume. The presentation consisted of a five-minute lecture on tape, accompanied by eighteen slides. Each role was identified as an animal. Each animal was shown on two slides: once in a cartoon sketch and once in a realistic picture. All 179 participants saw all of the eighteen slides, but a randomly selected half of the participants saw them in black-and-white while the other half saw them in color.

After they had seen the slides, the participants took a test (*immediate test*) on the material. The eighteen slides were presented in a random order, and the participants wrote down the character type represented by each slide. They received two scores: one for the number of cartoon characters they correctly identified and one for the number of realistic characters they correctly identified. Each score could range from 0 to 9, since there were nine characters.

Four weeks later, the participants were given another test (*delayed test*) and their scores were computed again. Some participants did not show up for this delayed test, so their scores were given the missing-value code, *.

The primary participants in this study were preprofessional and professional personnel at three hospitals in Pennsylvania involved in an in-service training program. A group of Penn State undergraduate students were also given the test as a comparison. All participants were given the OTIS Quick Scoring Mental Ability Test, which yielded a rough estimate of their natural ability.

Some questions that are of interest here are as follows: Is there a difference between color and black-and-white visual aids? Between cartoon and realistic depictions? Is there any difference in retention? Does any difference depend on educational level or location? Does adjusting for OTIS scores make any difference?

The data are stored in the Minitab worksheet called CARTOON.

Description of Cartoon Data

Variable	Description
1 ID	Identification number
2 COLOR	0 = black-and-white, 1 = color (no participant saw both)
3 ED	Education: 0 = preprofessional, 1 = professional, 2 = college student
4 LOCATION	Location: 1 = hospital A, 2 = hospital B, 3 = hospital C, 4 = Penn State student
5 OTIS	OTIS score: from about 70 to about 130
6 CARTOON1	Score on cartoon test given immediately after presentation (possible scores are 0, 1, 2, . . . , 9)
7 REAL1	Score on realistic test given immediately after presentation (possible scores are 0, 1, 2, . . . , 9)
8 CARTOON2	Score on cartoon test given four weeks (delayed) after presentation (possible scores are 0, 1, 2, . . . , 9; * is used for a missing observation)
9 REAL2	Score on realistic test given four weeks (delayed) after presentation (possible scores are 0, 1, 2, . . . , 9; * is used for a missing observation)

Furnace

Wisconsin Power and Light studied the effectiveness of two devices for improving the efficiency of gas home-heating systems. The electric vent damper (EVD) reduces heat loss through the chimney when the furnace is in its off cycle by closing off the vent. It is controlled electrically. The thermally activated vent damper (TVD) is the same as the EVD, except that it is controlled by the thermal properties of a set of bimetal fins set in the vent. Ninety test houses were used, 40 with TVDs and 50 with EVDs. For each house, energy consumption was measured for a period of several weeks with the vent damper active and for a period with the damper not active. This should help show how effective the vent damper is in each house.

Both overall weather conditions and the size of a house can greatly affect energy consumption. A simple formula was used to try to adjust for this. Average energy consumed by the house during one period was recorded as (consumption)/[(weather)(house area)], where consumption is total energy consumption for the period, measured in BTUs, weather is measured in number of degree days, and house area is measured in square feet. In addition, various characteristics of the house, chimney, and furnace were recorded for each house. A few observations were missing and recorded as *, Minitab's missing-data code.

The data are stored in the Minitab worksheet called FURNACE.

Description of Furnace Data

Variable	Description
1 TYPE	Type of furnace: 1= forced air, 2 = gravity, 3 = forced water
2 CH.AREA	Chimney area
3 CH.SHAPE	Chimney shape: 1 = round, 2 = square, 3 = rectangular
4 CH.HT	Chimney height (in feet)
5 CH.LINER	Type of chimney liner: 0 = unlined, 1 = tile, 2 = metal
6 HOUSE	Type of house: 1 = ranch, 2 = two-story, 3 = tri-level, 4 = bi-level, 5 = one and one-half stories
7 AGE	House age in years (99 means 99 or more years)
8 BTU.IN	Average energy consumption with vent damper in
9 BTU.OUT	Average energy consumption with vent damper out
10 DAMPER	Type of damper: 1 = EVD, 2 = TVD

Grades (Also GA, GB)

Scholastic Aptitude Tests (SAT) are often used as criteria for admission to college, as predictors of college performance, or as indicators for placement in courses. The data below are a sample of SAT scores and freshman-year grade-point averages (GPAs) from a northeastern university. (The university wishes to remain anonymous.) The sample of 200 students was broken down randomly into two samples of size 100 for ease of use in this handbook. These two samples are stored as GA and GB. The combined sample is stored as GRADES.

Description of Grades Data

Variable	*Description*
1 VERB	Score on verbal aptitude test
2 MATH	Score on mathematical aptitude test
3 GPA	Grade-point average (0 to 4, with 4 the best grade)

Lake

These lakes are all in the Vilas and Oneida counties of northern Wisconsin. Measurements were made in 1959–1963. The data are stored in the Minitab worksheet called LAKE.

Description of Lake Data

Variable	*Description*
1 AREA	Area of lake in acres
2 DEPTH	Maximum depth of lake in feet
3 PH	pH, a measure of acidity (a lower pH is more acidic; a pH of 7 is neutral; a higher pH is more alkaline)
4 WSHED	Watershed area in square miles
5 HIONS	Concentration of hydrogen ions

Peru

A study was conducted by some anthropologists to determine the long-term effects of a change in environment on blood pressure. In this study they measured the blood pressure of a number of Indians who had migrated from a very primitive environment, high in the Andes mountains of Peru, into the mainstream of Peruvian society, at a much lower altitude.

A previous study in Africa had suggested that migration from a primitive society to a modern one might increase blood pressure at first, but that blood pressure would tend to decrease back to normal over time.

The anthropologists also measured the height, weight, and a number of other characteristics of each subject. A portion of their data is given below. All these data are for males over 21 who were born at a high altitude and whose parents were born at a high altitude. The skin-fold measurements were taken as a general measure of obesity. Systolic and diastolic blood pressure are usually studied separately. Systolic blood pressure is often a more sensitive indicator.

The data are stored in the Minitab worksheet called PERU.

Description of Peru Data

Variable	Description
1 AGE	Age in years
2 YEARS	Years since migration
3 WEIGHT	Weight in kilograms (1 kg = 2.2 lb)
4 HEIGHT	Height in millimeters (1 mm = 0.039 in.)
5 CHIN	Chin skin fold in millimeters
6 FOREARM	Forearm skin fold in millimeters
7 CALF	Calf skin fold in millimeters
8 PULSE	Pulse rate in beats per minute
9 SYSTOL	Systolic blood pressure
10 DIASTOL	Diastolic blood pressure

Pulse

Students in an introductory statistics course participated in a simple experiment. The students took their own pulse rates (which is easiest to do by holding the thumb and forefinger of one hand on the pair of arteries on the side of the neck). They were then asked to flip a coin. If their coin came up heads, they were to run in place for one minute. Then all the students took their own pulses again. The pulse rates and some other data are given below. The data are stored as PULSE.

Description of Pulse Data

Variable	Description
1 PULSE1	First pulse rate
2 PULSE2	Second pulse rate
3 RAN	1 = ran in place, 2 = did not run in place
4 SMOKES	1 = smokes regularly, 2 = does not smoke regularly
5 SEX	1 = male, 2 = female
6 HEIGHT	Height in inches
7 WEIGHT	Weight in pounds
8 ACTIVITY	Usual level of physical activity: 1 = slight, 2 = moderate, 3 = considerable

Restaurant

The survey is described in Chapter 5. The data are stored in the worksheet called RESTRNT.

Description of Restaurant Data

Variable	Description
1 ID	Identification number
2 OUTLOOK	Values 1, 2, 3, 4, 5, 6, 7, denoting outlook, from very unfavorable to very favorable
3 SALES	Gross 1979 sales, in $1000s
4 NEWCAP	New capital invested in 1979, in $1000s
5 VALUE	Estimated market value of the business, in $1000s
6 COSTGOOD	Cost of goods sold as a percentage of sales
7 WAGES	Wages as a percentage of sales
8 ADS	Advertising as a percentage of sales
9 TYPEFOOD	1 = fast food, 2 = supper club, 3 = other
10 SEATS	Number of seats in dining area
11 OWNER	1 = sole proprietorship, 2 = partnership, 3 = corporation
12 FT.EMPL	Number of full-time employees
13 PT.EMPL	Number of part-time employees
14 SIZE	Size of restaurant: 1 = 1 to 9.5 full-time equivalent employees, 2 = 10 to 20 full-time equivalent employees, 3 = over 20 full-time equivalent employees, where full-time equivalent employees equals (number of full-time) + (1/2)(number of part-time)

Trees

People in forestry need to be able to estimate the amount of timber in a given area of a forest. Therefore, they need a quick and easy way to determine the volume of any given tree. Of course, it is difficult to measure the volume of a tree directly. But it is not too difficult to measure the height, and even easier to measure the diameter. Thus the forester would like to develop an equation or table that makes it easy to estimate the volume of a tree from its diameter and/or height. A sample of trees of various diameters and heights was cut, and the diameter, height, and volume of each tree was recorded. Below are the results of one such sample. This sample is for black cherry trees in the Allegheny National Forest, Pennsylvania. (Of course, different varieties of trees and different locations will yield different results, so separate tables are prepared for each variety and each location.) The data are stored in the worksheet called TREES.

Description of Tree Data

Variable	Description
1 DIAMETER	Diameter in inches at 4.5 feet above ground level
2 HEIGHT	Height of tree in feet
3 VOLUME	Volume of tree in cubic feet

Appendix B

Quick Reference for Commands

This appendix summarizes Minitab commands in Releases 7, 8, and 9.
Some specialized commands are omitted.

Commands are left-justified; subcommands are indented.
Text that Minitab uses is in bold. Non-bold text is just for annotation.

Notation

K	denotes a constant, such as 8.3 or K14
C	denotes a column, such as C12 or 'Height'
E	denotes either a constant or a column
M	denotes a matrix, such as M5
[ ]	encloses an optional argument

1. General Information

HELP	explains Minitab commands and subcommands
INFO	[**C...C**] gives the status of the worksheet
STOP	ends the current session

2. Input and Output of Data

READ	[to **'filename'**] data into **C...C**	
TAB	columns are separated with tabs	(Releases 8 & 9 only)
NONAMES	columns are not named (use with TAB)	
ALPHA	data are in columns **K...K** (use with TAB)	
SET	data into **C**	
INSERT	data [between rows **K** and **K**] of **C...C**	

READ, SET, and INSERT have the subcommands:

FORMAT	(fortran format)	
NOBS	= **K**	
FILE	**'filename'**	(Release 9 only)
SKIP	**K** lines	(Release 9 only)
END	of data	
NAME	for **C** is 'name', ..., for **C** is 'name'	
PRINT	the data in **E...E**	
WRITE	[to **'filename'**] the data in **C...C**	
TAB	separate columns with tabs	
NONAMES	omit column names (use with TAB)	
REPLACE	replace existing file	(Release 9 only)
NOREPLACE	do not replace existing file	(Release 9 only)

PRINT and WRITE have the subcommand:

FORMAT	(fortran format)	
SAVE	[in **'filename'**] a copy of the worksheet	
PORTABLE	worksheet can be used on other computer brands	
REPLACE	replace existing file	(Release 9 only)
NOREPLACE	do not replace existing file	(Release 9 only)
LOTUS	save in Lotus format	(DOS and Windows only)
RETRIEVE	the Minitab saved worksheet [in **'filename'**]	

PORTABLE	worksheet can be used on other computer brands
LOTUS	retrieve Lotus file (DOS and Windows only)
OUTFILE	**'filename'** put all output in this file
OW	= **K** output width of file
OH	= **K** output height of file
NOTERM	no output to terminal
NOOUTFILE	output to terminal only
NEWPAGE	start next output on a new page
JOURNAL	[**'filename'**] record Minitab commands in this file
NOJOURNAL	cancels JOURNAL
PAPER	output to printer (not for Release 9 Windows)
OW	= **K** output width of printer
OH	= **K** output height of printer
NOTERM	no output to terminal
NOPAPER	output to terminal only (not for Release 9 Windows)

3. Editing and Manipulating Data

LET	C(K) = K # changes the number in row K of C
DELETE	rows **K...K** of **C...C**
ERASE	all data in **E...E**
INSERT	(see Section 2)
COPY	**K...K** into **C**
COPY	**C...C** into **C...C**
COPY	**C** into **K...K**

These two versions of COPY have the following subcommands:

USE	rows **K...K**
USE	rows where **C = K...K**
OMIT	rows **K...K**
OMIT	rows where **C = K...K**

CODE	(**K...K**) to **K**, ..., (**K...K**) to **K**, for **C...C**, put in **C...C**
STACK	(**E...E**) on ... on (**E...E**), put in (**C...C**)
SUBSCRIPTS	put into **C**
UNSTACK	(**C...C**) into (**E...E**) and ... and (**E...E**)
SUBSCRIPTS	are in **C**
CONVERT	using table in **C C**, the data in **C**, put in **C**
CONCATENATE	**C...C** put in **C**

4. Arithmetic

LET = expression

Expressions may use
  arithmetic operators: +   -   *   /   ** (exponentiation)
  comparison operators: =   ~=   <   >   ≤ ≥
  logical operators: & (and) | (or)   ~ (not)
and any of the following: ABSOLUTE, SQRT, LOGTEN, LOGE, EXPO, ANTILOG,
ROUND, SIN, COS, TAN, ASIN, ACOS, ATAN, SIGNS, NSCORES, PARSUMS,
PARPRODUCTS, COUNT, N, NMISS, SUM, MEAN, STDEV, MEDIAN, MIN, MAX, SSQ,
SORT, RANK, LAG, EQ, NE, LT, GT, LE, GE, AND, OR, NOT.

Examples: LET C2 = SQRT(C1 - MIN(C1))
         LET C3(5) = 4.5

ADD	**E** to **E** ... to **E**, put into **E**
SUBTRACT	**E** from **E**, put into **E**

```
MULTIPLY        E by E ...      by E, put into E
DIVIDE          E by               E, put into E
RAISE           E to the power     E, put into E

ABSOLUTE value                  of E, put into E
SQRT                            of E, put into E
LOGE                            of E, put into E
LOGTEN                          of E, put into E
EXPONENTIATE                       E, put into E
ANTILOG                         of E, put into E
ROUND to integer                   E, put into E
SIN                             of E, put into E
COS                             of E, put into E
TAN                             of E, put into E
ASIN                            of E, put into E
ACOS                            of E, put into E
ATAN                            of E, put into E
SIGNS                           of E, [put into E]
PARSUMS                         of C, put into C
PARPRODUCTS                     of C, put into C
NSCORES (normal scores)         of C, put into C

COUNT the number of values in      C [put into K]
N (number of nonmissing vals)   in C [put into K]
NMISS (number of missing vals)  in C [put into K]
SUM             of the values in   C [put into K]
MEAN            of the values in   C [put into K]
STDEV           of the values in   C [put into K]
MEDIAN          of the values in   C [put into K]
MINIMUM         of the values in   C [put into K]
MAXIMUM         of the values in   C [put into K]
RANGE           of the values in   C [put into K]
SSQ (uncorrected SS)           for C [put into K]

RCOUNT          of E...E, put into C
RN              of E...E, put into C
RNMISS          of E...E, put into C
RSUM            of E...E, put into C
RMEAN           of E...E, put into C
RSTDEV          of E...E, put into C
RMEDIAN         of E...E, put into C
RMINIMUM        of E...E, put into C
RMAXIMUM        of E...E, put into C
RRANGE          of E...E, put into C
RSSQ            of E...E, put into C

INDICATOR       variables for subscripts in C, put into C...C
```

5. Plotting Data

These are character graphs, available in all versions of Minitab.

Users of Release 9 Windows, or of Release 9 Enhanced on VAX or UNIX workstations: If you want to type these commands for character graphs, first type GSTD. From that point on, Minitab will interpret all commands you type as character graphs. To type Professional Graphics commands again, type GPRO.

```
GSTD            enable character and G-graphs           (Release 9 only)
GPRO            enable Professional Graphics            (Release 9 only)

HISTOGRAM       C...C
  INCREMENT     is K
  START         at K [end at K]
  BY            C
```

```
       SAME              scales for all columns
STEM-AND-LEAF            display of C...C
   TRIM                  outliers
   INCREMENT             is K
   BY                    C

BOXPLOT                  for C
   INCREMENT             is K
   START                 at K [end at K]
   BY                    C
   LINES                 = K
   NOTCH                 [K% confidence] sign c.i.
   LEVELS                K...K

DOTPLOT                  C...C
   INCREMENT             is K
   START                 at K [end at K]
   BY                    C
   SAME                  scales for all columns

PLOT                     C vs C
   SYMBOL                is 'symbol'
MPLOT                    C vs C, and C vs C, and ... C vs C
LPLOT                    C vs C using tags in C
TPLOT                    C vs C vs C

    PLOT, MPLOT, LPLOT, and TPLOT have the subcommands:
       TITLE             = 'text'
       FOOTNOTE          = 'text'
       YLABEL            = 'text'
       XLABEL            = 'text'
       YINCREMENT        is K
       YSTART            at K [end at K]
       XINCREMENT        is K
       XSTART            at K [end at K]

TSPLOT                   [period K] of C
   ORIGIN                = K
MTSPLOT                  [period K] of C...C
   ORIGIN                = K for C...C [... origin K for C...C]

    TSPLOT and MTSPLOT have the subcommands:
       INCREMENT         = K
       START             at K [end at K]
       TSTART            at K [end at K]
GRID                     C [values from K to K] C [values from K to K]
CONTOUR                  C vs C and C
   BLANK                 bands between letters
   YSTART                is K [up to K]
   YINCREMENT            is K

WIDTH                    of all plots that follow is K spaces
HEIGHT                   of all plots that follow is K lines
```

6. Basic Statistics

```
DESCRIBE                 C...C
   BY                    C

ZINTERVAL                [K% confidence] sigma = K for C...C
ZTEST                    [of mu = K]        sigma = K for C...C
   ALTERNATIVE           = K   # K=-1 for LT, K=+1 for GT

TINTERVAL                [K% confidence] for data in C...C
TTEST                    [of mu = K] on data in C...C
```

```
     ALTERNATIVE        = K

TWOSAMPLE              test and c.i. [K% confidence] for samples in C C
   ALTERNATIVE         = K    # K=-1 for LT, K=+1 for GT
   POOLED              procedure

TWOT                  test and c.i. [K% confidence] data in C, groups in C
   ALTERNATIVE         = K    # K=-1 for LT, K=+1 for GT
   POOLED              procedure

CORRELATION           between all pairs of C...C [put into M]
COVARIANCE            between all pairs of C...C [put into M]

CENTER                the data in C...C put into C...C
   LOCATION           [subtracting K...K]
   SCALE              [dividing by K...K]
   MINMAX             [with K as min and K as max]
```

7. Regression

```
REGRESS               C on K preds C...C [put stand. res. in C [fits in C]]
   NOCONSTANT         in equation
   WEIGHTS            are in C
   SRESIDUALS         put into C  (standardized residuals)        (Release 9 only)
   FITS               put into C                                  (Release 9 only)
   MSE                put into K
   COEFFICIENTS       put into C
   XPXINV             put into M
   RMATRIX            put into M
   HI                 put into C (leverages)
   RESIDUALS          put into C (observed - fit)
   TRESIDUALS         put into C (deleted Studentized residuals)
   COOKD              put into C (Cook's distance)
   DFITS              put into C
   PREDICT            for E...E
   VIF                (variance inflation factors)
   DW                 (Durbin-Watson statistic)
   PURE               (pure error lack-of-fit test)
   XLOF               (experimental lack-of-fit test)
   TOLERANCE          K [K]

STEPWISE              y in C, predictors in C...C
   FENTER             = K    (F-to-enter, default is 4)
   FREMOVE            = K    (F-to-remove, default is 4)
   FORCE              C...C  (keep these predictors in all models)
   ENTER              C...C  (enter these predictors on first step)
   REMOVE             C...C
   BEST               K alternative predictors (print these)
   STEPS              = K (default depends on output width)

BREG                  C on predictors C...C
   INCLUDE            predictors C...C  (include these in all models)
   BEST               K models  (print info on these)
   NVARS              K [K]    (give models of these sizes)
   NOCONSTANT         in equation
   NOWARN             suppress warning for model with many predictors

NOCONSTANT            in REGRESS, STEPWISE, and BREG commands that follow
CONSTANT              fit a constant in REGRESS, STEPWISE, and BREG
BRIEF                 K
```

8. Analysis of Variance

AOVONEWAY	aov for samples in **C...C**
ONEWAY	data in **C**, levels in **C** [put residuals in **C** [fits in **C**]]
TUKEY	[family error rate **K**] (Releases 8 and 9 only)
FISHER	[individual error rate **K**] (Releases 8 and 9 only)
DUNNETT	[family error rate **K**] control group is **K** (Rel. 8 & 9 only)
MCB	[family error rate **K**] best is **K** (Releases 8 and 9 only)
TWOWAY	data in **C**, levels in **C C** [put res in **C** [fits in **C**]]
ADDITIVE	model
MEANS	for the factors **C** [**C**]
ANOVA	**C...C = termlist**
RANDOM	**C...C**
EMS	print expected mean squares
FITS	put into **C...C**
RESIDUALS	put into **C...C**
MEANS	for **termlist**
TEST	for **termlist / errorterm**
RESTRICT	use restricted model
MANOVA	[**termlist** [**/ errorterm**]] (Release 9 only)
SSCP	prints matrices H and E (Release 9 only)
EIGEN	prints eigen analysis (Release 9 only)
PARTIAL	prints partial correlations (Release 9 only)
NOUNIVARIATE	does not print univariate output (Release 9 only)
ANCOVA	**C...C = termlist**
COVARIATES	are in **C...C**
FITS	put into **C...C**
RESIDUALS	put into **C...C**
MEANS	for **termlist**
TEST	for **termlist / errorterm**
GLM	**C...C = termlist**
COVARIATES	are in **C...C**
WEIGHTS	are in **C**
FITS	put into **C...C**
RESIDUALS	put into **C...C**
SRESIDUALS	put into **C...C**
TRESIDUALS	put into **C...C**
HI	put into **C**
COOKD	put into **C...C**
DFITS	put into **C...C**
XMATRIX	put into **M**
COEFFICIENTS	put into **C...C**
MEANS	for **termlist**
TEST	for **termlist / errorterm**
BRIEF	**K**
TOLERANCE	**K** [**K**]

Also subcommands MANOVA, SSCP, EIGEN, PARTIAL, and NOUNIVARIATE,
listed with ANOVA.

9. Multivariate Analysis

DISCRIMINANT	groups in **C**, predictors in **C...C**
QUADRATIC	discrimination
PRIORS	are in **K...K**
LDF	put in **C...C**, coefs for linear discriminant function
FITS	put in **C** [**C**]
XVAL	cross-validation
PREDICT	for **E...E**

```
    BRIEF           K

PCA                 principal component analysis of data in C...C
    COVARIANCE      matrix, default is correlation matrix
    NCOMP           = K (number of components)
    COEF            put into C...C
    SCORES          put into C...C

FACTOR              [C...C]                                    (Release 9 only)
    CORRELATION     [M] factor the correlation matrix
    COVARIANCE      [M] factor the covariance matrix
    NFAC            K number of factors to extract
    ML              use maximum likelihood, default is PCA
    PRIORS          are in C, use as initial communality estimates in ML
    MAXITER         = K, maximum number of iterations for ML
    CONVERGENCE     = K, criterion for convergence in ML
    INLOADINGS      are in C...C, input these as loadings
    SORT            [zero out loading less than K] sort loadings
    VMAX            use varimax rotation
    QMAX            use quartimax rotation
    EMAX            use equamax rotation
    OMAX            = K, use orthomax rotation with specified gamma
    EIGEN           put in C M, store eigenvalues and vectors
    LOADINGS        put in C...C, store loadings
    COEFFICIENTS    put in C...C, store factor score coefficients
    SCORES          put in C...C, store factor scores
    RESIDUAL        put in M, store residual matrix
    ROTMAT          put in M, store rotation matrix
    BRIEF           K
```

10. Nonparametrics

```
RUNS                test [above and below K] for C

STEST               sign test [median = K] for C...C
    ALTERNATIVE     = K    # K=-1 for LT, K=+1 for GT
SINTERVAL           sign c.i. [K% confidence] for C...C
WTEST               Wilcoxon one-sample rank test [median = K] for C...C
    ALTERNATIVE     = K    # K=-1 for LT, K=+1 for GT
WINTERVAL           Wilcoxon c.i. [K% confidence] for C...C
MANN-WHITNEY        test and c.i. [K% confidence] for samples in C C
    ALTERNATIVE     = K    # K=-1 for LT, K=+1 for GT

KRUSKAL-WALLIS      test for data in C, subscripts in C
MOOD median test    data in C, levels in C [put residuals in C [fits in C]]
FRIEDMAN            data in C treat in C blocks in C [res in C [fits in C]]

WALSH               averages for C, put into C [indices into C C]
WDIFF               for C and C,   put into C [indices into C C]
WSLOPE              y in C, x in C, put into C [indices into C C]
```

11. Tables

```
TALLY               the data in C...C
    COUNTS
    PERCENTS
    CUMCOUNTS       cumulative counts
    CUMPERCENTS     cumulative percents
    ALL             four statistics above

CHISQUARE           test on table stored in C...C
```

TABLE	the data classified by **C...C**
MEANS	for **C...C**
MEDIANS	for **C...C**
SUMS	for **C...C**
MINIMUMS	for **C...C**
MAXIMUMS	for **C...C**
STDEV	for **C...C**
STATS	for **C...C**
DATA	for **C...C**
N	for **C...C**
NMISS	for **C...C**
PROPORTION	of cases = **K** [through **K**] in **C...C**
COUNTS	
ROWPERCENTS	
COLPERCENTS	
TOTPERCENTS	
CHISQUARE	analysis [output code = **K**]
MISSING	level for classification variable **C...C**
NOALL	in margins
ALL	in margins just for **C...C**
FREQUENCIES	are in **C**
LAYOUT	**K** rows by **K** columns

12. Time Series

ACF	[up to **K** lags] for series in **C** [put into **C**]
PACF	[up to **K** lags] for series in **C** [put into **C**]
CCF	[up to **K** lags] between series in **C** and **C**
DIFFERENCES	[of lag **K**] for data in **C**, put into **C**
LAG	[by **K**] for data in **C**, put into **C**
ARIMA	p=**K** d=**K** q=**K** [P=**K** D=**K** Q=**K** S=**K**] data in **C** & [put res in **C** [fits in **C** [coef in **C**]]]
CONSTANT	term in model
NOCONSTANT	term in model
STARTING	values are in **C**
FORECAST	[origin = **K**] up to **K** leads [put in **C** [limits in **C C**]]

13. Control Charts

These are character graphs, available with all versions of Minitab.

Users of Release 9 Windows, or of Release 9 Enhanced on VAX or UNIX workstations: You must use GSTD if you want to type these commands. See Section 5 for details.

XBARCHART	for **C...C**, subgroups are in **E**
MU	= **K**
SIGMA	= **K**
RBAR	use R-bar to estimate sigma, default is pooled s
RSPAN	= **K**
TEST	**K...K**
SUBGROUP	size is **E**
RCHART	for **C...C**, subgroups are in **E**
SIGMA	= **K**
RBAR	use R-bar to estimate sigma, default is pooled s
SUBGROUP	size is **E**
SCHART	for **C...C**, subgroups are in **E**
SIGMA	= K
RBAR	use R-bar to estimate sigma, default is pooled s

```
    SUBGROUP        size is E
ICHART              for C...C
    MU              = K
    SIGMA           = K
    RSPAN           = K
    TEST            K...K
MACHART             for C...C, subgroups are in E
    MU              = K
    SIGMA           = K
    RBAR            use R-bar to estimate sigma, default is pooled s
    SPAN            = K
    RSPAN           = K
    SUBGROUP        size is E
EWMACHART           for C...C, subgroups are in E
    MU              = K
    SIGMA           = K
    RBAR            use R-bar to estimate sigma, default is pooled s
    WEIGHT          = K
    RSPAN           = K
    SUBGROUP        size is E
MRCHART             for C...C
    SIGMA           = K
    RSPAN           = K
PCHART              number of nonconformities are in C...C, sample size = E
    P               = K
    TEST            K...K
    SUBGROUP        size is E
NPCHART             number of nonconformities are in C...C, sample size = E
    P               = K
    TEST            K...K
    SUBGROUP        size is E
CCHART              number of nonconformities are in C...C
    MU              = K
    TEST            K...K
UCHART              number of nonconformities are in C...C, sample size = E
    MU              = K
    TEST            K...K
    SUBGROUP        size is E
```

All statistical process control charts have the subcommands:

```
    SLIMITS         are K...K  (number of sigma limits)
    HLINES          at E...E   (draw horizontal lines)
    ESTIMATE        using just samples K...K
    TITLE           = 'text'
    FOOTNOTE        = 'text'
    YLABEL          = 'text'
    XLABEL          = 'text'
    YINCREMENT      = K
    YSTART          at K [end at K]
    XINCREMENT      = K
    XSTART          at K [end at K]
```

14. Exploratory Data Analysis

```
LVALS               of C [put lvals in C [mids in C [spreads in C]]]
MPOLISH             data in C, levels in C C [put res in C [fits in C]]
    COLUMNS         (start iteration with column medians)
    ITERATIONS      = K
    EFFECTS         put common into K, rows into C, cols into C
    COMPARISON      values, put into C
RLINE               y in C, x in C [put res in C [fits in C [coef in C]]]
```

```
    MAXITER        K (max. number of iterations)
  RSMOOTH          C, put rough into C, smooth into C
    SMOOTH         by 3RSSH, twice (default is 4253H)
  CTABLE           (coded table) data in C, row C, column C
    MAXIMUM        value in each cell should be coded
    MINIMUM        value in each cell should be coded
  ROOTOGRAM        data in C [use bin boundaries in C]
    BOUNDARIES     put in C
    DRRS           put in C
    FITTED values  put in C
    COUNTS         put in C
    FREQUENCIES    are in C [bin boundaries are in C]
    MEAN           = K
    STDEV          = K
```

15. Distributions and Random Data

```
  RANDOM           K observations, put into C...C
    BERNOULLI      trials p = K

  PDF              for values in E...E [put results in E...E]
  CDF              for values in E...E [put results in E...E]
  INVCDF           for values in E...E [put results in E...E]

    RANDOM, PDF, CDF, INVCDF have the subcommands:
    BINOMIAL       n = K, p = K
    POISSON        mu = K
    INTEGER        discrete uniform on integers K to K
    DISCRETE       distribution with values in C and probabilities in C
    NORMAL         [mu = K [sigma = K]]
    UNIFORM        [continuous on the interval K to K]
    T              degrees of freedom = K
    F              df numerator = K, df denominator = K
    Additional subcommands are BETA, CAUCHY, LAPLACE, LOGISTIC,
    LOGNORMAL, CHISQUARE, EXPONENTIAL, GAMMA, WEIBULL

  SAMPLE           K rows from C...C put into C...C
    REPLACE        (sample with replacement)
  BASE             for random number generator = K
```

16. Sorting

```
  SORT             C [carry along C...C] put into C [and C...C]
    BY             C...C
    DESCENDING     C...C
  RANK             the values in C, put ranks into C
```

17. Matrices

```
  READ             [from 'filename'] into a K by K matrix M
  PRINT            M...M

  TRANSPOSE        M into M
  INVERT           M into M
  DEFINE           K into K by K matrix M
  DIAGONAL         is C, form into M
  DIAGONAL         of M, put into C

  COPY             C...C into M
  COPY             M into C...C
```

```
COPY                M into M
  All three versions of COPY have the subcommands:
     USE            rows K...K
     OMIT           rows K...K
ERASE               M...M

EIGEN               for M put values into C [vectors into M]

In the following commands, E can be either C, K, or M:
ADD                 E to   E, put into E
SUBTRACT            E from E, put into E
MULTIPLY            E by   E, put into E
```

18. Miscellaneous

```
BRIEF               = K controls amount of output from REGRESS, GLM,  &
                        DISCRIMINANT, FACTOR, ARIMA, RLINE, FFDESIGN

RESTART             begin fresh Minitab session

OH                  = K number of lines for height of output
OW                  = K number of spaces for width of output
IW                  = K number of spaces for width of input

TSHARE              operate interactively
BATCH               operate in batch mode

SYSTEM              provides access to the operating system commands
CD                  [path] change directory              (DOS and Windows only)
DIR                 [path] list files                    (DOS and Windows only)
TYPE                [path[filename]]                      (DOS and Windows only)
```

19. Design of Experiments

```
FFDESIGN            K factors [K runs]
  BLOCKS            K or termlist
  REPLICATES        = K (replicates for each point)
  CENTER            = K (add these center points)
  FRACTION          = K (use this fraction of the design)
  FOLD              [on factors  fact...fact]
  ADD               fact = term, ..., fact = term
  ALIAS             info up to order K
  RANDOMIZE         [using base K]
  XMATRIX           put into C...C   (data matrix)
  LEVELS            are in K...K or C (levels for each factor)
  TERMS             put in C (terms that are in the design)
  BRIEF             K
PBDESIGN            K factors [K runs]
  REPLICATES        = K (replicates for each point)
  CENTER            = K (add these center points)
  RANDOMIZE         [using base K]
  XMATRIX           put into C...C
  LEVELS            are in K...K or C (levels for each factor)

FFACTORIAL          C...C = termlist
  BLOCKS            are in C
  COVARIATES        are in C...C
  EFFECTS           put into C...C
  COEFFICIENTS      put into C...C
  FITS              put into C...C
  RESIDUALS         put into C...C   (observed - fit)
  SRESIDUALS        put into C...C
```

```
TRESIDUALS        put into C...C   (deleted Studentized)
HI                put into C       (leverages)
COOKD             put into C...C   (Cook's distance)
DFITS             put into C...C
XMATRIX           put into M       (design matrix)
MEANS             for termlist
CUBE              for termlist     (two- and three-way interactions)
EPLOT             print a normal probability plot of effects
ALIAS             info up to order K
TOLERANCE         K [K]
```

20. Stored Commands and Loops

STORE and EXECUTE allow you to write simple macros (stored command files)
and loops. You can use STORE or an editor to create the command file.

```
STORE             [in 'filename'] the following commands
                      (Minitab commands go here)
END               end of storing Minitab commands
EXECUTE           'filename' [K times]
ECHO              printing of Minitab commands
NOECHO            printing of Minitab commands
YESNO             K, used to ask the user a question
```

The CK capability. The integer part of a column number or matrix
number may be replaced by a stored constant. The following example
prints C1—C5, since K1 = 5:

```
          LET K1 = 5
          PRINT C1 - CK1
```

21. High-Resolution G-Graph Commands

These commands are not available on Release 9 Windows.

Users of Release 9 Enhanced on VAX or UNIX workstations:
If you want to type these commands, first type GSTD. When you
want to type Professional Graphics commands again, type GPRO.

```
GOPTIONS          sets options for high-res graphics
  DEVICE          = 'device' graphics device (Rel. 7, except DOS, and Rel. 9)
  HEIGHT          = K inches
  WIDTH           = K inches
  PRINTER         send graphs to printer (default)     (Release 8 DOS only)
  PLOTTER         send graphs to plotter               (Release 8 DOS only)
```

G-graphs commands: GHISTOGRAM, GBOXPLOT, GPLOT, GMPLOT, GLPLOT, GTPLOT.
G-graph control charts: GXBARCHART, GRCHART, GSCHART, GICHART, GMACHART,
GEWMACHART, GMRCHART, GPCHART, GNPCHART, GCCHART, GUCHART.

All these G-graph commands have the same syntax and subcommands as
the corresponding character versions.

All G-graph commands except on Macintosh have the additional subcommand:
```
  FILE            'filename' to store graphics output
```

GPLOT, GMPLOT, GLPLOT, and GTPLOT have the additional subcommands:
```
  LINES           [style K [color K]] connect (x, y) points in C C
  COLOR           codes are in C
```

```
Styles for Lines                               Colors
0 solid           4 dots                       0 black (white on Mac)
1 large dashes    5 medium dash, dot           1 white (black on Mac)
2 medium dashes   6 medium dash, small dash    2 red      5 cyan
3 small dashes    7 medium dash, 2 small dashes 3 green    6 magenta
                                               4 blue     7 yellow
```

22. High-Resolution Graphs Commands for Release 9

We list here just a portion of the high-resolution graphics capability
that is available in Release 9.

If you want to type these commands, you must make sure the GPRO option is
in effect. When you want to type commands for standard graphs, type GSTD.

```
HISTOGRAM    C...C
  MIDPOINTS  K...K              # gives midpoints for each bar
BOXPLOT      C[*C] ... C[*C]    # second C in each pair is group variable
PLOT         C*C ... C*C
  JITTER                        # jitters points so you can see overlaps
TSPLOT       C...C
```

The control chart commands are XBARCHART, RCHART, SCHART, ICHART,
MACHART, EWMACHART, MRCHART, PCHART, NPCHART, CCHART, and UCHART.
These commands have the same syntax and subcommands as do the
corresponding character versions, with one exception. You can use
just one data column (i.e., do just one chart) with a command.

HISTOGRAM, BOXPLOT, PLOT, and TSPLOT have the subcommands described below.
Control chart commands have all these subcommands except SAME.
In AXIS, TICK, MINIMUM, MAXIMUM, and SAME, K is used to specify the axis.
Use K=1 for the x-axis and K=2 for the y-axis.

```
    TITLE      'text'           # adds a title to the graph
    FOOTNOTE   'text'           # adds a footnote to the graph

    AXIS       K                # K specifies the axis
      LABEL    'text'           # label for axis K

    TICK       K [K...K]        # tick marks, first K specifies axis
      LABEL    'text'...'text'  # a label for each tick mark on axis K

    MINIMUM    K K              # first K is axis, second K is min data value
    MAXIMUM    K K              # first K is axis, second K is max data value
    SAME       K [K]            # same scales on all graphs for listed axes
```

PLOT and TSPLOT have the subcommands described below.
If you use a group variable on SYMBOL or CONNECT, you can list one K
for each group on TYPE, COLOR, and SIZE. If you do not use a group
variable but give several graphs on the main command, you can list
one K for each graph on TYPE, COLOR, and SIZE.

```
    SYMBOL     [C]              # symbols for points, C is a group variable
      TYPE     K...K            # symbol type
      COLOR    K...K            # symbol color
      SIZE     K...K            # symbol size, default is 1

    CONNECT    [C]              # connects points, C is a group variable
      TYPE     K...K            # line type
      COLOR    K...K            # line color
      SIZE     K...K            # line size, default is 1

    LINE       C C              # adds a line, using (x, y) values in C C
      TYPE     K                # line type
      COLOR    K                # line color
      SIZE     K                # line size, default is 1
```

OVERLAY   # puts all graphs in one figure

Symbol Types	Line Types	Colors for Symbols and Lines	
0 none	0 null (invisible)	0 white	5 cyan
1 circle	1 solid	1 black	6 magenta
2 plus	2 dash	2 red	7 yellow
3 cross	3 dot	3 green	
4 asterisk	4 dash-dot	4 blue	
5 dot			

References

Chapter 1

The pulse data in Table 1.1 are from a classroom experiment by Brian L. Joiner.

Chapter 2

The trees data are from H. Arthur Meyer, *Forest Mensuration* (State College, PA: Penns Valley Publishers, Inc., 1953).

Chapter 3

The pulse data are from a classroom experiment by Brian L. Joiner.

The lake data in Exercise 3-6 were collected by the Wisconsin Department of Natural Resources and were provided to us by Alison K. Pollack and Joseph M. Eilers.

The furnace data in Exercise 3-19 are from a study conducted by Tim LaHann and Dick Mabbott of the Wisconsin Power and Light Company.

Chapter 4

The cartoon data in Exhibit 4.2 are from Stephan Kauffman, "An Experimental Evaluation of the Relative Effectiveness of Cartoons and Realistic Photographs in Both Color and Black and White Visuals in In-Service Training Programs," Master's thesis, The Pennsylvania State University, 1973.

The men's Olympic track data in Table 4.2 are from *The World Almanac and Book of Facts, 1993* (New York: Pharos Books, 1993).

The employment data in Table 4.4 are from R. B. Miller and Dean W. Wichern, *Intermediate Business Statistics* (New York: Holt, Rinehart & Winston, 1977). Copyright © 1977 by Holt, Rinehart & Winston. Reprinted by permission of CBS College Publishing.

The radiation data used in Exhibit 4.7 came from the Las Vegas Radiation Facility, Las Vegas, Nevada.

The marriage-rate data in Exercise 4-5 are from *The World Almanac and Book of Facts, 1993* (New York: Pharos Books, 1993).

The track-records data in Exercise 4-6 are from *The World Almanac and Book of Facts, 1993* (New York: Pharos Books, 1993).

The temperatures for various cities in Exercise 4-10 are from *The World Almanac and Book of Facts, 1984* (New York: Newspaper Enterprise Association, Inc., 1984).

The wind-chill data used in Exercise 4-11 were obtained from the U.S. Army.

The automobile accident data used in Exercise 4-12 are from Paul M. Hurst, "Blood Test Legislation in New Zealand," *Accident Analysis and Prevention* 10 (Pergamon Press, Ltd., 1978), pp. 287–296. Copyright © 1978, Pergamon Press, Ltd. Reprinted with permission.

The leaf-surface resistance data in Exercise 4-17 were provided by Thomas Starkey of The Pennsylvania State University.

Chapter 5

The Wisconsin Restaurant Survey was done by William A. Strang of the Small Business Development Center, University of Wisconsin, Madison, 1980. We also thank Robert Miller for his help with the analysis of these data and with the writing of Chapter 4.

The psychiatric commitment data in Exercise 5-4 were provided by a health care center in Wisconsin.

Chapter 6

The chest-girth data in Table 6.1 were obtained from Roger Carlson, "Normal Probability Distributions," in *Statistics by Example: Detecting Patterns*, F. Mosteller, W. H. Kruskal, R. F. Link, R. S. Pieters, and G. R. Rising, editors (Reading, MA: Addison-Wesley, 1973).

The grades data described in Exercise 6-5 were obtained from a university that wishes to remain anonymous.

The Penn State nuclear reactor data in Exercise 6-15 were obtained from testimony presented to the Atomic Energy Commission in 1971.

Chapter 9

The cholesterol data for heart-attack victims in Table 9.1 were collected by a Pennsylvania medical center in 1971.

The shoe-sole materials data in Table 9.2 are from G. E. P. Box, W. G. Hunter, and J. S. Hunter, *Statistics for Experimentation* (New York: Wiley & Sons, 1978). Copyright © 1978 by John Wiley & Sons, Inc. Reprinted by permission.

The Parkinson's disease data in Table 9.3 were obtained by Gastone Celesia in a study at the University of Wisconsin.

Chapter 10

The fabric flammability test in Table 10.1 was conducted by the American Society for Testing Materials. Related analyses are "ASTM Studies DOC Standard FF-30-71 on Flammability of Children's Sleepwear," *Materials Research Standards*, May 1972, and John Mandel, Mary N. Steel, and L. James Sharman, "National Bureau of Standards Analysis of the ASTM Interlaboratory Study of DOC/FF 3-71 Flammability of Children's Sleepwear," *ASTM Standardization News*, May 1973.

The driving performance data in Table 10.2 are from David C. Howell, *Statistical Methods for Psychology*, Third Edition (Boston: PWS-Kent, 1992), p. 418.

The pancake data in Table 10.3 were given to us by Gregory Mack and John Skilling.

The billiard-ball data in Table 10.4 are from Albert Romano, *Applied Statistics for Science and Industry* (Boston: Allyn and Bacon, Inc., 1977), p. 300.

The plywood data in Exercise 10-2 are from a study directed by Frank J. Fronczak of the U.S. Forest Products Research Laboratory.

The pendulum data in Exercise 10-3 were obtained from John Mandel. These data also appear in his article "The Acceleration of Gravity," in *Statistics by Example: Detecting Patterns*, F. Mosteller, W. H. Kruskal, R. F. Link, R. S. Pieters, and G. R. Rising, editors (Reading, MA: Addison-Wesley, 1973).

The data on manganese in cow liver in Exercise 10-5 were obtained from the National Institute of Standards and Technology.

The potato-rot data in Exercise 10-10 were collected at the University of Wisconsin.

The alfalfa data in Exercise 10-11 are from an experiment conducted at the University of Wisconsin.

The meat-loaf data in Exercise 10-12 were obtained in a study at the University of Wisconsin conducted by Barbara J. Bobeng and Beatrice David.

Chapter 11

The test score data set in Table 11.1 was collected in a statistics class at The Pennsylvania State University.

The stream data in Exhibit 11.5 and Exercises 11-24 and 11-26 were gathered as part of an environmental impact study.

The SAT scores data set in Exercise 11-8 was obtained from *The Digest of Educational Statistics: 1983–1984 Volume*, U.S. Government Publication, and from *The World Almanac and Book of Facts, 1993* (New York: Pharos Books, 1993).

The track data in Exercise 11-9 are from *The World Almanac and Book of Facts, 1993* (New York: Pharos Books, 1993).

The maple tree data in Exercise 11-18 were provided by Erik V. Nordheim of the University of Wisconsin.

The Peru data in Exercise 11-22 are from P. T. Baker and C. M. Beall, "The Biology and Health of Andean Migrants: A Case Study in South Coastal Peru," *Mountain Research and Development* 2(1), 1982.

The regression data in Exercise 11-30 are from F. J. Anscombe, "Graphs in Statistical Analysis," *The American Statistician* 27(1), February 1973, pp. 17–21. Reprinted by permission of the American Statistical Association.

The magnifying glass data in Exercise 11-34 were collected by a student in an introductory physics class.

Chapter 12

The penny-tossing data in Table 12.1 are described in W. J. Youden, *Risk, Choice and Prediction* (North Scituate, MA: Duxbury Press, 1974).

Further discussions of the ESP data in Table 12.2 are given in Marvin Lee Moon, "Extrasensory Perception and Art Experience," Ph.D. thesis, Pennsylvania State University, 1973. An additional article by Moon appeared in *American Journal of Psychical Research*, April 1975.

The accidental-deaths data set in Exercise 12-1 is from *The World Almanac and Book of Facts, 1984* (New York: Newspaper Enterprise Association, Inc., 1984).

The precipitation data in Exercise 12-2 were collected in State College, Pennsylvania.

The data on migratory geese in Exercise 12-3 are from R. K. Tsutakawa, "Chi-Square Distribution by Computer Simulation," in *Statistics by Example: Detecting Patterns*, F. Mosteller, W. H. Kruskal, R. F. Link, R. S. Pieters, and G. R. Rising, editors (Reading, MA: Addison-Wesley, 1973). Copyright © 1973 Addison-Wesley Publishing Company, Inc.

The Harrisburg, Pennsylvania, infant mortality data in Exercises 12-4 and 12-5 are from Vilma Hunt and William Cross, "Infant Mortality and the Environment of a Lesser Metropolitan County: A Study Based on Births in One Calendar Year," *Environmental Research* 9, 1975, pp. 135–151.

The car defects data in Exercises 12-6 and 12-7 were obtained from L. A. Klimko and Camil Fuchs, "An Analysis of the Data from Wisconsin's Dealer Based Demonstration Motor Vehicle Inspection Program," Statistical Laboratory Report 77/7, Department of Statistics, University of Wisconsin, Madison, 1977.

The Mark Twain data in Exercise 12-8 are from Claude Brinegar, "Mark Twain and the Quintus Curtius Snodgrass Letters: A Statistical Test of Authorship," *Journal of the American Statistical Association* 58, 1963, pp. 85–96.

Chapter 13

The acid data set in Exercises 13-3 and 13-4 was collected in a chemistry class.

The presidents data set in Exercise 13-8 is from *The World Almanac and Book of Facts, 1984* (New York: Newspaper Enterprise Association, Inc., 1984).

The bird-fat data in Exercise 13-15 are from Jack Hailman, "Notes on Quantitative Treatments of Subcutaneous Lipid Data," *Bird Banding* 36, 1965, pp. 14–20.

The azalea data in Exercise 13-16 are from a study done at The Pennsylvania State University by Professor Stanley Pennypacker, Department of Plant Pathology.

The wine data in Exercise 13-17 are from W. Kwan, B. R. Kowalski, and R. K. Skogerboe, *J. Agric. Food Chemistry* 27, 1979, p. 1321.

Chapter 14

The ignition-key data in Exhibit 14.1 are from N. Farnum, *Modern Statistical Quality Control and Improvement* (North Scituate, MA: Duxbury Press, 1994), p. 185.

The plating data in Exhibit 14.7 are from *Statistics for Business: Data Analysis and Modelling*, J. D. Cryer and R. B. Miller (Boston: PWS-Kent, 1991), p. 317.

The cap torque data in Exercise 14-1 are from *Statistics for Business: Data Analysis and Modelling*, J. D. Cryer and R. B. Miller (Boston: PWS-Kent, 1991), p. 292.

The documents data in Exercise 14-4 are from N. Farnum, *Modern Statistical Quality Control and Improvement* (North Scituate, MA: Duxbury Press, 1994), p. 245.

Chapter 16

The plywood data in Exercise 16-5 are from a study directed by Frank J. Fronczak of the U.S. Forest Products Research Laboratory.

The driving performance data in Exercise 16-6 are from David C. Howell, *Statistical Methods for Psychology*, Third Edition (Boston: PWS-Kent, 1992), p. 418.

The poplar data in Exercise 16-8 are from P. R. Blankenhorn, T. W. Bowersox, C. H. Strauss, and others, "Net energy and economic analyses for producing *Populus* hybrid under four management strategies," final report to U.S. Dept. of Energy, Washington, DC, ORNL/Sub/79-07918/1, 1985.

INDEX